Edward Jones
333 Orange Ave. #34
Coronado, CA 92118

DEKE!

DEKE!

U.S. MANNED SPACE: FROM MERCURY TO THE SHUTTLE

Donald K. "Deke" Slayton
with Michael Cassutt

A TOM DOHERTY ASSOCIATES BOOK
NEW YORK

DEKE!—U.S. MANNED SPACE: FROM MERCURY TO THE SHUTTLE

This book is printed on acid-free paper.

A Forge Paperback
Published by Tom Doherty Associates, Inc.
175 Fifth Avenue
New York, N.Y. 10010

Forge® is a registered trademark of Tom Doherty Associates, Inc.

Design by Brian Mulligan

Library of Congress Cataloging-in-Publication Data

Slayton, Donald K., 1924–1993.
Deke! : U.S. manned space: from mercury to the shuttle / Donald K. "Deke" Slayton with Michael Cassutt.
p. cm.
"A Tom Doherty Associates book."
ISBN 0-312-85918-X
1. Slayton, Donald K., 1924–1993. 2. Astronauts—United States—Biography. I. Cassutt, Michael. II. Title.
TL789.85.S55A3 1994
629.45'0092—dc20
[B] 94-2463
CIP

Printed in the United States of America

0 9 8 7 6 5 4 3 2

For Bobbie,
and for our children and grandchildren

As soon as Gus scrambled out of the hatch, he had begun swimming for his life. The goddamned capsule's going under! . . .

Deke! . . . Where was Deke! . . . Surely Deke would be here! . . . He had done as much for Deke. Somehow Deke would materialize and save him . . . Deke! . . . Or somebody! Deke!

—Tom Wolfe, *The Right Stuff*

Every time Pruett thought of Deke Slayton, and the orders grounding the solidly built pilot from space flight, he wanted to go somewhere and be sick. Deke could manhandle a fighter in a skilled fashion that few pilots in the world ever approached. When he flew as a test pilot at Edwards, the supply shop ran out of G-suits, which helped a man's body to fight off the punishing effect of plus-gravity forces. So Deke went ahead and flew anyway, and without the suit. And he built up a tolerance to plus-G that made even old-time veterans blink their eyes. He could pull more than nine G in a fighter, without that suit, and still not black out. He pulled more pressure on the centrifuge than any other member of the astronaut team. But he had something in his heart that kicked every now and then, and a panel of doctors knew fright, and they grounded Slayton.

—Martin Caidin, *Marooned*

1

BEGINNINGS

I guess when it comes to space and aviation, I've seen and done a lot in fifty years. My name isn't the first one to come to mind when somebody says the word *astronaut*, but I was one of the original Mercury guys—the one who got screwed out of a mission for medical reasons. I hung in there and wound up running the Astronaut Office. Neil Armstrong became the first person to walk on the moon because I selected him. I eventually got into space, however, on Apollo-Soyuz—thirteen years after I should have.

When I was four years old, growing up on a dairy farm in Wisconsin, I was fond of running across the country road to the neighbor's place. There wasn't much traffic on that road, but my mother was terrified that I'd get hit by a truck. With three other children younger than me, she had her hands full and her eyes elsewhere.

So whenever she let me out into the front yard, she tied me to a tree with a rope. I was tethered like a puppy. I could run around, but only so far. I certainly couldn't reach the road.

Eventually I convinced my mother that I wasn't going to go running into the road, and I was set free. But I can make the case that ever since I was young I have wanted to explore . . . and people have tried to stop me.

It's not as though there was anywhere to run to. The Slayton farm was a mile north of Leon, Wisconsin, which itself wasn't much more than a wide spot in Highway 27, which runs south of the city of Sparta. Sparta was a small town in those days—I remember the population well, because it was the same as the number of feet in a mile—5,280. I think it's now grown almost to a nautical mile, 6,010 or something.

In 1959, the year I was selected as a Mercury astronaut, Leon's population was 150. It probably still is, for all I know. It had a general store named the Farmer's Store, a feed mill, a couple of filling stations, a garage, and two farm implement dealers.

I had an older half sister, Verna, and an older half brother, Elwood. My dad, Charles Sherman Slayton, was married once before, but his first wife died when Elwood was born. Elwood was farmed out to an aunt to raise, so he was more like a cousin than a brother . . . he didn't live with us.

Verna, who was four years older than Elwood, did live with us. She eventually died of multiple sclerosis. Elwood is still up in Wisconsin.

My dad remarried to Victoria Larson, and I'm the oldest from the second marriage. I've got a brother, Howard, one year younger. A sister, Bev, a year younger than he is. A sister, Marie—she's two years younger than Bev. My youngest brother, Dick, is nine years younger than I am. So that's the family.

I should point out that to my family, and to the rest of the world until I was in my thirties, I was always Don Slayton. Nobody called me Deke until I became a test pilot at Edwards in the 1950s.

Slayton is supposed to be English, but three of my grandparents came from Norway. The English one we've traced back to New England somewhere. My dad used to say he was probably a cabin boy or horse thief. There are a lot of Slaytons around the country, I've discovered. I go someplace and somebody pops up to say he's my relative via this channel, forty times removed.

My dad was an avid reader, even though all he had was an eighth grade education. He took the *Saturday Evening Post* and the *Farm Journal*. When I was a little older he even became the township tax collector, but managed to stay liked by everyone.

He took jobs because farming was a tough way to make a living, especially in the thirties. Nobody had money in those days and by most standards we were poor. We reused Christmas and Valentine's Day cards to save a few pennies. But, fortunately, being on a farm, we never worried about starving to death. We always had plenty of good, healthy food to eat. We didn't have fancy clothes, but for farming all you need are old overalls.

My dad used to go out and run a road grader for the county, building roads. That's how we got cash. Verna's husband really ran the farm.

Entertainment was a luxury. I do recall being taken into Sparta once to see my first movie, which turned out to be some Nelson Eddy–Jeanette MacDonald thing. My parents must have liked it, but I didn't. I've hardly been to a movie since.

We didn't even have a radio until I got to high school, because we didn't have electricity. For a while we had a Delco battery plant set up in the basement of the house. That gave us electricity for lights, but not much. It was only after I left in 1942 that we got 110-volt electricity.

No indoor plumbing, either. That didn't come in until we got electricity.

(Growing up the way I did, I'm still amazed at the number of things that are electrical these days. Once I had a conversation with a guy about nuclear war and nuclear winter. He was telling me it wouldn't bother him: he'd just jump in his car and head out to the country. I pointed out that his car probably had an electronic ignition and if he tried to put gas in the car the pumps were also electrical.)

Fishing was something we did all the time. There was a stream down in back, half a mile from where we lived. All you had to do was cut a cane pole and get a piece of string; hooks were cheap. You could get a lot of entertainment for very little.

I also started hunting when I was very young: I was using a gun from the time I was six years old, and bought an old sixteen-gauge shotgun—single shot—when I was eight or nine. I think I paid a dollar and a half for it. It didn't work half the time. It had an outside hammer. I can't remember how many times I'd find a rabbit and pull to shoot the son of a gun and the gun would misfire. You could say I did a lot of unintended game conservation in those days.

We also learned how to deal with animals. These days everybody's got pets, and we're probably giving more feed to them than we are to farm animals. When I was young, people would usually have a family dog around, maybe a few stray cats, but nobody paid much attention to them. My wife, Bobbie, and I have some little Lhassa apsos—it gets down to forty degrees and you've got to get those little guys inside or they'll freeze. But those old farm dogs, it would get down below zero and they'd just dig a hole in a hay bale—or in the snow—and curl up, spend the night. Animals are pretty tough if they're on their own.

We also had some sheep. I remember grinding an old mechanical shaft that runs the shearing scissors. You'd usually cut chunks out of the hide here and there, grab a slab of pine tar, slap it on the wound, and off they'd go.

We sold a few eggs, anything that would get a penny here or there. Today my brother doesn't have any chickens . . . no pigs . . . no sheep, nothing. He's got dairy cows and that's it.

The problem is, he's got about half a million dollars invested in machinery. You need about five to six hundred acres to make any financial sense out of it, and he's only got a couple hundred. So when he was more active (and I guess his son-in-law still is) he did a lot of job-shopping for other people. He'd do his own combining, then he'd go combine for some other people.

You can't find farm labor; you've got to have the machinery. And you've got to have a big base of land to make the machinery pay. Howard's got three tractors, for example, which is just a hell of a capital investment.

Lot of people are just selling off the land or moving out of the farm. There are bigger and bigger chunks with less people. The old family farm is a thing of the past. It happened pretty fast, too.

I always thought the Russians could reset to where we were about 1939. They've got the people and that style of farming is manpower intensive.

I had to pump water and carry wood before I even started going to school. I started milking cows, first thing morning and night, when I was probably six years old. With our cows in the wintertime, you had to pitch hay and feed in front of them, and shovel manure out from behind them, every day. Summertime they ran in the pasture, so it was a little easier.

The crops we raised were mostly to feed the cattle. Corn and hay and oats. Of course, we didn't have a tractor, we had four horses. The oats were raised for the horses.

Harvesting was just plain manual labor. Mow the hay down and let it dry, windrow it. Then it was just hand work. Get a pitchfork and stack it . . . come out with the wagons and pitch it onto the wagons, haul it into the barn. Then you'd lift it up to the hay mow and pitch it around up there. It was hot, dirty work, but it was good for you. Today they just chop it and blow it into a silo or bale it. One guy can do a hell of a lot of hay in a day.

We had threshing parties where all the neighbors would get together and travel from one farm to the other. You had one guy who was running the threshing machine, which he'd rented out, and he'd come to our farm for the day. All the surrounding neighbors would come in with their wagons and haul in the bundles of oats and corn. Some people would do that while others would carry the grain to the granary. When you finished that guy's crops, you headed down the road to the next guy's.

The family that was getting the work was always obligated to feed the crew. Ladies would chip in with cooking, too. Some of the biggest meals I've ever had were threshing meals.

That happened three times a year—one was the grain threshing, another was when they were filling the silos with green corn. And the last time would be late in the fall, when the corn had matured and you were shredding the corn off the stalks. But it was basically the same format for all of them.

These days each guy does his own haying with a combine and a truck and a trailer. It's all automated and not very manpower-intensive, which is good, because there isn't a lot of manpower around. In the 1930s and 1940s the standard working wage for a hired guy on a farm was a dollar a day, a ten- or twelve-hour day, whatever hours were worked. When I got in high school, I used to do some day work for people, and that's the pay I got, too.

My dad tried to be diversified. We had chickens and sheep, but our income was from whole milk. My uncle over the hill used to sell cream. He had one more step to go through besides milking the cows: he had to run the milk through this old cream separator, which is a hand-cranked thing. He sold nothing but the cream, and that went to a butter factory in the area. Our milk went to the Pet Milk factory in town, where it was converted to condensed as opposed to fresh milk.

We did have one cash crop, and that was tobacco, which was really labor-intensive. You had to grow tiny plants under cloth in beds, then pull those out and hand plant them. Tobacco planters are still in use today because nobody's ever figured out a better way to do the job. You had a big water barrel mounted on a pair of wheels and two little seats back behind it, right next to the ground. A guy sits on each seat, and then there's a little blade right in between them that digs a furrow. You've got a clicker to time it, and you just sit there with your lap full of fresh tobacco plants and stick them in there one at a time. First it's my turn and then it's your turn. You go click-click-clicking across the field. The water tank dumps a shot of water in with every plant as you stick it in there. It's just a mechanical timing device. Having horses trained to walk at the right speed was a challenge in itself. You had to have exactly the right team of horses. With a tractor today it's easy; you set the throttle any place you want it.

That was the first major job. Then, when the stuff started to grow, you had to hoe it continually. You always got big green worms like tomato worms that grow on tobacco. You'd have to go through the plants constantly, drag these big worms out, and slam them on the ground to kill them. We didn't put insecticide on anything because we didn't have any—nothing that was safe.

Finally, when the tobacco was mature, you'd chop it down, string it up on laths, and hang it up in a shed to dry. This was all just a hell of a lot of work.

You used to do that in September, then leave it there until Christmas time, when you usually got a warm spell where it was above freezing. The tobacco was now dry, but there was enough moisture in the air that if you touched the leaves they wouldn't break up. This was casing season. You'd take down the plants and spend the next month in the barn stripping the leaves off and putting it in bales. Finally you'd sell it in the spring.

It was really hard work. One of the things I tell my friends who still have the bad habit of smoking . . . if you ever saw people fixing tobacco, you'd give it up. We'd take all the really nice leaves with no wormholes and put them in one bale. They went up to the cigar manufacturers to make cigar wrappers. Tobacco in Wisconsin was mostly for wrappers.

But any trash that had wormholes in it went into another bale, and that went to cigarette companies. They just chopped it up and converted it to cigarettes.

You'd think with all this I'd never have taken up smoking, but at that time smoking was the adult thing to do. My dad smoked and my uncle smoked—everybody I knew smoked. In third or fourth grade we'd take shavings out of the hand pencil sharpener and try to roll a cigarette. That didn't taste too good.

Howard and I tried snuff once. This was in the middle of winter—we were up in a pile of straw, and we got so damn dizzy we couldn't stand up.

It's amazing to me that people still smoke today, with all the current knowledge. In our day nobody knew it was that bad for you. We knew it was a smelly habit. I never wanted to get too close to my dad when I was young because he always smelled like smoke.

My dad lived till the age of eighty-three and he was still smoking. But he died of lung cancer. If he hadn't been out in the fresh air on a farm his whole life, he'd have probably died about twenty years earlier. I think in an outdoor career you can probably tolerate it better than you can sitting in an office smoking.

The thing about farming then—and it's not a lot different now—is that when you talk about a farmer, you're defining an all-purpose person. A farmer has to be a small businessman, an entrepreneur, a bookkeeper. He has to know how to deal with animals; he's sort of a half-assed veterinarian. He's a horticulturalist. These days he's even got to be an electrician. He's got to be a carpenter, a mechanic, it's all these things rolled into one. It's a very demanding occupation.

A farm is a good place to be raised. I support the idea of compulsory farm rearing; it would be a great character builder for everyone. But it's hard to create that environment. . . . My first wife used to say, you can't simulate poverty.

One thing happened to me on the farm that affected me for a long time: I cut off the ring finger on my left hand.

I did it when I was five, before I started school. I was following my dad around on his horse-drawn hay mower. We had an old mower with a sickle bar, two horses on it. I'd go out there and follow him around, like little kids do.

The bar would get clogged up with hay and you'd have to back the horses up a step and get off and clean the sickle bar off. Then go on and mow again.

So I was going to be helpful, and I reached down there to clean the bar off and about that time the horses took this one step . . . it went click and zipped that finger right off.

I was luckier than hell because I could have lost all of them. It didn't hurt, but of course, my dad was pretty upset about it. He took me uptown, where they trimmed it off. First time I ever had anesthesia . . . one of the few times I ever did.

It never particularly bothered me physically. When I took up boxing, I found I could hit just as hard with my left as I could with my right. But I also think it made me self-conscious. I was probably in my twenties before I got over the idea that the missing finger was the first thing anybody noticed about me.

I had to walk to grade school in Leon except in wintertime, when we used to ski across the fields to school. In the spring and fall we'd follow the road around, which was a mile and a half walk. Rain, shine, or whatever, you went to school. They didn't close it because it was raining or snowing.

The Leon School was an old two-room schoolhouse, grades one through four in one room and five through eight in the other. One teacher for each room, six to twelve kids per class. A lot of the kids showed up for school having heard nothing but Norwegian all their lives; they couldn't speak any English. (In my family, where my mother and her relatives spoke Norwegian, it was the other way around. They'd use Norwegian whenever they didn't want us to know what the hell was going on.)

So we all lost an opportunity to learn another language at an age when it would have been very easy to do. I've always regretted that. (It took me months, when I was pushing fifty, to learn enough Russian to get by.)

My first four years were with Miss Ovelia Melby, who made me her teacher's pet. I didn't mind too much, though it was kind of awkward, since she was going out with a cousin of mine. The second four years were with Miss Martha Schamens and with a young guy named Ernie Betts. Betts later became the assistant secretary of agriculture.

The only mischief I got into was tossing a snowball through one of the classroom windows. I was ashamed of myself and confessed. I always did well with the schoolwork, but I had problems with presentations. I didn't like to get up in front of people and talk.

When I was chosen class valedictorian for eighth grade graduation, my self-consciousness kicked in: I froze onstage.

High school was five miles farther away than Leon, in Sparta itself. I'd ride the school bus sometimes. I was interested in sports, but farm kids really couldn't play football, since the season conflicted with harvests. I tried out for track, but I turned out to be too slow. I had a cousin in town who was on the boxing team, so I tried that, under Coach Eddie Blewett and his assistant, Killer Kane (later a coach at the University of Wisconsin).

I found that there were some other farm kids taking boxing, and one of them had his own car, an open Model T. So I used to ride back and forth with them during the boxing season. A lot of the time I'd simply run home; it was more productive than running around the goddamn floor of the school gym. By that time I also had a bicycle, and sometimes I rode that to school.

(Some of those older guys are still around and into Formula One airplane racing. In fact, that's how I got into it when I retired from NASA in the early 1980s. It's pretty unusual that you've got about twenty-five active racing planes and three of them are guys from Sparta, Wisconsin.)

I don't think I expected to become a world welterweight champion boxer, but I held my own. I stayed in my weight for the next forty-some years.

Howard and I belonged to the Future Farmers of America—FFA—in high school, and we started raising animals to show at the county fair. I raised Shropshire sheep and he raised Berkshire pigs.

I also went out for the band and learned the trombone. The director, Beldon LaBansky, was also a lieutenant in the Army Reserve. He was the kind of feisty guy you couldn't ignore.

This was the big band era; it also happened to be the time we got a radio and I was exposed to music for the first time. I became a big fan of the Dorseys, Glenn Miller, Benny Goodman, Les Brown, all of the big bands. Sparta had a city band in addition to the school group, and the school band would play concerts on Sunday evenings sometimes.

Here's where I had my first experience with alcohol. I tried drinking some wine, naturally drank too much, and threw up. That ended those experiments for a while.

One day in November the school band was out on the field at marching practice. It was a pretty nice day for that time of year—you didn't even need your jacket in the morning. About midday snow started to fall. So much snow fell just that afternoon that some of the country kids were trapped at school.

This became the Armistice Day blizzard—November 11, 1941—and killed hunters all over Minnesota and Wisconsin.

My mom used to tell a story of how I got all worried about wars when studying geography. I was eleven and the Italians had just invaded Ethiopia. So I went to her for reassurance that there wouldn't be any more. When I was just starting my sophomore year at Sparta High School, the Germans invaded Poland, and by God, here was another war. It was still pretty far from Sparta, however. I don't recall anybody being too worried that we would be involved.

But over the next few months I began to notice airplanes going in and out of Camp McCoy, an Army base a few miles outside Sparta. I began to think that flying around looked like a great thing to be doing—it beat working on a farm. I got so well known for this that in the yearbook my motto was, "Keep 'em flying!"

So in my junior year I transferred out of all my ag courses and loaded up with physics, chemistry, and math.

Transferring like that caused one of the few real arguments I ever had with my dad. We always got along. Working on a farm with each other every day of our lives, we had to. But I was the oldest boy in the family, you see, and I was supposed to take over the farm. So my dad was pretty disappointed when I told him I was more interested in becoming a pilot.

Fortunately, my brother Howard stayed in ag, and we all just assumed he would take over the farm.

The only mark on my school record was an F I got in chemistry my junior year. It was particularly surprising because I had gotten an A the quarter before. But the teacher, Clarence Simonson, had caught me smoking and decided to teach me a lesson. Unfortunately, it didn't take.

Then, midway through my senior year, on Sunday, December 7, 1941, the Japanese attacked Pearl Harbor. The next day we were at war.

Like all landlubbers, I was interested in the Navy. Partly this was because all my older friends who were entering the service all wanted to join the Navy, and partly because I wanted to be a Navy pilot. I knew that to be a pilot, you needed some college education, but I wasn't going to let that stop me.

It so happened that in March 1942 the Army Air Corps changed its rules. They would now accept high school graduates as candidates for flight school. So, on April 8, 1942, I signed up with the old Army Air Corps as a private unassigned.

2

FLYING

Things being what they are with the military, I graduated from high school in May 1942 all ready to go and fight the Japanese or the Germans . . . and sat on my butt for the next three months. The problem was, the war had just started and they had all these people signed up. They couldn't cope with it all.

I did get called up once that summer and got as far as Milwaukee before they rescinded the orders and I went back home again.

One thing I did during that trip, however. Here I was wanting to be a pilot, and I'd never been in an airplane. I figured I should really see if I would like it. So while I was in Milwaukee I went to Lake Michigan, where there was a guy who who had a little Luscombe with floats. He charged me five dollars and took me up, flew me around awhile, then came back. I decided flying was pretty neat.

I finally got my call in August.

I was shipped down to the aviation cadet center at San Antonio, Texas. They were oversaturated; they didn't have uniforms to issue people, so we went around in the clothes we had. We lived in tents and when we walked around, we were up to our knees in sand. Then it rained. We wound up building little wood platforms to keep out of the mud.

I had one big hurdle to get over when I got to San Antonio—the physical exam for pilot training. I came through in perfect health, except for my missing finger. For a moment it looked as though I was going to be disqualified on the spot. But they checked the regulations and discovered that the ring finger on your left hand (if you were right-handed; reverse it if you were left-handed) was the only finger you could have missing on either hand and be qualified as a pilot. Your ring finger, they decided, is the most useless finger on your hand.

Don't ask me who the hell wrote those regulations.

I did some boxing at San Antonio and found myself going up against a guy who had been a sparring partner for Joe Lewis. I didn't know that when I started, and I was lucky I didn't get hurt.

Like everybody else in the Army, for lack of anything better to do, I also took up smoking. I kept at it for the next twenty years or so.

We got issued uniforms and went through the standard preflight training. Part of that was a battery of tests. They were picking people for three different categories—pilot, bombardier, or navigator—and they decided right then and there where you were going to go for training. As usual, they'd ask you what you wanted to do . . . then have you do the opposite.

Obviously I said I wanted to be a pilot. I was approved and assigned to Class 43E. Then they moved me up a class, because they ended up with some vacancies in the earlier class and they wanted them all filled. So they backed me up to the D class. Which was a break for me.

Except that since I was still assigned to 43E, no one had expected me to go anywhere for a couple of weeks. So they loaded me up with shots. Of course, being an old farm boy, I'd never had a shot in my life. Suddenly I had yellow fever and typhoid and smallpox all working on me. There I was on the train from San Antonio up to Vernon, which is where the primary school was, and I was so goddamn sick from all these shots that I couldn't stand. I got over it, but I've never forgotten that train trip. Never had another quite as bad.

Primary training took place at Vernon, Texas, up near the Panhandle. Vernon was one of those places where you could stand in the mud and have dust blowing in your eyes. North Texas in the wintertime.

We had several types of planes to learn on: the Stearman, a biplane trainer, and several single-wings—the Fairchild PT-19 and the North American AT-6 Texans. I liked the PT-19; I didn't have to be concerned about liking flying, having made that flight at Milwaukee, and I thought it was a lot of fun. As usual, there were a few guys didn't take to it. We had some guys who were prone to get airsick. But I never had that problem. Within the first ten hours I was flying around upside down. Everybody else was still trying to figure out how to fly right side up.

My instructor was an old country guy, a civilian, like all the instructors in primary. He was a hunter and he'd take me out to the Red River, between

Oklahoma and Texas. We'd fly around dueling, do a lesson in the morning, then run down the river to see where all the ducks were. This was in the fall, so the ducks were flying. Then it was back to Vernon, where I'd drop him, go get lunch, then go fly again solo. He would go out and duck hunt. And I'd be out there above the river, flushing ducks for him.

Out on the Red River was the first time I came close to killing myself in an airplane. When you fly low enough to flush these ducks, they all come up off the water right in front of you. It shows how stupid I was: I had them through the prop and through the wing and everything else. Fortunately, I never hit one, but once I ducked my head when all these ducks were coming around the canopy . . . and hit the surface of the river with the wheels of the plane.

I could just as easily have flipped ass over teakettle. That would have been the end of the day right there.

I think my instructor got a lot of ducks that fall, though. He had a couple of other students, and I think he had everybody chasing for him.

Primary training was pretty straightforward. Just a lot of flying, which suited me fine.

After three months I was promoted from primary to basic flight training and was shipped from Vernon to Waco. One of the differences here was that now we flew off concrete runways, though we still had some auxiliary grass fields scattered around the town.

I started flying a new plane, too, the Vultee T-13. It had a fixed landing gear and looked a lot like the T-6, except the gear didn't retract. The T-13 was called the "Valiant," but we called it the "Vibrator."

The Confederate Air Force, the guys who restore and fly old warplanes, has still got a few of them. Fred Haise, one of the Apollo 13 crew, crashed in one of them in 1973 and got himself all burned. That was in a Confederate Air Force T-13 they had rigged up to look like a Japanese torpedo bomber.

It wasn't a bad airplane, but it shook like mad. It was sturdy, though. You could do snap rolls all day with that damn thing.

Being an instructor in those days was pretty challenging. You were probably as likely to be killed by one of your students as shot down by the Germans. In some cases guys would freeze up on the controls. Get scared, panicked about something . . . just lock up.

Even communicating from instructor to student was tough: all we had was an old gosport tube from the front to the back cockpit, and that wasn't much. If that didn't work, you had to chop the throttle and yell.

So it was risky for instructors. We had military instructors in Waco, and one day this guy took two of us up for formation training.

"All right," he said, "let's break up and do a little dogfighting. Just give you a little feel for that."

Well, being green as grass, I came in making a high side pass on him. I think he thought I was going to run right over him. *I* didn't think I was going to, but *he* did. I got close enough, anyway, and it really scared him pretty bad. "Slayton," he told me, "don't *ever* do that again."

At that time there were also some Royal Air Force—British—pilots going through basic. A couple of them were instructors. One day this T-13 came back with an RAF instructor, and there was a great big damn gouge through the engine cowl, just a couple of inches below the spinner on the prop. The canopy was all smashed up. Worst of all, the whole tail, from a few inches behind the leading edge, was just torn up.

Everybody stood around, scratching their heads, trying to figure out how in the hell you could do that kind of thing. And this guy wouldn't talk. He wasn't going to tell anybody what had happened.

Well, the instructors finally got it out of him. He and the other pilot had been flying down the Bosque River, trying to go under this bridge. At the last minute they realized they weren't going to make it. They pulled up to go over it and hit a bridge cable. Luckily, the cable hit the cowling while the prop was *horizontal*. Props are going pretty fast, so you can imagine the odds against that. It was the first impact with the cable that caused the gouge in the cowling.

Then, of course, the cable broke, and the end snapped around, smashing the canopy and tearing most of the tail off the plane.

There's an awful lot of just pure horseshit luck in flying. By the time a pilot gets to the test pilot level, if nothing else, he's proved he's lucky.

Coming out of Waco, I faced another decision point. I knew I was going to be a pilot if I managed to get through the next level, which was advanced training. Most of the people who were going to wash out already had by this point.

Now I had to choose whether I wanted to be a single-engine or multi-engine pilot. Well, that wasn't a choice: I wanted to be a single-engine fighter pilot. And my best friend wanted to be multi-engine pilot. The military way, again: we each got the wrong orders.

Being dumb little cadets, we went in to see the commander. I said, "Sir, he wants to go to multi-engine and I want to go to single-engine, so there must be a slot available either way. Why can't you just switch us?"

The commander kind of shook his head. "Boys, you don't understand the Army."

So my friend went off to single-engine and I never saw him again.

Multi-engine advanced training happened to be located in Waco. The transition from small, single-engine flying to multi-engine was pretty painless on the technical level. I was a little frustrated because I couldn't do aerobatics. Flying multi-engine meant you just ground along. No snap rolls. You did a lot of cross-country flying and some instruments, though we didn't get a lot of instruments in those days. We got a little time in the old Link trainers. But I don't think anybody came out of advanced flight school proficient in instrument flying. Of course, the instruments weren't that great to begin with.

Night flying was even worse. In those days they had the night lines—a series of beacons—all over the country. You could follow the night lines until visibility got so bad you couldn't; then you were really in trouble. Night flying still kills more people in airplanes than any one thing. At least in Texas you didn't have to worry about running into a mountain.

The airplanes we flew included the Beechcraft AT-10 and the Cessna AT-12. They were multi-engine tail draggers.

The most challenging airplane we had there was a Curtiss 18-I, one of the first all-metal airplanes. It would take off, land, and cruise at 120 miles an hour, which isn't very fast, but showed fairly high wing loading for an airplane in those days. That is, the wings were designed to be smaller and still give sufficient lift to the airplane. Theoretically a plane with high wing loading wasn't supposed to fly. According to legend, a bumblebee isn't supposed to fly, either, if you use aerodynamic rules.

The Martin B-26, which also had high wing loading, was just coming into use at that point, and it was killing people left and right. Pilots called it the "Flying Prostitute," because it had "no visible means of support." Because the 18-I had about the same wing loading as the B-26, it was hoped one would serve as a good trainer for the other.

I never got to find out how effective the Curtiss was. When advanced training was over, there was yet another decision point—you had to pick your favorite multi-engine airplane, the one you wanted to fly from there.

Well, my favorite multi-engine airplane was the Lockheed P-38, because that was the only multi-engine fighter. So that was my number one. The A-20, which was a pretty fast airplane in those days, was my number two. The Martin B-26 was my third, and the North American B-25 was my last. I damned well didn't want to get anything bigger than that. I didn't want to fly cargo planes and I didn't want any four-engine bombers. I picked all the small multi-engines there were.

As you might expect, I got my last choice, the B-25. On April 22, 1943, I graduated from advanced. I was officially a pilot in the Army Air Corps. I had been out of high school for almost one year.

B-25 combat training took place in Columbia, South Carolina. I decided pretty quickly that the B-25 wasn't a bad airplane. Its only drawback was that it was a bomber.

Most of these planes were built for medium-altitude bombing and were equipped with a couple of fifty-caliber machine guns. We had a few of the G model, which had a 75mm cannon up in the nose. There were also some that were designed for ground attack; they had that 75mm in them, plus some wing-mounted fifties. The H model had eight fifties in the nose.

We did a combination of medium-altitude and low-altitude combat training. In the Pacific they were doing quite a bit of low-altitude bombing, but not in Europe. They had had one group that had tried low-altitude bombing over there, and they had a hundred percent turnover in three months. The enemy had too much firepower.

I was just getting used to the airplane, learning gunnery and bombing, and especially formation flying. These days in formation flying, you get into some lousy weather, you snuggle in tighter than hell and go. In those days, when you had to go through a cloud deck, the procedure was that the lead airplane kept on the same heading, while the wingmen went off at thirty-degree angles, flying for a minimum of thirty seconds and doing what they call a timed heading change before coming back. The plan was that everyone would climb up through the clouds and join up again on top.

This worked great if all you had was a nice thin deck of stratus. If the weather didn't accommodate—if you had a thicker layer of clouds—you'd have airplanes scattered all over the place. But I can see why they did that in those days, because some people weren't too damn swift at flying formation.

One day while I was at Columbia I took off as a single and joined up on a guy's wing at a thousand feet. Another B-25 coming in on the leader's left wing came too far right and tore the tail right off the lead plane with his wing. Well, nothing flies too well without a tail. The lead plane just flipped over and went right into the ground. The crew had parachutes, but at a thousand feet there was no chance to use them.

You see a couple of those and you understand why people aren't too prone to want to sit up there and lead the formation.

It was interesting to me that the guy flying the airplane that got crunched was terrified about going into combat. He didn't want to go into combat;

he wanted to stay and instruct. The message I got out of that was death can get you anywhere, any time, so don't worry about it.

I survived three months of combat training and finished it in August. I was assigned to 340th Bombardment Group, which was then based in North Africa. Some guys got lucky enough to fly replacement airplanes over to Europe. The rest of us were simply assigned as replacement crews; the Army didn't have enough airplanes.

So we got loaded on a Liberty ship convoy out of Newport News, Virginia. It was a long, miserable trip. We had nothing to eat but C-rations, which was only slightly better than having nothing to eat at all. There was hardtack—bread—and that was it.

Liberty ships made about eight knots, rocking and rolling the whole time. There were a lot of ground troops down in the hold, where they had these little bunks about two feet wide stacked eight deep. The air crews weren't near as prone to get as seasick as the ground troops, but those ground guys were throwing up all over everybody. It was a mess, so I stayed up on deck most of the time, where I could breathe. I told myself it was a good thing I hadn't wound up in the Navy.

It took us thirty-five days to go from Newport News to Zerni, North Africa. At the time we left, of course, the bomb group was still in North Africa. By the time we got there, they'd moved to Italy.

We stopped in Zerni for three days, but they never let us off the ship. We were headed for Naples.

While we were crossing the Atlantic there were a lot of submarine alerts, so the convoy was noodling around, changing course, zigzagging around, and we had depth bombs going off regularly. We didn't lose any ships.

As we were coming into the Mediterranean, however, through Gibraltar and just off the coast of Morocco, a couple of subs attacked our convoy at dusk. The subs were followed by some German bombers, and three of our ships were sunk.

Some American P-39s joined the fight, flying from the coast of Africa. They came charging in and jumped those damn bombers. They didn't get the Germans before the Germans bombed us, but they got a couple of them later.

So before I got to Italy I had already seen my first aerial combat . . . from underneath.

3

COMBAT

Naples is a dirty place even now, but it was a lot worse in September 1943, when the Germans had just departed. We moved into a big old set of barracks that they had just ransacked and abandoned. It was shot up and torn up. So we unrolled our sleeping bags and slept on concrete floors. At least there was a floor to sleep on.

The first night we were there, we got bombed. That was my second air attack from the wrong perspective. The Germans knew exactly where those barracks were, obviously, because they'd just left them. They came in there and strafed and bombed. I never knew how many people got hit.

We spent a couple of days in that barracks, then went off to an airport where we were supposed to get shipped off to a bomb group down in the heel of Italy, to San Petrazio. Well, somehow the C-47 we were supposed to leave on wasn't available, and we ended up that night sleeping under the wing of the airplane on the airfield. Then the airfield got bombed that night.

In retrospect it was kind of stupid to stay under the wing of that airplane. The shrapnel was falling all over the place from the antiaircraft fire. If it had knocked a hole in that aircraft and set it on fire, we'd have been in real trouble.

So I got bombed three times before I ever got to a damned combat unit. I decided I didn't like it too well from the receiving end.

One thing I did from that point on: no matter where I was living in Europe, either in a tent or a house, I had a nice deep foxhole somewhere handy. The only problem was keeping other guys out of it long enough for me to get in.

* * *

I finally reached the 340th Bombardment Group, which was stationed in an olive grove outside San Petrazio. Since the group had just come out of North Africa, all they had for shelter were double-walled desert tents. The idea was to give you some insulation, which I'm sure worked great in the desert. But it rained like hell in Italy, and as soon as that top level of the tent got wet enough, it bent down, rubbing the bottom level and putting a hole in it. When you got those two levels together at night, you just had water streaming in all over the place. Those desert tents weren't much better than having nothing.

I flew my first combat missions out of the mud at San Petrazio. Before we left the States, we were assigned into crews, and we went to Europe as a full B-25 crew of six. I had just turned nineteen and I was copilot. The pilot was a first lieutenant: he was an old guy, probably twenty-one or twenty-two. The others were a bombardier-navigator and three gunners, one of whom doubled as a radio man.

We got to Italy figuring we were going to fly as a crew. Of course, the first thing that happened was they split us all up. I don't think they ever flew replacement crews intact. I never did fly a mission with my original crew. The pilot wound up flying six or seven missions, ended up with a nervous breakdown, and had to be sent home.

Our bombardier got killed. He was flying on another crew, and they came back in an aborted mission carrying a full load of bombs and crashed on landing. Blew the whole thing all to hell. It was one of those accidents that probably shouldn't have happened . . . just some pilot coming in to the landing field heavier than he should have been flying, winding up short. We never heard any great amount of detail.

So I flew my first combat mission as copilot with a flight commander—a captain—who'd been in Europe for a while. I don't think he was even twenty-five years old.

(It amazes me to remember how young everybody was. Some months later, when we stole some Italian houses and moved into them, I lived with a guy who was twenty-seven. I thought he was an old son of a bitch.)

Whoever said you've got nothing to fear but fear itself was exactly correct, because on my first combat mission I was scared spitless. I didn't know what to expect, but I was really scared.

We went in there on this bomb run . . . we're leading the flight . . . and I kept looking around and wondering, when was all this goddamn stuff supposed to start to fly? But I never saw a burst of flak. We unloaded a full load of bombs—and I don't even remember what the target was, a bridge, maybe, somewhere in Italy. It was what we call a milk run. So after that I kind of relaxed.

At that time the bomb line had just moved through Naples and up into the mountains, to the abbey at Monte Cassino, and there it stopped. Most of the time I was in San Petrezio we bombed north of that line, trying to knock out bridges or troop concentrations, or went off to the Balkans.

At one point, three days in a row they sent us over Athens at the same time, same place, same direction! We were too young and too stupid to question orders in those days. They'd give you the intelligence briefing telling you about flak concentrations and what you could expect for fighters. We all thought it was kind of stupid going to the same place the same time, three days in a row. But nobody stepped up and said so.

We didn't get shot up bad the first flight, but it got worse. The last trip over Athens we only got half the airplanes back. I felt pretty damn lucky, because after the third flight I counted over three hundred holes in my airplane. Lot of them were pinholes, like a 20mm one in the right wheel well that blew out the tires. Of course, we lost all hydraulics and had to crank the landing gear down by hand. But the airplane was still flying. A B-25 would take a lot of hits without being too vulnerable.

The Germans had certain techniques. They would hit you with flak for a while, but they would never send fighters when they were shooting at you with antiaircraft. Now, antiaircraft flak was always black from our vantage point. When you saw a red burst, it meant the flak was going to stop. That's when the fighters hit you.

We had P-38 escorts, which was good, because those guys would stay in there with you. We faced Junks—German Junkers—a few times. Boy, I hated those bastards. You'd never know they were there until you got hit. They'd be at about twenty-five thousand feet, and they'd come diving through the formation. By the time they got there they'd shot the hell out of you.

I don't believe that damned bomb line ever moved all through the winter of 1943–44. There was so much natural high ground for the Germans that they had a pretty good defensive position and we just couldn't move them. They had a lot of 88s—which were normally artillery—and were using them as antiaircraft guns. We hit one bridge where they had *four hundred* heavy guns concentrated. You get a lot of flak from four hundred guns.

Can you imagine how much weaponry the factories in Germany must have turned out? They weren't just supplying the Italian theater, but the Russian front, too. The weapons were of good quality, too. That German 88 was a very effective artillery piece, and I think the Germans even mounted them on some tanks, like the Tiger. I know for sure that it made a hell of a good antiaircraft gun.

So we learned to play a kind of calculated game. We computed that a shell was traveling a thousand feet a second from the time it fired. So if you were at ten thousand feet, you had about ten seconds from the time it was fired until the thing was going to hit you.

The theory was you needed to maneuver fairly regularly if you were going to try to outguess the enemy. It was always a guessing game whether to break left or right or go straight ahead. It took a lot of guts to go straight ahead, but most of the time that was the right thing to do, because sometimes your instincts ran you right into the middle of that shit. It could be brutal, no matter how you guessed.

Take the beachhead at Anzio, where we had something like fifty thousand German troops pinned down for weeks. We were putting everything we could into the air over that place—I eventually made thirteen flights over that damned beachhead. We were using all the air power we had, even taking heavy bombers like the B-17 and B-24 down to lower altitudes where they didn't belong.

We went in there one day behind a group of 24s and I saw *seven* of them going down all at once.

Bad as Anzio was to fly over, it was worse being on the ground. At one point there was a move to give the air guys a little experience on the ground and get some of the ground guys into the air. This probably came from Mark Clark, the Army commander over there at that time. Some of his recent experiences may have been behind this decision.

There were two major rivers between Naples and Rome, running into the sea, the Voltruno and the Rapido. The bomb line was on one of them and Mark Clark had his headquarters on the other. One day some Air Corps planes picked the wrong river and bombed the hell out of Mark Clark's headquarters. So he wasn't very happy with the 12th Air Force, his air arm, at that point.

Worse yet, he was once off the coast at Anzio in a PT boat and some Junks came in unopposed and strafed the hell out of him.

So for a while we wound up with an artillery guy flying with us to keep us honest. This guy had a great way of explaining the difference between the bomber and and the bombed: "When you're up here, all the bombs are addressed, 'To whom it may concern.' When you're down there, you're only worried about the one with your name on it."

We were in San Petrazio for six weeks when command moved us up to Foggia. There was quite a complex up there—fifteen different airfields. We moved onto an old B-24 field.

The runway was made of PSP—pierced steel planking—but everything else was mud. Here I was one day, parked on the grass with a full load of bombs on, mired in this goddamn mud with the power almost full on.

All of a sudden a Jeep comes running down right in front of me. At that moment the airplane breaks loose out of the mud. Suddenly here's a Jeep wheel flying up through the air. I'd gotten the vehicle with the left prop. I don't remember if I flipped it upside down or not, but I'll never forget that goddamn spare tire flying up through the air.

Anyway, Foggia was a muddy mess. Then they cut out an airfield for us on the south slope of Mount Vesuvius, out of the grape vineyards. The ground there was all lava and made a great runway. They had just taken a bulldozer to it, knocking the grapes off. Even when it was wet, that stuff gave you a good, hard surface. Our runway was so wide, in fact, we could take off eight abreast.

Here's where I had one of my closer calls. We were doing one of those eight-abreast formation takeoffs. I was on the downwind side, and I got caught in the prop wash of the airplanes upwind just as we broke from the ground. We were maybe fifteen or twenty feet in the air, and all of a sudden I'm sitting there with the wheel and rudder hard over. Damn near upside down, but not quite. Fortunately, we came back over before we hit, but I remember looking back at the gunner, who was sitting between the pilot and me, and he was just white as a sheet.

I didn't have time to get scared. It was just too quick.

Then command hauled us up to a C-47 field on the north side of Vesuvius to Pompeii, to another dirt field. There was an A-20 group about a mile and a half down the road from us that had flown their planes up to this field the afternoon before. And they had them lined up wingtip to wingtip down the side of the field, all loaded with bombs.

A flight of C-47s began takeoff . . . and the same thing happened to the guy on the outside that had happened to me. He got lost in the dust went right through that row of A-20s. He took the nose off one, smacked through the middle of the second one, knocked the tail off a third. . . . He ended up with nothing left but the cockpit. And he survived. I went to see him in the hospital a couple of days later, which convinced me that the old 47 was a pretty solid airplane.

We stayed in Pompeii for a while, flying every day, with Mount Vesuvius rumbling and rolling, throwing up big clouds of smoke. It wasn't bothering us, particularly. We'd take off, fly around it up north to do our missions.

The area was Italian farm country, and some of us had moved in with Italian families in old stone farmhouses. Vesuvius would roar every now and then, but I didn't pay much attention to it.

Late one night we woke up because there was noise on the roof of the house. There was a big roar and about three minutes later big rocks start falling down out of the clouds! It got to be daylight, and every once in a while there'd be a big bunch of rocks thrown like that. . . . We all put on our helmets and decided to make a dash for the mess tent, which was a dumb thing to do. The tent wasn't going to protect us, either.

The ash and rocks were coming from the north to the south. There was a cloud cover at the time we were getting hit with all this stuff, which was probably good. It was wet, and that prevented a big cloud of dust and dirt, which would have made some real problems.

It didn't take long to decide that the airplanes were all knocked to hell with rocks. We had forty-eight airplanes on that airfield. As far as I know, they're still sitting there, because we just abandoned them. They loaded us into trucks and just hauled ass out of there. We went around the mountain to Naples.

That was Vesuvius. With everything else going on in the world, I don't know if it even got publicized.

I do have a beautiful painting by a guy in California, which shows a group of B-25s flying over Naples with Vesuvius erupting in the background.

Other Voices

MARIE SLAYTON MADSEN

I don't want to cry poor or anything, because we always had clothes and food, but this was still the end of the Depression, and nobody had much.

Don was always very conscientous, just a great big brother. When he went off to the service he was very faithful about writing home—and sending money. One day I remember all of the kids at home got twenty dollars each from him! It doesn't sound like much now, but back in 1943 that was really something.

I flew fifty-six combat missions in Europe between October 1943 and May 1944. That wasn't the only flying I did, of course, but that was the bulk of it. It averaged out to two or three flights a week. I had my first dozen missions by Christmas 1943, which qualified me for a visit to a rest camp on the island of Capri.

Take the week of February 9–16, 1944. On the ninth I flew over Rome, hitting an enemy troop concentration. The next day, the tenth, Anzio: clouds over target. The thirteenth, back to Anzio: lots of flak and two planes down. Later that day we did a practice bombing run. On Valentine's Day we hit Perugia and saw some flak; there was another training flight later that same day.

On the fifteenth we were over Monte Cassino. My notes say we "blew off hill." On the sixteenth we were headed to Orte, but had to turn back because we had no fighter escort. So later that same day we hit Anzio again; one plane down and several emergencies—that is, other planes heavily damaged.

That was a typical week, several missions in which nothing happened combined with a couple that were as busy as you could possibly want. They ranged in length from one and a half to two and a half hours, no more, but sometimes they seemed longer.

The Cassino flight on the fifteenth was the only mission I ever ran on the abbey itself. They had stood people down for three days around the system to get ready to hit that, because we were trying to annihilate the damn place. I don't know how many bomber groups were sent in there. B-24s went in ahead of us. They were trying to do away with the place, and I guess they did pretty well.

All this time we were confident we were going to win the war. But the Germans were pretty resilient. You'd go in there and knock the hell out of a bridge, and you'd think, boy, they won't get anything across there for a month . . . and two days later they're up and running again. I have a lot of respect for what the U.S. and U.N. forces did in Iraq in such a short time. Of course, they had smart weapons. We were just dropping iron bombs in patterns, and it's harder than you'd think to knock out bridges and roads and marshaling yards that way.

After the eruption of Vesuvius we were a bomber group without any bombers, so we moved down to Salerno, south of Naples, where we were given a whole set of brand-new B-25Js.

Our older Mitchells, the Gs, had been modified in the field. About the time I got to Europe somebody had discovered that they were getting the hell shot out of them from the rear. So they cut the tail off and stuck a single fifty-caliber machine gun in there. That worked pretty well.

Then when the new planes came out, they had a tail gunner's position with two fifties and an enclosure. They also had improved firepower elsewhere; open waist guns now were enclosed. So it was a nicer airplane for combat.

The new J models were shiny aluminum. Command wasted no time getting them painted in camouflage.

Next we went from Salerno to Corsica. By the time we were set up on

Corsica, the bomb line had moved toward Rome; on the east coast it was moving even faster. Command decided it was time to get in some low-level bombing, and they asked for volunteers to begin training for missions like that.

By this time I was one of the older guys—I was twenty years old now—and I'd been promoted from copilot to pilot. It was actually kind of foolish, I guess, but I volunteered to start doing low-level, because I'd done some of that before and it was kind of fun. But I ended up running out of missions about that time.

Every air force group had its own way of defining a combat tour. The 8th Air Force in England was running twenty-five-mission tours, and the 12th—the one I was in—said fifty missions was it for medium bombers. It was probably shorter for heavy bombers. I just happened to get six beyond the limit.

One reason I left Europe is the same old story. The whole time I was in Italy I was still trying to get into fighters, with no luck at all. Finally they said, "If you want to get into fighters, the thing to do is go back to the States and go through rehabilitation. Combat veterans can do anything they want to do. All you've got to do is tell them what you want and it's yours."

I figured I had it made.

4

THE PACIFIC

In May 1944 I sailed back across the Atlantic in another transport, but this trip was a lot less uncomfortable, since I wasn't crammed into a Liberty ship with seasick GIs.

The East Coast rehabilitation center for troops returning from Europe was Miami Beach, and that's where I wound up for three weeks of R&R as the Allies invaded France on D-Day. I was already itching to go back to Europe, or even to the Pacific, as a fighter pilot. I was a combat veteran and combat veterans could do anything they wanted, right?

They just laughed at me. "Slayton," they said, "you've got two choices. You can either become an instructor for B-25 advanced training in Texas, or you can be an instructor for B-25 combat training in South Carolina. What's it going to be?"

The best choice was the combat training assignment. So I was ticketed for my second tour at Columbia.

I also got a three-week leave to go home to Wisconsin. It was only my second visit since I had been called up two years earlier; the first visit was when I finished flight school. I had changed. I was a smoker and a drinker now; I was a pilot; I'd been in combat; I'd been all over the United States and even parts of Europe. Everybody seemed a little behind the times.

By then I had heard about a new twin-engine plane that was supposed to be faster than anything else around. It was the Douglas A-26, and they were putting a squadron together up at Selfridge Field, Michigan.

I arrived at the Columbia Army Air Base at six P.M. one evening in July 1944. By eight the next morning I had a request on the CO's desk asking for transfer to the A-26 outfit. It took six weeks, but it got me out of there.

I spent the rest of 1944 at Selfridge Field, on Lake St. Clair, about thirty miles outside of Detroit. While I was there, they moved one of the Martin B-26 outfits from Italy back to the States, to Columbia, where I'd just come from, with the idea of converting them to the A-26. So then they took a bunch of us, me included, and sent us down there to transition them. We knew before we left that we were getting them ready to go to the Pacific.

The A-26 wasn't a P-38, but it was close to what I'd wanted to begin with. You couldn't do aerobatics with it, but you could outrun anything at that time, low-level. And it had more firepower than anybody else, too: fourteen forward-firing fifties in addition to fourteen five-inch rockets and a bomb bay full of bombs. Crew of two, pilot and gunner.

We had two remotely controlled turrets, one on the top and one on the bottom, both of them operated by the gunner. We'd lock the top one forward and use it for strafing. If anybody was still shooting at you when you went by a target, the gunner would hit them them with the bottom turret. It was pretty effective.

On May 13, 1945, we started our move to the Pacific, nine planes taking off from Hunter Airfield in Savannah, Georgia, headed for Sacramento, California. It was supposed to be an eight-hour flight.

Weather prediction in those days wasn't much better than guesswork. We hit a front early on and the group got scattered all over the place. (A couple of the planes didn't even leave the East Coast.) It was getting dark and I knew I wasn't going to make it to Sacramento, so I started looking around for a place to set down.

I wound up at Muroc Army Air Field in the desert north of Los Angeles. It was a pretty desolate-looking place in the middle of nowhere, and I was happy to get out of there the next morning. I didn't realize, of course, that eventually Muroc would become Edwards Air Force Base, and that I would wind up spending six years living there.

Sacramento to Hawaii . . . Hawaii to Christmas Island . . . Christmas to Canton, China . . . Canton to Tarawa . . . Tarawa to Eniwetok . . . Eniwetok to Guam . . . Guam to points west . . . points west to Okinawa. It was a roundabout trip; we got there on July 13.

I flew seven combat missions over Japan with the 319th Bombardment Group, commanded by General Randy Holzapple. The group flew twenty-five missions total.

This was completely different from fighting Germans in Italy. There wasn't really a lot of opposition. For example, we went over Nagasaki in

early August, eight abreast, firing bombs, rockets, and guns. W
as we went across the target because some clouds moved i
around, came back again, and *then* bombed the son of a bi
lost an airplane in spite of the fact that the enemy had a lo
In Germany you'd have lost half a formation.

With the Japanese, you'd see all kinds of flak ahead as you were coming in. But as soon as you got into your bomb run, it all quit. We figured they were all diving for their holes. Because God almighty, we should have been dead ducks.

I saw one Japanese fighter the whole time. Bedcheck Charlie would come around once in a while at night. Woke up one morning and there was a big goddamn shrapnel hole in the tent, but I hadn't even heard it.

The group left the States with ninety-six airplanes, lost one at Eniwetok on the way over, and still had ninety-five of them left at the end of the war.

On August 5, 1945, we hit Taramigu, a target on the island of Kyushu.

Everything on Okinawa was aimed at the coming invasion of Japan. We were strafing the hell out of the coastal towns, trying to get things cleaned out so we could invade. Command was bringing people up out of the Philippines to Okinawa.

At that time there were a lot of rumors going around. We came back from one mission to find everybody saying that the war was over, that Truman had agreed to some kind of truce. Everybody had a couple of bottles of booze saved for some event—like the end of your tour—and when we got that goddamn message, all those bottles came out. Everybody dumped them into one big pot. Naturally, we all got drunker than skunks.

Three o'clock in the morning, they ring everybody out of bed: we're going on another mission. The war is *not* over.

That was probably the toughest mission I flew the whole war. We were all drunk on our ass and just feeling horrible.

Then on August 6, 1945, we heard that some big bomb had been dropped on Hiroshima, but we didn't think that much about it. Nobody was telling us the war was over. Two days later I was on another raid, to Tsuiki Airfield on Kyushu. The day after that, the ninth, we heard about the second bomb. I never thought the war was over until then. (And I still flew another mission, on the twelfth. My last combat mission, as it turned out.)

I'm not one of those who ever criticized Harry Truman for dropping the A-bomb, because the invasion of Japan would have been a bloody goddamn mess. We'd have lost half a million people if that had happened. And you'd have had at least a million Japanese casualties.

* * *

The war was over, so everybody wanted to get home. Naturally, everybody couldn't go home at once, so the Army developed a point system—you got so many points for missions, time overseas, that kind of thing. Those with the most were sent home first.

I ended up spending almost two months in Okinawa before I got rotated. Shortly after the war a damn typhoon hit Okinawa.

We had a coral runaway; the tent I was in with four guys was down at the edge of it, below an embankment.

They clocked the winds at a hundred and thirty-five miles an hour before the wind meter finally blew away. We'd look up and see the airplanes, which were all chocked and tied down, sitting there with their props spinning. It was really blowing.

Fortunately, tents are pretty resilient. It's like they're made of bamboo . . . they bend but they don't break. So we weren't in any real danger. But some of our guys had gotten ambitious and had scrounged wood and corrugated sheet metal to try to fancy them up, putting on tin roofs, and that stuff was flying all over.

(Our squadron leader, Fred Morris, had appropriated some of this lumber—which we had wanted for an officer's club—for his own cabin. I think he regretted it.)

We came out of it okay, but the Navy really caught hell. They had a whole bunch of ships sunk. In fact, the boat I later returned to the United States on, a small Navy flattop, had gotten caught in the typhoon and its deck had been lifted and broken loose. Luckily, the ship didn't have any airplanes on it; the Navy was using it strictly for troop transport.

We didn't have a lot to do. No missions to fly. Even though the war was technically over, you still had to be careful: the Japanese were still setting up some mines and trip wires. They were very fanatic.

The Marines had originally taken the north end of Okinawa while the Army took the south end. The Marines fought their way up and they fought their way back, and they said we just took it. But most of the Japanese were still there.

The Army was more methodical than the Marines. When they took something, they held it. But there were a hell of a lot of Japanese there at the end of the war who didn't want to give up yet. Maybe they didn't believe that the war was really over. They were still holed up in caves. For entertainment we'd go out with infantry patrols and scrounge around in those caves, trying to flush out a few Japanese. Never did find any live ones, though we found a lot of dead ones. We fired into a few caves, threw a few grenades in there. It was mostly trying to find something to keep yourself occupied.

When World War II started, the Air Corps consisted of a few thousand regular officers and reserves. Almost all of the hundreds of thousands of people brought in to the service after Pearl Harbor, like me, were what they called "AUS"—members of the Army of the United States. So I wasn't a regular officer; I wasn't even a reserve. This meant that I was in the Army for the duration of the war, no more. I had no military career.

But I wanted one. I wanted to keep flying planes. So I applied for a regular commission.

While I was waiting, my turn at rotation came up, and I boarded the carrier USS *Pookie* headed for Seattle. I wasn't traveling alone—I had a pet.

Ned Parsekian, my navigator on Okinawa, had adopted a little white dog. Since he hadn't been in Europe and he didn't have a lot of points built up, he was going to be get stuck on Okinawa for months or get shipped to Japan for occupation duty. So he wanted me to take that little dog back to the States for him.

Of course, there weren't any pets allowed in those days, so I stuffed that little dog in my duffel bag to smuggle him on board. We had landing ships hauling us from the shore out to the carrier, where we had to climb up rope ladders while all our luggage got piled in a big rack. I threw my duffel bag, including this poor little dog, in there.

I had been up on deck for about an hour, waiting to recover my luggage, when all of a sudden I heard over the loudspeaker: "Will the owner of a little white dog please report to the bridge?"

I went. Fortunately, I was an officer instead of an enlisted man, so I was able to talk them into letting me keep the little guy.

So I had to take him out and walk him on the deck. Every time he'd shit, I'd have to clean it up. It was pretty cramped quarters and a lot of people weren't too happy having that dog on board. I got him back to the States and took him up to the farm, left him with Howard. Ned never did come back for him, and the poor dog got hit by a milk truck.

I spent Christmas 1945 in Seattle, waiting for a troop train to Camp McCoy, Wisconsin. We loaded up on Christmas Day, then rode across the Dakotas down into Camp McCoy. The whole way the temperature was twenty below zero; after my awful train trip from San Antonio to Vernon, this was my least favorite.

Camp McCoy was my separation center. You had to decide whether you wanted to get out or whether you wanted to stay in and join the regular Army. I had already applied for a regular commission, and no decision had come back. I wound up reenlisting for one year.

I did take a look at the potential for flying for the airlines, but all the cargo pilots were becoming airline pilots. It looked like kind of dull flying, anyway. I didn't have any enthusiasm for that.

The Army was also recruiting A-26 pilots to go into China, where a civil war was still going on at that time. They were organizing a Flying Tiger-like outfit with A-26–qualified guys. I was interested in that, but so were a lot of other people.

I got sent from McCoy down to Albany, Georgia, as a B-25 instructor, and maybe spent a month there. Then on down to Boca Raton, Florida, to an Army air base there. (I tried to find that base a few years ago. The whole area is nothing but a big shopping center. No evidence it was ever an airfield.)

Down at Boca Raton I had two assignments. First off, I was an instructor pilot, though I didn't do a hell of a lot of flying. For my secondary duty, however, they made me an assistant information and education officer.

My job was to put together educational material for the troops and give them lectures, keep them educated. It turned out that among the many places we were getting literature from was the Russian Embassy. Glossy documents, pamphlets. I was a dumb first lieutenant at that time and I presumed this was coming from a legitimate source—they were still our allies at the time. I merrily spread it all over the base.

I got called in by CIC—military intelligence—and they wondered what I was doing ordering all this Russian propaganda and delivering it to the troops. I was really in trouble there for a while. We finally established that the guy who had the job before me had ordered all this stuff and gotten on the subscription list. I managed to convince them I was an innocent bystander. Just a little stupid.

That was my major success as an information officer . . . spreading Russian propaganda all over Boca Raton.

It was fun being a grown-up back in the States. Fred Shayne, a friend of mine from Okinawa, had gotten transferred down to Boca Raton with me. I drove down from Albany to Boca Raton with him, and we dogged around—bachelors having a ball.

Fred had an old Ford Model V, the first Model A they made with a V-8 engine, back in 1932. So that was our transportation. But I had never owned a car in my life. Whenever I'd gone back home on leave, my dad still didn't think I was competent to drive a damn automobile. The Army trusted me with an airplane, but he didn't trust me with the family car. Until then I'd never had enough money to own a car, anyhow, and running around like we were, there wasn't any point in owning one in the first place.

Well, Fred happened to go by a used car lot one day and saw this big Packard Super 8. He came running home and said, "Don, here's a car you gotta have. Look how cheap it is! Only eight hundred dollars. You can take that anywhere in the country and double your money on it." It had a blue book value of $1,200.

So I bought the damn thing and never had anything but trouble with it. The radiator was all screwed up and overheated constantly. I finally drove it up to Wisconsin on leave and blew out all the tires. I was shopping for tires on the go. I left it up there with Beldon LaBansky, who was also running a gas station. He took pity on me, took it over on consignment for me, and sold it. I think I was lucky to get $500 for it.

That was my first experience with high finance and automobiles.

Before I got rid of it, though, I'd had a couple of adventures. I'd drive this big four-door Packard up to the base gate, with Fred sitting in the backseat. The guards would think I was hauling a general and they'd snap to and salute.

Fred and I would drive down to Miami and spend the night down there drinking and chasing girls, then drive back up.

There weren't any superhighways then. They just had that one beach road with all those circular turnabouts in the middle of these towns. We were driving back just about sunrise one day, Fred driving. I'm sitting there asleep . . . woke up and said, wait a minute. When I went to sleep the sun was on my right arm. Now it's on my left arm. We were just buzzing along the road . . . he'd done one of those turnarounds and gone all the way around it, heading back toward Miami!

We'd have probably ended up there if I hadn't woken up. Anyway, we had a lot of those kind of misadventures.

When I signed up, it was for one year, at which point I had to make another decision. I was still trying to get a regular commission, but I had discovered that the only guys getting it had college degrees or two years of college. So I decided to bite the bullet. On October 21, 1946, I made my last Army flight. I had logged almost 1,100 hours of flying time as a pilot and 331 as a copilot, including sixty-three combat missions. I was twenty-two years old and I was going back to school.

5

COLLEGE

I got out of the Army in November 1946 and enrolled at the University of Minnesota. I picked Minnesota because I wanted to take aeronautical engineering, and Wisconsin didn't have it in those days. In fact, I think they still don't.

I went up there on the GI Bill, which was one of the better things the country did—I never would have made it otherwise—starting the second quarter of 1947. They had looked at my background and given me half a year of credits based on my military experience. Since I was going to attend full-time and get through as quickly as possible, I took a pretty heavy load.

The GI Bill covered tuition at school, but not much else. I still had to support myself, so I tried all kinds of part-time jobs. I was a total failure at all of them. The first thing I tried was selling encyclopedias—Collier's had a deal for students to sell the things door-to-door. You had to wear a tie and a hat and look like a professional businessman. There I'd be, walking the streets of Minneapolis and St. Paul at nights in the middle of a Minnesota winter, hauling a satchel full of books. I don't know that I sold any. A few people took pity on me; they'd bring me in for a cup of coffee. I probably wasn't even smart enough to cash in on the fact that I was a veteran.

Then I went to work in a ladies shoe store. And I didn't do much better at that. Finally I got a job at the Montgomery Ward warehouse, unloading freight cars at night. That was something I knew how to do, since it was just plain old manual labor.

Everything in the world came into that store: linoleum, refrigerators, and soft goods and hard goods. I learned the hard way that you could put an amazing amount of stuff in a boxcar. It really is a very efficient way to transport stuff. It also takes a long time to unload.

That was all right with me. I was getting paid by the hour, and also getting a discount at Montgomery Ward.

When I first moved up to Minneapolis to start school, I couldn't find housing. That was pretty typical of most college towns in the late 1940s. All these veterans were back from the war and going to school on the GI Bill.

They had a whole bunch of Army cots set up underneath Memorial Stadium and that's where I lived for the spring of 1947. There was a rooming house right next to it, and I finally got a room in there. It was only a block from the stadium, at 629 Washington. The whole area has been torn down and rebuilt, including the stadium.

Since I had boxed in high school, I decided to take it up again in college. I figured it was one way to stay in shape. We did our training under the bleachers at Memorial Stadium.

Aside from making sure I had food and shelter, I found academic life pretty easy. The first couple of quarters were the toughest, especially English classes, where I occasionally had to make presentations. My old self-consciousness came back and I had a tough time.

Being on the GI Bill at school you had to carry a certain number of credits. I was carrying overload up until the next to the last quarter. Suddenly I discovered I wasn't going to have have enough courses to take in the summer to stay qualified for the money. (I was trying to get my courses and my credits to match and it wasn't coming out.)

It turned out I was taking a course in engineering from a strange guy who would come in once a week to give you a quiz. He gave you one question and you had five minutes to answer. You could use any book you had, any notes, it didn't matter. The whole thing was wide open. When you got through, you turned the paper in and he'd give you a grade.

The tricky part was this: you started out with a grade of fifty percent no matter what you did. Whether or not you turned in a paper each week was up to you. If you decided your answer was wrong and didn't turn in the paper, you got the fifty. If you turned it in and were wrong, you got a zero. If you were right, you got a hundred. You had to have seventy-five to pass the course. It was a pretty interesting philosophy.

That was the course I decided to retake in the summer. I showed up enough so that I wouldn't get dropped, and never turned in a paper. The teacher thought I was crazy until I explained my situation to him; then he understood. (I had the same teacher that summer.)

While I was at Minnesota, I had got to know one of the Picard brothers. August and Jean were two pioneers in high-altitude research using balloons.

August was a physicist and Jean was an aeronautical engineer. (Jean's wife was Geodet, who as it turned out was a friend of Bob Gilruth, another Minnesota graduate who later became my boss at NASA. Bob had Geodet on as a consultant for a few years down at the Manned Spacecraft Center.)

But old Jean was the classic absent-minded professor. He was bearded when that wasn't too common, and he'd walk around on a summer day with his overshoes on, carrying his umbrella. He was just wrapped up in thought.

But he was the smartest guy I ever saw. We had a seminar with him, and there wasn't any subject that he didn't understand intimately. Of course, I may not have been smart enough to make a good judgment, but he sure impressed the hell out of me.

Minnesota was a pretty good school, but I was anxious to move on. There were a lot of people around who wanted to become professional students, but I wasn't one of them. My motivation was to get a degree and get out of there.

Most of the guys I knew were mature students—veterans—who didn't have time to get wrapped up with fraternities and that kind of bullshit.

While I was at Minnesota, I flew with the Air Force Reserve. (The Army Air Corps had become its own service in October 1947.) They flew T-6 trainers out of Wold Chamberlain Field, the main airport in Minneapolis. There was no money to speak of, but it was still flying.

I had this friend from my days as an air cadet, Truman Anderson, who was originally from Minneapolis. We had gone down to San Antonio together in 1942. I wasn't up in the Twin Cities for long when I discovered that Truman was a captain in the Air National Guard unit there. (As a reserve you technically belong to the Air Force and are subject to call up by the president and secretary of defense. National Guard and Air National Guard units report to the governor of a state.)

This Minnesota ANG unit was flying P-51 Mustangs, the best fighter planes in the world. They also had a utility flight squadron with a couple of A-26s. So I got in touch with Truman. "Sure," he said, "I can get you in the Guard. You're an A-26–qualified guy and we're flying A-26s. The only problem is, you're a captain in the reserves, and the only slot I've got is for a second lieutenant. You'd have to go back to second lieutenant."

I said, "I don't give a damn about rank as long as I'm flying."

In addition to having some decent planes to fly, the ANG had another drawing point: they were paying $40 a month, which was big money to a guy living on the GI Bill. So I dumped out of the reserves and went into the Guard. It was starting all over again at a rank I'd had five years in the past, but that was fine with me: I'd spent the whole damn war trying to get into fighters and never made it. Here was my chance.

The Guard was based at Holman Field in St. Paul, down there in the Mississippi River bottom, surrounded on three sides by bluffs. You could drive along the river and look down on the field. I suppose it was okay for landing helicopters, but the only safe way to take off an airplane was to the southeast, over the river itself. Going the other way, you never got above the top of that first big bank building in downtown St. Paul. We probably scared a lot of people in those days. That field is still there, only now it's for general aviation.

Getting into the Minnesota ANG was a big break for me. I got checked out in 51s and did a bunch of 51 flying. I also flew the 26s.

One drawback was that the utility squadron also had a C-47 transport, a Gooney Bird, a state airplane they used to haul the governor or other dignitaries around. Since I was in utility flight, they insisted I check out in the Gooney Bird. It was good if you wanted to meet dignitaries—and I met Hubert Humphrey—but I didn't really like to fly the airplane. You had to fly once every ninety days to stay current, so I developed a Gooney Bird allergy. I'd miss a turn deliberately, so I wouldn't be current.

Of course, this never worked. A week after I was noncurrent, the governor would want to go someplace. Since I was a student and everybody else had jobs, I'd get the call to get my ass over to Holman. Only now I'd have to get requalified first. I went through this constantly.

The primary mission of the 51s was to be prepared to intercept Soviet bombers coming into U.S. airspace. One time the Air Force sent a couple of B-36s toward Minneapolis and St. Paul, and we got scrambled to intercept those guys. We also got warned not to get too goddamn close to them. I think they were worried because the B-36 was such a big airplane. With real big airplanes, people are prone to get into problems if they get too close to them.

That's how they lost the XB-70 at Edwards in 1966. Joe Walker got too close to that thing without realizing it and got caught in tip vortices and whap! He smacked right into it. That's happened to a lot of people.

We flew all the time, not only on weekends, but also during the week. The unit had a permanent cadre of maintenance people and a full-time commander. There was a weekly unit meeting where they'd call everybody in and you had drill and military type lectures. Two weeks a year the whole group would go out on maneuvers up to Camp Ripley, up in northern Minnesota near Bemidji.

Summer camp was fun. I flew an A-26 pulling targets for the artillery. We had a cable drum in the back of the airplane with a mile of cable on it. We'd climb up to about ten thousand feet, the guy running the winch would then reel the cable out about half a mile as we flew over the antiaircraft guys and let them shoot at us. It was a little dangerous if you remember that the artillery guys were basically weekend warriors.

We never got hit, though we did have a couple of cables shot off. I guess that meant they hit the target.

In boxing I had my nose broken six or seven times in a fairly short period of time. It got to the point where I couldn't wear an oxygen mask. One day I was going out to fly when the unit commander discovered the situation. He gave me an ultimatum: "You can either box or you can fly with me, but you ain't gonna do both. We can't have you running around here not able to wear an oxygen mask."

So that was the end of my boxing career. It was a pretty easy decision to make.

I graduated in with a bachelor's degree in aeronautical engineering, but nobody needed engineers. A few companies came around interviewing, and one of them was Boeing. To my surprise, they hired me. Since I graduated in a summer class, there were only about ten of us in the class: only two ended up getting jobs.

It was August 1949. I had gone through a four-year course at the University of Minnesota in two and a half years.

6

BOEING

My starting salary as a junior engineer at Boeing was $200 a month, which was about what I had made as a second lieutenant in the Air Corps in 1943, but it was better than the GI Bill, and it was a job. I was happy to have it.

I moved into a rooming house in Seattle and started riding to work with a couple of other guys from the University of Minnesota who had gone to work for Boeing, too.

It was a miserable time. I didn't know anybody at first, and didn't find the people in Seattle to be particularly friendly. My social life was zilch. Eventually I became friends with Ralph and Vivian, a couple who lived in the rooming house for a while. Then they split up. I went out with Vivian a few times, but it was nothing serious.

Some time later during my stay in Seattle I dated a girl who was eighteen years old. Pretty soon she broke up with me; she decided that I was too old for her. Well, I was twenty-six by then. (My buddies bought me a cane as a consolation prize.)

Seattle wasn't all bad. I did a lot of fishing on the weekends, up in the mountains. There was good hunting, too. I also learned to ski. Now, I had skiied when I was a kid in Wisconsin, but that was just a matter of pointing yourself down a hill.

Two hours outside Seattle there were real mountains. I figured, hell, I know how to ski. My first run down a real ski slope I zipped right off the trail into the trees. I was lucky I didn't get hurt, but I realized I'd better get serious and learn how to turn and stop.

I kept on skiing for years. It's the closest I've ever come to flying without a plane.

* * *

Another problem with Seattle was that there was no flying. And I sure wasn't happy with my job.

Boeing in those days was the biggest aircraft manufacturer in the United States, probably the world. They had a massive facility, the biggest thing in Seattle, and thousands of people working there: engineers, assembly technicians, secretaries. It was easy to get lost in the crowd.

The first thing that happened to me there was I got assigned to what they call a change group. I guess it was a typical way to break in a junior design engineer, because all the new guys who got pulled in the same time I did got put in this group, too.

Design changes had already been made and put on detailed aircraft plans in pencil, and now they had to be transcribed and changed to ink drawings. In those days all that work was done in India ink on a special paper, vellum.

My problem was that I was the worst draftsman in the world. Dealing with India ink, which is sloppy and permanent, is bad enough if you like what you're doing, and I didn't. I spent six weeks in the change group and decided right then that this was not the life for me.

I wasn't seeing a lot of other alternatives. Here I had a piece of paper that said I was qualified to become an aircraft design engineer; I wondered when I was going to stop this drafting bullshit and start designing aircraft.

Eventually I got transferred out of the change group to a team working on the B-52 bomber, which was still in the design stage then. The kind of work I was supposed to do there was called structural design, which was closer to what I wanted. The drawback was that you actually did your work in a medium that was even trickier than ink and vellum . . . with little inscribing pens right onto metal templates. It was very meticulous work.

My particular piece of the B-52 was the bulkhead between the crew compartment and the aft body of the airplane. I spent maybe the next year working on that damned bulkhead. It didn't take a lot of brainpower. People would come in with the electrical group and say, put this size a hole here for this cable to go through. Then this other guy comes in and you've got to put this structural support for this box and this bracket. Then we need three holes here . . . it was that kind of stuff. It wasn't like, for example, doing analyses on structural deformation—it was being a glorified draftsman.

I bought another car, because it was obvious I couldn't be around this place without transportation. I hadn't shopped around too long before I finally bought an old Nash for about $150.

Seattle is a hilly place. Well, I bought this Nash at a lot down in the valley; I lived up on one of the hills. And before I got up the hill, the goddamn transmission went out on it. On the way home.

I got a friend of mine who had a car to tow the Nash the rest of the way home, and stuck it in the garage at the rooming house. I spent the next two months rebuilding the transmission at night, and finally got it going again.

Finally I accumulated enough money to buy a new car. (My experiences with used cars had me leaning that way.) I bought a bullet-nosed Studebaker Club Coupe, a six-cylinder, which was a pretty nice little car. I kept that until I went back to Minneapolis in the Air Guard. If I was smarter, I'd have kept the son of a bitch for the next ten years.

After about a year of the change group and the B-52 bulkhead, I got moved again, this time to the team doing modifications on the KC-97 transport. Fortunately for me, this was in June 1950. The Korean War broke out.

I figured I was through punching a clock. I wanted to go fly again.

In moving to Seattle, I'd had to drop out of the Minnesota Air National Guard. There was no Guard outfit in Seattle for me to transfer over.

Somebody had told me you automatically reverted to your reserve status if you left the Guard, so I decided to become a reservist. Active reservists flew on weekends and had a two-week obligation in the summer, but there was a class called inactive reserve, which is what I thought I was. And I didn't do a damn thing during the year and a half I worked for Boeing.

When the war broke out, I drove down to McChord Field, which was the nearest Air Force base to Seattle, and said, "I'm tired of this goddamn engineering. I never liked it anyway. Put me back in the cockpit."

They started checking and they said, "You're not a reservist, you're nothing. You're a civilian."

"You've got to be kidding."

They told me I was required to reapply for inactive reserve status within six months of the time I dropped out of the Guard. I hadn't done that, so so therefore I was nothing.

Well, I sat down and wrote a letter to Hoyt Vandenberg, who was chief of staff of the Air Force, and said, "All my friends are getting called up to go to Korea. None of them want to go, but I do. I want to go, but I can't get back in the Air Force." I never got an answer.

I called John Dolny, my old squadron commander at the Air Guard in Minneapolis, and said, "John, if you guys ever get called up, let me know. Hold me a slot. And I'll guarantee I'll be there."

It was back to the KC-97 for me for the next few months. Finally, in February 1951, I came home from work one night and saw I had a wire sitting there . . . from John Dolny. He said, "We're being called up first of March. If you can get back here in forty-eight hours, I've got a slot for you."

So I just dropped everything and jumped in my car. I got the hell out of Seattle and Boeing. And I was sure happy to leave.

Off I went to Minneapolis, figuring I was back in the Air Force, or at least the Minnesota Air Guard.

Well, the first thing I found out was that I couldn't pass the physical! I'd been on a drawing board for so long that my eyes had gone to hell.

I had some airline friends who had had the same problem, and they knew a doctor downtown. So they sent me down to see him. For about three months I went through this eye therapy . . . exercises. After all, your eye is nothing but a muscle, and you've got to exercise it. I looked through different lenses at different patterns. And he got me back up to twenty-twenty and I got cleared to fly.

I was pretty happy being back in Minneapolis-St. Paul again. I was a full-time captain in the Guard, on my way to active duty, and I was living high on the hog. I thought I'd improve my social standing.

I liked my little six-cylinder Studebaker, but then the company came out with a new V-8. Better yet, it was a convertible.

Now, in the middle of a Minnesota winter nobody needs a convertible. They had a hell of a good price on it—too good for me to pass up, for a new car. So I traded in my nice little six-cylinder for this new V-8, which became the biggest lemon I've ever bought in my life. Even worse than my first two cars. Before I drove it five hundred miles, it needed a new valve job.

I ended up driving that thing down to Maxwell Air Force Base in Alabama in 1952, to command and staff school, where it rained every day for six weeks. Whenever I'd put the top down, I'd get rained on. Every morning it was a big decision—do I put the top down and let it get dry or do I put it up and leave it wet? I don't think that car was ever dry after that.

I ended up getting orders to go to Germany about that time. So there I am with this lemon, but I didn't know what to do about it. So I had it shipped to Germany. It got over there about three months after I did and I went up to Bremen from Wiesbaden to drive it back. That night I parked it down by the bachelor officers' quarters I was staying in. I went out the next morning and the top was slashed wide open.

There I am with an almost brand-new car with the top slashed open, and a motor that wasn't worth a shit, in Germany where they don't have a lot of parts for brand new Studebakers. I wish I had some of those letters I wrote to Studebaker in those days. I made a career out of writing them nasty letters.

None of it did me any good.

While I was waiting to get medical clearance to fly, I became maintenance officer for the Air Guard. We had four squadrons in the group at that point: the two Minneapolis ones, one in Duluth, one in Sioux Falls, South Dakota.

One day early in the summer of 1951 the whole Sioux Falls squadron was in Minneapolis on its way to maneuvers. They had all their airplanes out there parked on our ramp. We had our whole complement of airplanes, a lot of them in the hangar, some on the ramp.

We were over in the O-club that night, having a couple of drinks and dinner, when a big storm hit the area. You could feel the pressure change in the O-club. Suddenly your ears pop. So we knew something had happened pretty close by. When things settled down, we dashed out of there and ran over to the hangar.

The top was blown off and airplanes were sitting upside down all over the place. They had just been flung out. About ninety percent of those planes were on top of each other like toys.

So I had my work cut out for me for the next few months. As maintenance officer it was my job to get those airplanes in the air again—and eventually most of them did.

Since I was flying again, I also did all the flight tests on them. Every time we got one ready to fly and checked out properly and inspected, I'd take it up, wring it out, bring it back, and sign it off. Then go get another one and run through that cycle again.

I'd normally do a couple of spins with them, one left and one right, just to make sure they were rigged properly. That was the last thing I did, then I'd bring it on back.

It was when I was doing a maintenance test flight like this that I had my one and only encounter with an unidentified flying object.

I was up about the middle of one afternoon—a nice sunny day—wringing out this particular 51. I had just come out of a spin at around ten thousand feet over the Mississippi River, near Prescott, where the Mississippi and the St. Croix meet, about twenty-five miles from the Twin Cities. I was heading back to Holman Field when all of a sudden I saw this white object about my altitude, at one o'clock.

I didn't think anything about it. My first thought was that it looked like a kite. But logic said nobody's flying a kite at this altitude. So I started kind of watching it to see what it was.

I was closing on it, but I still didn't think too much about it. The closer I got, the more it looked like a weather balloon, and I'm thinking, that's what it's gotta be. Then I flew past it a little high, about a thousand feet off. It still looked like a three-foot-diameter weather balloon to me.

My guess on the dimensions couldn't have been too far off. I had plenty of gas, so I figured I'd make a pass on it. Burn some gas and have a little fun. I pulled into a turn.

But when I came out of that turn and headed straight at it, all of a sudden it didn't look like a balloon anymore. It looked like a disk on edge!

I thought, that's strange. Then I realized I wasn't closing on that son of a bitch. A P-51 at that time would cruise at 280 miles an hour. But this thing just kept going and climbing at the same time at about a forty-five-degree climb. I kept trying to follow it, but he just left me behind and flat disappeared.

I wondered what that was, but I never saw it again. I turned around, headed back, and landed, and didn't tell anybody about it for two days. I was afraid they'd think I'd lost my mind.

A couple of evenings later I was over in the O-club with my boss, a full colonel, and after I couple of beers I thought I'd better tell him, and I did.

He said, "Get your ass over to Intelligence in the morning and give them a briefing." So I did. They sat there and nodded and took notes.

Then they told me: Just for your information, the day you saw this object a local company was flying high-altitude research balloons. They had a light airplane tracking it, and a station wagon on the ground. Both observers were watching this balloon and had seen this object come up beside the balloon. The object appeared to hover, then it took off like hell.

The guys on the ground tracked it with a theodolite, and they'd computed the speed at four thousand miles an hour.

I guess they were trying to tell me I wasn't exactly crazy: somebody else had seen something unusual, too. But I never heard another thing about it.

Karl Henize, one of the scientist-astronauts I hired for NASA years later, had worked at Northwestern University with Professor J. Allen Hynek, who did studies of unidentified flying objects. Hynek contacted me once about my story, and I've also talked to a number of UFO-type people.

My position is, I don't know what it was: it was unidentified. Maybe what I saw was that company's weather balloon—maybe the object going four thousand miles an hour to these guys on the ground was me. Maybe there was something about the environment and the setup that confused me. I don't know. Or it could have been something unknown. (I don't automatically presume that it came from Alpha Centauri, just because I can't identify it.) It's still an open question to me.

I've heard of other stories like this. There had been a lot of stuff in the press in the late 1940s—Kenneth Arnold and his "flying disks" near Mount Rainier, for example. There was a report of a Guard guy down in Kentucky who got scrambled to chase something he couldn't identify. Airline pilots have seen things. But no one's said anything like that to me personally . . . and, of course, I didn't run around telling my story.

Sidney Sheldon published a novel in 1989 called *The Doomsday Conspiracy*, which dealt with UFOs, and got Gordo Cooper to write an introduction to it. But even Gordo never mentioned anything like that to me. There have been two or three space program reports that have gotten picked up by the UFO people, but those weren't legitimate.

I sort of wondered if my story wound up in Project Blue Book, the Air Force's official investigation of UFOs. I know people have been saying for years that a UFO crashed out in New Mexico in 1946, and it's been hushed up ever since.

In my experience it's pretty tough to keep a secret that big that long. Of course, I was pretty surprised at the way they were able to keep the lid on the F-117A Stealth fighter and B-2 bomber and some of those airplanes for as long as they did. But you're not going to do it forever. They'd probably have done even better with the 117 if they hadn't had a couple of them crash.

7

AIR FORCE

I had gone back to Minneapolis in February 1951 because Dolny said a call-up to active duty was "imminent." Naturally, it was more than a year later, March 1, 1952, before anything like that actually happened.

The war was still going on in Korea, though it had deteriorated into what eventually became a stalemate. There was still fighting going on—dogfights between F-80s and F-84s and Russian MiGs. I wanted to go, unlike most of the guys I knew. A lot of them had stayed in the reserves or the Guard after World War II, never thinking they'd get called up. They were married, they had children, they had careers. I had none of those things. All I wanted to do was fly.

While waiting for this call-up, which looked for some months like it would never happen, I applied for a regular USAF commission for the third time. This time, with my degree in aeronautical engineering, I was accepted. I went from being a captain in the Minnesota Air National Guard to a captain in the U.S. Air Force.

When our Guard unit was finally called up, everyone got scattered in different directions. A few guys actually did wind up flying missions in Korea. But I wasn't one of them.

I got assigned to attend the Air Command and Staff School at Maxwell Air Force Base in Montgomery, Alabama. I guess the Air Force figured a brand-new twenty-eight-year-old captain needed some additional military education before they could trust him in a squadron. I arrived in May 1952 for a six-week course. When I finished that, I wanted to be on my way to Korea.

By that time a slot had opened up in the inspector general's office at headquarters, 12th Air Force, for a maintenance inspector. That just happened to be what my MOS—military officer's skill—still said I was. Since I was available, that's where I was assigned. Great. Except that HQ 12th Air Force was in Wiesbaden, Germany. It was a typical military decision.

I protested: Germany's a nice place, but it wasn't where I wanted to go. It didn't make any difference. So off to Germany I went.

My feeling was that if I wasn't going to get to Korea, I wanted to be in flight test. That was my ultimate goal, anyway. I wasn't in Wiesbaden a week when I filed an application to get the hell out of there and off to Edwards Air Force Base in California.

That didn't take. They said, you've got a three-year tour. When you've done your three-year tour, then we'll talk. So there I was, stuck in Germany.

At least Wiesbaden had more to offer socially than Seattle had. The bachelor officers' quarters were in a fancy hotel right in the heart of the city. Anything interesting that was going on, I was close to it.

One night there was a volleyball game on the base. Some of the players were women who had civilian jobs there, including Marjorie Lunney, this pretty brunette from Los Angeles who was the secretary to the inspector general.

Marge broke her wrist during the game, and I volunteered to carry her to the infirmary. That's how we met and started going out.

She was a couple of years older than me and had been born in Toronto. Her parents were in the process of emigrating from Ireland to America. She had grown up in L.A. and started working as a civilian for the Air Force. By the time we met, she had been stationed in Japan and was on her second tour in Germany.

My job involved going around to every Air Force base in France and Germany doing inspections. The idea was we'd hit them once a year to be sure their planes and weapons were up to snuff, that their paperwork was in order. This was almost as bad as being a junior design engineer in the change group. On the plus side, there was some interesting travel, to Ireland and Austria, for example. I remember taking a train to Zermatt, Switzerland, for Christmas 1954 . . . and having the train stalled by an avalanche.

But I didn't want to spend three years doing inspections. As soon as they decided not to ship me back to the States, I started trying to get into a fighter outfit. Even that took me eighteen months.

My boss in the inspector general's office was a colonel named "Pop" Gunn. Being an old fighter pilot, he was sympathetic to my situation. But he wasn't going to let me out of HQ as long as he was there.

He knew I wanted to join a fighter squadron with the hottest planes.

There were three in Germany that were just getting the new F-86s about the time Gunn finally got transferred back to the States. So he did me a favor and tipped me to an opening in one of the three F-86 units.

I moved to Bitburg as a squadron maintenance officer in the 36th Fighter Day Wing.

One of the many things I hadn't liked about Wiesbaden and HQ 12th Air Force was that the only planes we had available to fly, to maintain proficiency, were C-47 Gooney Birds. I still had my Gooney Bird allergy, but it was either take one of them out and fly it around or don't fly. (I had managed to get a ride in a T-33 trainer once, which was my first experience with a jet.)

At Bitburg they had F-86 jet fighters, having just switched over from F-84s. So I had to get qualified in a T-33 before I'd be ready to go.

My first problem was that I'd been flying those damn heavy-handed Gooney Birds for so long that when I got into a T-33 without any tip tanks, it was so sensitive I couldn't keep the son of a bitch going straight. I finally learned to cope with that.

Then I got in the 86, which I found easier, at first, until I found myself at altitude, flying formation, and discovered the 86 could be pretty sensitive, too.

But it's like everything else: once you get used to it, it's really nice. I really learned to fly with a light touch, just using my fingertips.

Marge had stayed behind in Ramstein, which was the new base HQ 12th Air Force moved to in 1954, when I transferred to Bitburg. But we still saw each other whenever we could.

I had applied again for test pilot school and, with jet pilot time added to my engineering degree, had been accepted. I was scheduled to leave Germany by June.

So Marge and I got married in Ramstein on May 18, 1955. It was another of my notable accomplishments . . . the last marriage performed by German civil authorities under Occupation laws, which meant that the justice of the peace ceremony was the actual marriage while the church wedding was all for show.

It was a show, too. Neither of us had family present, of course. But a bunch of my squadron buddies showed up wearing black armbands. Marge didn't think that was too funny.

8

TEST PILOT

I'd visited Edwards when it was still called Muroc that one time during World War II, on my way to Okinawa when I had refueled a plane and gotten out of there. There was nothing impressive about it then, and it didn't look a hell of a lot more developed ten years later. It had expanded, of course. In 1945 the flight test center was still located in Dayton, Ohio, at Wright-Patterson. A lot of those offices had moved west in 1949, so there were new hangars and runways. But it was spread over so much barren desert, you could hardly tell.

I was pretty familiar with what was going on at Edwards. You couldn't escape it in the Air Force. Not only had Yeager broken the sound barrier in the X-1 back in October 1947, but all sorts of new designs were being tried out every year, pushing past Mach 1 to Mach 2 and even Mach 3: the Douglas D-558-2 Skyrocket; the X-2, which was just going into powered flight trials while I was at the test pilot school.

A lot of this supersonic research was conducted by the National Advisory Committee for Aeronautics, N.A.C.A., which had its High Speed Research Station at the base, too.

Exciting as all this cutting-edge stuff was, the primary job of the Air Force Flight Test Center was to qualify new aircraft for operational use. That is, to get new fighters and bombers and cargo planes from the drawing board to the squadron. It was a painstaking process that usually took years, and in those days also killed some guys.

For example, that F-86 jet fighter—which had been the hot new plane in 1953—had first flown at Edwards in the same month, October 1947, that Yeager broke the sound barrier. (At that time the F-86 was known as the XP-86. Before 1950, Air Force fighters were designated P, for *pursuit*. That was later changed to F, for *fighter*.)

Following test flights by the contractor, North American, what we call Phase I testing, the plane was turned over to the Air Force for Phase II tests. Advanced testing, Phase III, followed before production of the F-86 could begin. The A model of that plane was available in June 1948, a very short time as this process goes. And it was still *five years* before I got hold of one of the F models.

The test pilot school was designed to train pilots for this kind of work. It wasn't anything fancy, just some classrooms in an old barracks on the original Muroc base, located next to the officers' club and a little arcade. These were the only real source of entertainment in the area. Pancho's Happy Bottom Riding Club had burned down by then.

I arrived in June 1955 and spent the next six months going to school, studying and flying. (They expanded the course to a year just after I graduated.)

The guy in charge of the school was a Major Folvy. There were a dozen guys in my class, including three who had already been picked to become future instructors at the school; they weren't even pretending to become test pilots. One, a Lieutenant Colonel Lenhart, took over as commandant of the school as soon as he graduated. Two other guys, Harry Stohrs and Ralph Matson, also stayed and became instructors. Those guys all had a fairly strong engineering background; they'd been through the Air Force Institute of Technology at Wright-Patterson. It was good for the rest of us. There were still nine of us competing for maybe three slots in fighter test.

In those days we did all our academic work—running all our performance numbers—on little hand calculators or slide rules. Today guys in test pilot school are sitting there with laptops. They can do more real-time calculating in five minutes than we could do in a six-month course.

Flight test instrumentation was archaic as hell. You didn't have real-time telemetry: you were putting stuff on a tape recorder, where the data rate was pretty low. In the cockpit they had photo panels, meaning there was a 16mm camera looking at your instruments the same time you were.

I suppose it was risky. You didn't know what performance you really had until you got back and looked at the data. Today they do everything real-time with telemetry and computers. You do a test point and you can do analysis on it and say, obviously we're going to the next one, and proceed. We had to lay out a flight plan and say, okay, I think I can go this far this time. If you guessed right, you were okay. If you didn't . . .

I uncorked a TF-102 twice in roll coupling, which wasn't ever supposed to happen. The contractor, General Dynamics, was supposed to have certified the damn thing, and I was just going to do the spot-check certification on them. I went through three test points and hit the fourth, which turned out to be one point too far. Whammo! The son of a bitch unwound on me.

We found we had a little 1.5-G lateral instability on that aircraft, the kind of thing that could have broken the tail off. Of course, I could have bailed out, but I never actually had to do that. (Bailing out is more dangerous than people think.)

When I graduated from school in December 1955, I was lucky enough to get assigned to fighter test at the Flight Test Center. So did Donald Sorlie, who had been one of my classmates. One other guy went back to Wright-Patterson, where they did all-weather testing—a later phase of the process.

Another one of my classmates went to Lockhead, as the Air Force guy assigned to acceptance flight test at the office of the contractor. (After Phase III, once manufacturing has begun, each aircraft off the line has to be run through acceptance test before it can be delivered to the Air Force.) He was the first guy killed in an F-104. He had a flameout and couldn't make himself believe, like a lot of other people, that the plane wasn't going to glide to the lake bed. And he was damn near over the lake bed when it happened. What everybody ultimately discovered, if they lived long enough, was that the 104 was going to hit where the nose was pointing when it lost the engines. Usually the nose is pointing straight down. This guy thought he had plenty of altitude to glide down onto the lake bed. He was about seven hundred feet short.

Of course, the early models of the F-104 had a feature that made ejecting from the plane at low altitude a bad idea: the seat fired down through the bottom of the plane, not up. If you were at low altitude, you were suppose to roll the plane so you wouldn't be ejecting straight into the ground. The worry was that a pilot ejecting at, say, five hundred miles an hour in flight was going to fly into the F-104's high tail. Eventually everyone realized that you weren't likely to be going that fast if you had a real problem, like a flameout, and later models let you eject upward safely, even at zero altitude.

But the 104 wound up killing a whole bunch of people first.

There were thirteen pilots in the fighter section of test ops at the time. Horace "Duke" Hanes, a colonel, was the overall chief of flight test, which including not only the test pilots but also the engineers and analysts.

Pete Everest, who was just getting ready to set a world speed record in the X-2 rocket plane, was the head of flight test operations—the test pilots—

which had sections for fighters, bombers and cargo planes, and helicopters. He doubled as the section chief for bombers and cargo planes. Stu Childs was the head of the fighter test section under Everest; he had just succeeded Robert "Silver Fox" Stephens.

The overall base commander was Brigadier General J. Stanley Holtoner. I thought he was a jerk: he kept showing up in fighter test wanting to fly planes he had no business flying. One time I flew chase while he took a ride in the F-107, which was the first plane to have an advanced avionic flight control system. That was because the F-107 had more movable control surfaces than your standard fighter.

Well, Holtoner was flying along, and I was pacing him, when suddenly I look over and he was gone! Just disappeared! There had been a little glitch in the avionics system and it had flung him straight up in the air. He was pretty annoyed; I could hardly stop laughing.

In 1957 another one-star, Marcus Cooper, replaced Holtoner. He was a little better.

Fitz Fulton, who I later worked with in the shuttle approach and landing program twenty years later, was at Edwards then. At that time, he was an Air Force bomber test pilot. He also flew the B-50 bomber that carried the X-2 rocket plane to altitude. Fighter test was only one of the things going on: my old Boeing engineering project, the B-52, was in test at that time. Then the B-58 came along. There were new cargo planes, the C-130 and KC-135. And they were still flying some heavies for research, a three-engine Martin XB-51, the B-66, and the B-47.

We also had some people off working part of the time on secret stuff on North Base. I learned later that the U-2 was what was being flight tested. It was one of those dumb things where they told us, "If you see this funny little airplane, pretend you didn't see it and don't go near it." Well, they had a good reason. The U-2 was so damned unstable, they didn't want anybody in another plane coming in there and throwing any turbulence around.

Some of the other guys in fighter test in early 1956 were Iven Kincheloe, Lou Schalk, Bob White, and Mel Apt. Several of them got to be quite famous, as test pilots go.

There were two guys named Don—me and Don Sorlie—and that turned out to be confusing in air-to-ground chatter, so they started calling me "D.K." Pretty soon it just became "Deke." So in one sense you could say Deke Slayton was born at Edwards.

Everybody asks pilots, who's the best one you ever saw? Of course, the right question to ask any pilot is, who's the *second* best pilot you ever saw? James Wood, who later became chief test pilot for the X-20 Dyna-Soar and the F-111, was good. So was Chuck Yeager, though we didn't overlap at Edwards.

But of all the guys I know, Bob Hoover was probably the best goddamn fighter pilot in the world. He had been in World War II and Korea as a P-51 pilot, then gone into test work as Yeager's backup on the X-1. He's still flying air shows.

We had some good guys. By the time you got to Edwards, you'd been through the operational experiences and flight test school, and into fighter test. I didn't think there was anybody in there who didn't belong, which is more than I can say about the space program; we definitely had a couple who didn't belong there. But that was not true of fighter test. Everybody could do the job.

As usual with the new guys, the first thing you do is get stuck in flying chase. I say stuck, but actually it was great experience. You'd get up in a T-33 or F-86 or F-100 and chase the contractors around. I enjoyed it because it was a lot like dogfighting. For example, say there's an F-104 test point in which the pilot's got to tumble the plane. And you're flying an F-100F with a photographer in the back. You've got to stay close enough to the 104 that the photographer can document what's going on. Nobody knows—it's just totally out of control. It could get pretty squirrelly.

Other Voices

BOB DREW

In the late 1950s I was a civilian test pilot for Douglas in Los Angeles. Douglas primarily built fighters for the Navy, but given our proximity to Edwards, we did a lot of our day-to-day testing up there.

The whole business of qualifying any military plane for acceptance is to prove that it has met certain performance goals—points. The points have to be certified with instrumentation and witnessed by a military test pilot. The Navy had an arrangement with the Air Force to allow people from Edwards to do that certification for us. So we were constantly looking for Air Force test pilots to fly chase.

We had a tough time finding them, since it wasn't thought to be glamorous work. Even when we did, sometimes it wasn't worth it. I remember putting one aircraft into a flat spin with the cockpit smoking—in a bit of trouble—and calling to my chase pilot that day to tell me which way I was going. "I don't know," was the answer I got.

Deke wasn't like that at all. He loved to fly so much he was always happy to help out, and very conscientious when he did. I would give him a call and tell him, say, that I was going to be up checking stability control or stick force per G on an airplane, and he would be right there where I wanted him.

I think he liked flying chase because it exposed him to all kinds of aircraft—Chance-Vought, Douglas, Grumman. Every one of them was a little different, and Deke was ready to learn them all.

One of my earliest assignments was flying chase in a T-33 for Pete Everest in the Bell X-2. He had made four or five powered flights already, going all the way up to Mach 2 and becoming, as his book title later had it, *The Fastest Man Alive*, but on my flight he had a premature engine shutdown. Well, a T-33 is hard to slow down. I went shooting right past the X-2, and Everest got pretty pissed off. "I'm supposed to be leading this parade!"

Not to be outdone, Iven Kincheloe set an altitude record in the X-2 about the same time, reaching 126,200 feet. So he became *The First of the Spacemen*.

The X-2 crashed a couple of months later. Mel Apt took over from Iven Kincheloe because the Air Force wanted to have a few more pilots experienced in rocket planes. (The veteran rocket pilots—Yeager, Kit Murray, and Everest—had either been reassigned or were about to move on.)

So Apt took the X-2 up for his first time on September 27, 1956. He started well and wound up setting a world speed record by hitting Mach 3.

But then something went wrong. Maybe he wasn't as familiar with the plane as he should have been, or maybe he got fooled by the instruments. (As I said, they were pretty crude and tended to lag.) Anyway, he got into roll coupling at a very high speed, and the X-2 came apart. The front end of the X-2 was an ejection pod, but Apt was never able to get the parachute open.

I missed out on flying chase that day because I was in the hospital. A few days before, I had been doing braking tests in the F-105. This meant coming in for a landing, then slamming on the brakes to see how they would hold up. Well, they started burning. So I shut down the engine, which was a dumb thing to do, because all that did was dump fuel on the flames. So I started it up again as I slid down the runway, came to a stop, opened the canopy, and hung over the edge with the engine still running until the fire truck could get there.

Then I chopped the throttle and fell off the plane. I guess it was the weight of my parachute and the distance I dropped, but I gave myself a hernia. So I was in the hospital getting sewed up when the X-2 made its last flight.

We had more planes under development at Edwards at one time than anyone has had since. There were the subcentury series planes like the F-89H and the F-86H. Then the 100F, in addition to 101s, 102s, 104s, 105s, and 106s.

There were usually two primary pilots on each airplane. During the time I was there, each of us was cleared to fly five airplanes. They changed the rules later on, knocking it down to two or three. I guess they thought five was too many for people to stay really proficient.

A test pilot now is lucky to have even one new plane in his entire career. At Edwards in the late 1950s I'd jump into a T-33, at one end of the spectrum, go out and fly photo chase on somebody, then jump in a 104 and go off and do an engine test. Then do a test point on a 105. Then to a 101, all in the same day. I flew five, sometimes six days a week.

It was certainly the golden age for fighters. When I left Germany, there were two F-100s sitting on the ramp, the first F-100s to get overseas. I never got a chance to fly one there, of course. But when I got to Edwards, I discovered that the 100 was considered to be an old airplane, best suited for chase. We were flying 101s, 102s, 104s, 105s, and 107s. There are a lot of stories to go with every one of them.

Like the 101A, which was a McDonnell plane originally intended to be a long-range bomber escort and later aimed at being some kind of interim interceptor. It was an improved model of the XF-88 Voodoo. Like the F-104, the 101 had a high horizontal tail. Both planes were prone, if you got them too slow, to lose pitch control. Then they'd just go ass over teakettle, and you wouldn't know what they were going to do.

The first guy who got killed when I got into fighter test was flying a damned 101. He got into a pitch-up, just like the 104, then made the classic mistake of a test pilot. Instead of trying to get out of the thing, he stayed with it too long. When he finally decided to eject, he was too late. That happened regularly.

That was one problem with the 101. The other was that it was so goddamn heavy that you'd be taxiing out to the runway and keep blowing out tires! They put a mod on it where they effectively strapped on another set of wheels for taxi and takeoff. When you got off the ground, they dropped off and went rolling across the desert.

Sometimes we'd sit around complaining about some of these stupid ideas. But it was an innovation that worked as an interim measure, until they finally got better tires and wheels.

Basically, however, Edwards was so advanced as far as the rest of the world was concerned that in 1958 Andy Anderson and I got sent to England for a month to help them test their first supersonic fighter, the P.1. They simply had no experience with something like that.

The P.1 was a terrific plane, with the easy handling of an F-86 and the performance of an F-104. Its only drawback was that it had no range at all. This was pretty typical of English fighters, I guess.

Looking back, however, I'd have to say that the P.1 was my favorite all-time plane.

The Convair F-102A Delta Dagger was the first plane I was assigned to as project pilot. It was a big delta-winged plane intended to be a supersonic interceptor. During its early development they'd found that it wouldn't go supersonic except in a dive. They had solved that problem by the time I came along, but with all the modifications required, they eventually came up with a whole new airplane, the F-106 Delta Dart.

But there was still a lot of performance test to do on the 102A, and on a two-seat version called the TF-102. Bob White had been the project pilot, and he trained me because he was about to shift over to the F-105. (I kind of followed in Bob White's trail. When he succeeded Kincheloe as the Air Force's prime pilot for the X-15 in the summer of 1958, I took Bob's place on the 105.)

One of the things we tried to do was figure out how to stop the 102 on a short runway, the kind of thing you might very well find in a forward combat zone.

We added a tailhook to the 102. I was supposed to simulate a landing and rollout, going over an arresting chain at the right speed and dropping the hook.

Well, hitting that chain at precisely the right speed was harder than we thought. On one of the tests I was going too fast . . . the hook caught all right, but it picked up the whole damn chain and took off with it. I'm flying down the runway with a huge chain following me. Part of the chain broke; the part that was left smacked into the 102.

White and I took two F-102s down to the last national air show there ever was, in Oklahoma City in 1957. They had one 101 and a couple of 100s, but it was the first time they had two 102s. It was also the first time they ever took an F-104 off the base at Edwards. Lou Schalk flew that one down there. Now, Lou Schalk loved daiquiris, and he put his daiquiri mixer and all his makings in the missile bay of my 102, because he didn't have room in the 104 for it. We spent about a week drinking at night and taking go pills all day long.

Looking back, I'm amazed at the amount of drinking and flying we did. Part of it was just that it was the thing everybody did. Being the age we were, we could get away with it . . . just like we could get away with flying without G-suits, which happened frequently. It wasn't as though we were exercise fanatics or anything like that. I might run a little bit, just to stay in some kind of shape, but I don't know anyone who had a real exercise program. In some ways, it's a miracle we survived.

We were all young guys, mostly, in our early thirties, all married, all with small children the same age. My son Kent Sherman was born at Edwards on April 8, 1957. I have a picture taken a few months later showing about five infants lined up on a couch. They look like quints.

One of the things that needed to be done in service testing was finding out how the F-102 was as a weapons platform. So Fox Stephens and I spent six months at Holloman Air Force Base in Alamogordo, New Mexico, on temporary duty doing weapons tests on the 102. Fox stayed out there and a couple of years later wound up as the first Air Force pilot on the A-12, which was an early version of the SR-71.

Our mission was to shoot down nine Matador missiles. The 102 was armed with an infrared-sensing Falcon missile and a radar one called Genie. We were supposed to shoot down Matadors in the daytime and at night, using different combinations of missiles.

For the first test, we took off from Holloman and set up in a racetrack pattern at forty thousand feet, waiting for the first Matador to be fired out of White Sands, all ready to shoot the thing down.

They launch the Matador out of White Sands . . . we're communicating with the guys on the ground . . . and we never see the goddamn thing!

The controllers finally admitted they had a problem. They couldn't make the Matador respond to command, couldn't make it turn. So here it was, climbing straight out and heading north. They didn't have destruct packages on anything in those days, so they sent Fox and me off to chase her down.

We hit the burners and ran as fast as we could run. The best a 102 would do was about Mach 1.25. A Matador would do about .95, and it had a head start. We never did catch it: we ran out of gas and had to come back.

We heard the thing finally crashed just short of Santa Fe. After that, anytime they got ready to launch, we went down to the pad early in our planes and picked them up in the count, just like we used to fly chase. We'd fly formation on the Matador until we got it where it was supposed to be . . . then we'd back off and shoot it down.

I don't know how many people ever knew they'd had some wild missile running out on that range. And I'm sure that wasn't the only one that got away from them.

From the F-102 and TF-102 I moved into the F-105 as overall test project manager. They also split the fighter test section in two, and made me chief of one section.

The 105 was built by Republic and was called the Thunderchief. Later on everybody called it the Thud. It had been conceived as a high-speed (Mach 1.5), all-weather interceptor back in 1950. Not only was it supposed to be able to handle itself in a dogfight, it was also supposed to be able to carry a nuclear bomb, if necessary. This might have been one of the reasons a lot of the initial 105 tests were secret. There wasn't even a picture of the thing published until it had been flying for quite a while.

I had another one of my trickier episodes with the F-105. I was at thirty-eight thousand feet doing a test to see how it would turn at low speed. I started out supersonic, then put it in a tight turn, slowing down so I was subsonic.

Suddenly, wham! The airplane snapped over on me, rolling four or five times, winding up with me in a flat, inverted spin. That is, I was upside down going around and around like a spinning plate.

Nobody had spun an F-105 before, so I had no procedures to fall back on. So I decided to try the usual spin recovery technique, and all that happened was that I wound up going around in the opposite direction. I tried it a few times; same thing.

By now I'd fallen to twenty-seven thousand feet—nine thousand feet in a few seconds—and I was dropping fast. I tried popping open the speed brakes on the 105—these were petallike pieces on the tail that you could open when you wanted to add some drag. The idea was that this might throw me into a vertical spin, which I knew how to recover from.

While I was waiting to see if that would work, I checked the altimeter again. Ten thousand feet. I tried to reach my ejection handle . . . just in case . . . but couldn't. The G-forces were pulling me out of the seat. I found that if I touched the stick, I would be forced back into the seat. So I knew I could eject if I had to.

Throwing the speed brakes had finally aimed the nose toward the ground. I tried to break out of the spin . . . didn't quite make it, though I noticed a little hesitation in the plane before it reversed direction. . . . So I gave it one more try . . . neutralized the controls . . . wham! Just as suddenly as I'd gone into a spin, I was out of it.

The whole business took forty seconds. I didn't have time to worry. And

I'd managed to save a $15 million airplane long enough for that particular problem to get fixed.

I eventually flew some tests in the F-106 Delta Dart, which was the highest number I reached. It was coming out at the same time the 105 was and they were running parallel tests. I also did quite a bit of the later 104 stuff, though Lou Schalk and Bill Gordon were the primary guys.

About the time I left in 1959 the F-4 Phantoms were coming in. They ran a competition between the F-103 and the F-4H, and the F-4 won. I flew chase on that a couple of times, but I never got a chance to fly one until long after that. In fact, Dick Scobee took me on an F-4 flight when we were running approach and landing tests on the shuttle at Edwards in the 1970s.

Scobee was like Fitz Fulton, one of the few cargo pilots who moved to fighters. Fitz was primarily a guy who flew heavies, but he was flying 104s damn near all the time he was at Edwards. There's still just a little bit of prejudice against guys with that background in the astronaut office. We used to have an old saying: P-3 pilots never make fighter ops.

One of the other things that was warming up when I was at Edwards was the space race. I was down in Los Angeles at a meeting of the Society for Experimental Test Pilots when somebody came in and told us the Russians had launched a satellite called Sputnik.

The fact that the Russians had tested an intercontinental ballistic missile had been in the papers before that. No one knew what they were up to. Some people in the Air Force had a proposal for a Project Man-in-Space-Soonest, Project MISS, which basically involved sticking a man in the nose cone of an Atlas rocket and shooting him into space. That looked to me like a stunt more than anything.

What was more promising was the X-15 rocket plane, then in development from North American. This was a joint N.A.C.A. and Air Force program, with a little bit of Navy involvement, too, aimed at reaching altitudes of fifty miles and speeds of Mach 5 or greater. If we were going into space, this was the way to go.

In April 1958 they created a new branch for "manned spacecraft" in the flight test group. Iven Kincheloe, Bob White, and Bob Rushworth were assigned here.

Kincheloe had flown the X-2, which made him the prime X-15 pilot. He was clearly the guy somebody higher up in the Air Force had chosen to become the first man in space. Kinch was a good pilot, but there were better stick and rudder guys around. But he came out of Korea as an ace and knew a lot of high-ranking people in the Air Force. He was also a good politician and had a real outgoing personality. So he became the golden boy.

Unfortunately, just a few months after this, on July 26, 1958, Kincheloe took off from Edwards in an F-104, flying chase on Lou Schalk. Kinch's 104 flamed out before he reached two thousand feet. Because the 104 still had that downward ejection seat, he had to roll the plane before he could get out, but he was too low. His chute never opened fully.

Bob White took over the X-15, and that's when I took over the 105. So I got busy at that time and didn't worry too much about what else was going on with Sputniks or Project MISS for the next few months.

One day in January 1959, however, I got sealed orders: I was requested report to Washington, D.C., for a classified briefing.

9

PROJECT ASTRONAUT

Ever since Sputnik's launch in October 1957, there had been a regular war going on in Washington. Space was the new "high ground," and everybody wanted to take it.

The Air Force thought it was the logical service to have some kind of space force. It had already won a big battle with the Army over control of ballistic missile technology—the Army wasn't going to be allowed to have any missiles with a range greater than two hundred miles. (This decision, by Secretary of Defense Charlie Wilson, hadn't stopped the Army's Ballistic Missile Agency from continuing its "research." The ABMA, whose key player was Wernher von Braun, still managed to launch America's first satellite in January 1958 aboard a version of the Redstone missile called the Jupiter-C.)

The Air Force's Western Development Division, under General Bernard Schreiver, had a huge program for developing intermediate and intercontinental ballistic missiles. They had three different missile systems in the works, the Atlas, the Thor, and the Titan. (It seemed like a duplication of effort at the time, but thirty-five years later all three of those missiles were still in active use as space launchers.)

They had formed the Advanced Research Program Agency, which was where the Man-in-Space-Soonest idea was located.

The flying Air Force, of course, was involved in the X-15 and had even begun to develop its own winged spaceplane, the Dyna-Soar. This struck me as the most logical next step and where I wanted to work.

What happened is that President Eisenhower, apparently because he didn't want to distract the Air Force from its missile- and bomber-building programs, and also because he wanted to keep things small, had nixed the idea of a big military manned space program. He had asked a panel to find some existing government agency to handle the project, and they had turned to the National Advisory Committee for Aeronautics.

N.A.C.A. had been founded in 1917 to do aeronautical research. Over the years it had grown to where it had eight thousand engineers at five different centers—Langley Field near Norfolk, Virginia, was the oldest; Ames in Palo Alto, California; Lewis in Cleveland; the High Speed Station out at Edwards; and also a launch facility at Wallops Island on the Virginia coast. N.A.C.A. had a highly respected bunch of people. They performed research into the design of supersonic vehicles, for example, which led directly to the design of the X-1.

On October 1, 1958, N.A.C.A. became the National Aeronautics and Space Administration, absorbing all the old N.A.C.A. centers in addition to the Army's Jet Propulsion Laboratory in Pasadena. The first administrator was the president of Case Institute of Technology in Cleveland, Keith Glennan, a former member of the Atomic Energy Commission.

Even before N.A.C.A. became NASA, some of its engineers at Langley Field started working on an idea for a manned satellite. They had designed a blunt-edged cone that would fit on top of a Redstone or Atlas missile. It was to be called Project Astronaut.

The first thing that happened was that NASA was told to lose the Project Astronaut name: it put too much emphasis on the man in the spacecraft. Abe Silverstein came up with the name Mercury, and that was announced by Keith Glennan, the NASA administrator, on December 17, 1958, the anniversary of the Wright Brothers' flight at Kitty Hawk.

Other parts of NASA had spent the last few weeks of 1958 trying to decide just who ought to be the candidates for Project Astronaut. A selection team was formed that included three military medical guys: Lieutenant Colonel Stanley White, USAF; Captain William Augerson, U.S. Army; and Lieutenant Robert Voas, USN. Working with them were Warren North, a NASA test pilot; Charles Donlan, a NASA engineer who was deputy for the Mercury program; and Allen Gamble, an industrial psychologist.

They came up with a whole bunch of professions from which to get astronauts, including test pilots, Navy submarine officers, high-altitude balloon researchers, even mountain climbers. They were going to issue a public call for volunteers, run 150 of them through some examinations, and pick a dozen. On December 9, this notice was actually published in the *Federal Register*.

Even as it was published, however, the committee was having second thoughts. Getting some of these people trained in a reasonable time was going to be the real problem. Bob Voas, for one, suggested that NASA should simply choose test pilots. They went a step further and decided that the best source of test pilots was the military. (Military pilots' records were detailed and available, and you wouldn't have to be negotiating salaries with active duty officers who volunteered.)

A week after Christmas 1958 this idea was briefed to President Eisenhower, who approved it in about five minutes, thus killing the idea of a national call for volunteers.

So the committee screened the service records of 508 military test pilots and came up with 110 who met the initial qualifications, including me.

These 110—58 Air Force, 47 Navy, 5 Marine—were divided into three groups of about the same size. Invitations were sent out to each group in turn, asking them to come to Washington for a briefing. The NASA committee had no idea how responsive test pilots would be; when most of the guys agreed to at least hear the proposal, NASA didn't bother going beyond the first two groups, so only 69 of those 110 even got an invitation.

Obviously quite a few guys from Edwards got called in there with me—like Howard Lane, who had been a class behind me at the test pilot school and was now the head of the fighter branch in test ops. The invitations weren't limited to those of us doing flight test, since the records included whoever had graduated from the Air Force and Navy test pilot schools, guys who were doing test pilot work at Wright-Patterson or Eglin or Patuxent River, even guys who had already rotated out of test by that time.

They even had some guys from flight test engineering, people who had been through the school but had been shuttled off to the engineers. They weren't really test pilots.

I had no particular interest in spaceflight—some in astronomy. I had picked up a book about Edwin Hubble, the guy who first figured out that the universe was billions of light-years across, at one point. And while I suspected this mystery summons to Washington might have something to do with spaceflight—NASA announced in January 1959 that McDonnell Aircraft in St. Louis was going to build its Mercury capsule—I wasn't sure.

It turned out I was in the second group scheduled for briefings. Some of the guys in the first group came back to Edwards and spilled the beans, so I knew what to expect.

NASA was so new, its whole headquarters staff fit inside the Dolley Madison House, facing Lafayette Park in downtown Washington. On Monday, February 9, I joined thirty-some other guys for the second in a series of briefings on Project Mercury. Making the presentation was Charlie Donlan, Bob Gilruth's number two guy at NASA at that time. With him was Warren North, the NASA test pilot and engineer, and Bob Voas, the Navy shrink. Gilruth himself was there, as were some other doctors—White and Augerson, and from the surgeon general's staff, Bill Douglas. An air force colonel named Keith Lindell was there, too.

The briefing was pretty straightforward. NASA was planning a series of manned and unmanned flights using the Redstone and Atlas missiles beginning in about a year. The Redstone flights would be suborbital; the Atlas would be orbital. Highest national priority. Volunteers were needed. NASA hoped to select twelve test pilots.

The unmanned capability in the Mercury turned some guys off. Not me. I figured that as aircraft flew higher and faster, ultimately you had to have a system that had an unmanned capability.

I also knew that a test pilot was the best person for this—not just for the flight itself, but to help in the development of the program.

So after hearing the briefing, I agreed to continue with the process. I told NASA I was still interested in the program, then went back to Edwards to wait. I think the enthusiastic reception surprised NASA. They had been worried that test pilots wouldn't stick around for a program like this, which is why they'd figured on hiring a dozen. But right about that time they cut the proposed number to six.

(One idea that was raised in the briefings was that NASA wanted its astronauts to resign from their military service and become civilians. I won't say it was shouted down, but it was dismissed pretty quickly. Given the state of NASA and Project Mercury in February 1959, you'd have had to be an idiot to give up your Air Force or Navy career to join them.)

Other Voices

THOMAS McELMURRY

I was in that second briefing with Deke. At the time I was assigned to Holloman Air Force Base in New Mexico, having gone through the test pilot school in Class 56-B, the year after Deke.

I was one of those guys who was pretty skeptical about NASA and Project Mercury. First off, they wouldn't promise that if you joined the program, you'd get a flight. Second, when asked about staying in the cockpit—being able to fly airplanes—one of them actually said, "We can get you excused from that." Well, that was the last thing any of us wanted!

I figured since I had come all this way, I would at least go through the first interviews, and I must have done pretty well. I had pretty good qualifications, including a master's degree from the Air Force Institute of Technology, in addition to test work.

But I was deaf in one ear! I had gotten myself into the National Guard back in 1940 without having to pass a real stringent medical, and one way or another had managed to get waivers to keep active.

At one point during the first discussions, the matter of my ear came up, and I was told not to worry about it. As the day wore on, though, I was feeling a little less willing to put up with any aggravation. I decided that if I got one more question about it, I was walking out. They asked, and I did. I guess I didn't want to join Project Mercury very badly.

I thought that was it for me and NASA. Imagine my surprise when five years later, in 1964, I wound up going to work for Deke at NASA.

I was invited to go to the Lovelace Clinic in Albuquerque in a couple of weeks for the first round of medical tests.

All the services were itching to turn their medical people loose on the astronaut candidates, though obviously the Air Force and the Navy, with their experience in aviation medicine, had the inside track. Brigadier General Don Flickinger of the Air Force, a flight surgeon high up in the Systems Command, was the one who suggested Lovelace, which was a private civilian clinic even though its director was a retired general named A. H. Schwichtenberg. They had done some of the early Man-in-Space studies in addition to tests on the U-2 pilots. "Send them to Lovelace," Flickinger said. "They're used to keeping secrets."

It wasn't only national security stuff they were supposed to be keeping secret: as a private clinic, Lovelace wouldn't have to turn over medical data to the services, so a guy could go through the tests somewhat secure in the knowledge that he wasn't going to screw up his flying career.

A total of thirty-six guys were invited, but four of them turned it down at that point. That wasn't unexpected. For some people the timing wasn't right to go into Mercury: they might have just been transferred to some new job, for example.

The commander at Edwards, Brigadier General Marcus Cooper, had advised those of us who'd gotten the NASA invitations to avoid this astronaut stuff and to stay right where we were. The Air Force didn't have any confidence at all that Mercury was going to work. If Mercury had been an Air Force program, you might have gotten a little different edge on it, too.

But I think they viewed this as being a disruption of a good Air Force officer's career. After all, the X-15 was in the planning and development stage. That was the real future in space, they felt.

I might have felt the same way if I'd had anyplace else to go. One reason I got interested in Mercury is because I was already worried about getting shipped out of Edwards to somewhere, and I didn't want to go anywhere. I was happy right where I was.

Air Force tours at that time were nominally five years, but it was just being cut down to four. Counting the time I had spent at test pilot school, I had been at Edwards for three and a half years, so I was due for a PCS, permanent change of station.

I only had two choices. I could either go down to Eglin and do some armament testing or go to Wright-Patterson to do all-weather testing and maybe get some more schooling. Those were the only other test jobs around in the Air Force. I wasn't very enthused about either one of them. I had the best job in the world right where I was.

So Mercury hit me just about right. If it had happened in another time frame, I'd have probably said the hell with it.

The thirty-two of us were divided up into five groups of six each, with two guys bringing up the rear, and fed through the Lovelace Clinic in Albuquerque, New Mexico, for a medical testing program that would take seven days and three evenings.

I was in the third group to go through Lovelace, beginning Saturday, March 6. With me was a Navy guy named Scott Carpenter plus candidates Frazier, Crandall, Solliday, and Iddings.

If you didn't like doctors, it was your worst nightmare. Take the standard medical examination, where they check your heart and they check your blood and they stick things in your body, and multiply it by ten. They had a captive group, and they exploited it. I did pretty well on the treadmill, as I recall. I also remember watching Scott Carpenter blow into a tube of mercury for about three minutes, twice as long as anybody else.

Ultimately all thirty-two made it through Lovelace, except for a Navy test pilot named Jim Lovell, who had had some minor liver ailment show up that disqualified him.

The next phase of testing was at Wright-Patterson Air Force Base in Dayton, Ohio. We went through there directly from Lovelace, five groups of five and one of six, each for one week.

While I thought the Lovelace physical examinations were excessive, I could at least see the point. But the idea behind all these so-called "stress" tests at Wright-Patterson escaped me completely. I'd flown combat missions and done operational and test flying for seventeen years by that point, like just about everybody else in the process. The fact that I had *survived* should have told them all they needed to know about stress. What were they supposed to learn from hooking me up to an idiot machine with flashing lights? Or asking me what I saw on a blank piece of paper? Or baking me in a chamber to 130 degrees? At least by putting me in a blackout chamber they let me catch a nap.

Eighteen candidates made the final list after Wright-Pat, and these were submitted to the NASA astronaut selection panel, which consisted of Gilruth, Donlan, Warren North, and a gaggle of doctors: White, Douglas, Augerson, and Voas. I guess they just couldn't get the number reduced to six, so they agreed on seven names. Gilruth passed them on to Abe Silverstein, who got final approval from NASA administrator Glennan.

I was back at Edwards by this time, of course, dealing with a pile of paperwork. Then came the call from Charlie Donlan. It was pretty terse: "You've been selected to join us, if you're still interested."

"Yes, sir, I am."

Then he told me the other guys who were selected. Three Navy guys: Scott Carpenter, Alan Shepard, and Walter Schirra. One Marine: John Glenn. Two other Air Force guys: Virgil Grissom and Gordon Cooper.

When I heard Gordo's name, my first reaction was, something's wrong. Either he's on the wrong list or I am. Gordo was an engineer at Edwards. As far as I was concerned, he wasn't even a test pilot. But I figured NASA had its reasons.

Donlan's call came on Friday, April 2, 1959. I went down the hall and told my boss, Stu Childs, that I'd been accepted, then went home and told Marge. She wasn't surprised, since I had had the attitude from the beginning that I was going to be selected. I called my parents in Wisconsin, warning them not to tell anyone else.

Early the next week I flew to Andrews Air Force Base, outside Washington. On Wednesday, April 8, I met a few of the NASA people, like Walter Bonney and Shorty Powers. It wasn't until the morning of the press conference, Thursday, April 9, that I met the other six astronauts.

My first impressions changed over the years, but this is what they were. Al Shepard seemed kind of cold and standoffish. John Glenn was trying to be nice to everyone. Wally Schirra was telling a joke. (I later learned that Wally remembered the first joke he'd ever been told.) I couldn't help but like Gus Grissom, who was, like me, another Air Force test pilot. Scott Carpenter I had met at Lovelace. Gordo I knew by sight from Edwards.

They hauled us into the auditorium at NASA headquarters and introduced us to the press. "Ladies and gentleman," Bonney said. "The astronaut volunteers."

There was this brief introduction, then a pause for reporters on afternoon newspapers to rush out with this information. Which left us facing the cameras with nothing to say.

I've never seen anything like it, before or since. It was just a frenzy of light bulbs and questions . . . it was some kind of roar. I know I stumbled through a couple of answers. Somebody asked me what Marge thought of my becoming an astronaut. I said I thought she was more worried about finding a baby-sitter.

We were sitting in alphabetical order, so I was at the far left end of the table next to Al Shepard. He was as surprised as I was.

What was the real surprise was watching John Glenn. He ate this stuff up. Somebody asked if our wives were behind us. Six of us said, "Sure," as if that had ever been a real consideration. Glenn piped up with a damn speech about God and family and destiny. We all looked at him, then at each other.

That night we had a big dinner with General Thomas White, the Air Force chief of staff, and Admiral Arleigh Burke, the chief of naval operations. This was in addition to NASA administrator Keith Glennan plus an editor from *Life* magazine named Ed Thompson and a fellow by the name of Leo D'Orsey.

Shorty Powers, our public relations officer, warned us we could expect to have a lot of people bugging us for our personal stories. That was one of our first clues that this was going to be something different from just being a test pilot. We were all green as grass and didn't know what he was talking about. He said it was recommended—and we later learned it wasn't just top NASA management, but President Eisenhower himself—that we band together as a group and accept one of the proposals that had already been made.

D'Orsey, who had represented Eisenhower on a book and magazine deal for his memoirs, offered to act as our representative for no fee. He told us that *Life* had made an offer giving them exclusive rights to our personal stories and coverage of our family lives. We were to get $500,000 spread out over four years.

Everyone recommended that we accept the offer, and we did, signing a contract with *Life* in which everybody participated equally.

Then we were turned loose and told to be back at Langley in three weeks.

10

MERCURY

I flew back to Edwards to tell Marge we were moving by the end of the month. Two weeks later we packed up Kent, now two years old, and started to drive cross-country, figuring to stop in Leon for a few days on our way to Virginia. After the usual misadventures—we were driving an old station wagon and had four flat tires on the way—we got there on Monday the nineteenth and wound up spending the week there.

It wasn't particularly relaxing: reporters kept calling my folks (who were living in town by then, having turned the farm over to my brother Howard) and showing up to talk to me. I realized that this was something I was going to have to deal with from now on.

In Virginia, Gus and Wally and I made plans to move into Stonybrook, a housing development at the south gate of Fort Eustis, which was just up the peninsula from Langley. Gordo and Scott lived at Langley Air Force Base, which was where the NASA center was located. Al had been assigned to the naval air station at Oceana, which was about twenty-five miles south of Langley, just across the bay, so he kept on living there. John was in Arlington, Virginia, outside Washington, D.C. He left his family there and moved into the bachelor officer's quarters at Langley, going home on weekends.

Gus and I wound up living next door to each other. Wally was just a few doors the other side of Gus. (Wally was standing there when we drove up in our broken-down wagon, and said we looked like a bunch of Okies.) So we usually commuted to Langley together. It's probably no surprise that they became my closest friends in those early days.

* * *

On Monday morning, April 27, 1959, the Mercury astronauts went to work. Since we had already won over the competition—the guys who didn't get selected—we weren't immediately competitive with each other. We did everything as a group. Our first offices were on the east side of Langley Field in a building that dated back to World War I. Our first job was to get familiar with NASA and with Mercury.

NASA's administrator was Keith Glennan; his deputy, an old-style N.A.C.A. guy, was Hugh Dryden. Below Dryden was the Space Task Group, whose director was Bob Gilruth. Working with Gilruth were Chuck Mathews (head of the Mercury Operations Division), Max Faget (Flight Systems), and Jim Chamberlin (Engineering/Contract Administration). A few months after we arrived, Gilruth brought Walt Williams from Edwards to add his expertise to the Operations Division. Williams and one of his engineers, Christopher Columbus Kraft, invented the job of flight director.

What was great about NASA in those days was that it was small. And, in a bureaucratic sense, short. An engineer working on a Mercury system didn't have to go through a bunch of layers to get to Gilruth or Faget. By the summer of 1960, a year into Mercury, there were still only seven hundred people working on the thing. That's not a lot as projects like that go.

The organization was also exempt from a lot of civil service rules at the start, meaning it could just hire people who were good without having to go through the time-consuming business of publishing a "register" and having to consider everybody else inside the government first. (We wound up hiring a couple of dozen guys from a canceled Canadian fighter project, including Jim Chamberlin, who became one of the principal designers of the Gemini spacecraft. That never would have happened in another U.S. government organization, and it wouldn't happen in NASA today.) I wound up working with Gilruth, Faget, Mathews, and Chamberlin for years. Hell, I'm *still* working with Max Faget.

We reported directly to Bob Gilruth at Langley. He was a Minnesotan, big and bearlike, but with a reedy little voice that probably made people underestimate him. He was in his mid-forties when we arrived, and had been working on transonic research—speeds a little below and a little above Mach 1—as early as World War II. More recently he had headed a bunch of young engineers working in the Pilotless Aircraft Research Division (PARD) at N.A.C.A. There was nothing in his background to suggest that he would become one of the key people in getting a man to the moon, but he did it.

Bill Douglas became our personal physician. He wound up going through a lot of training and medical tests with us. When they ordered pressure suits, they didn't just order seven . . . they got eight, one each for the seven of us, and one for Bill.

We also had two sort-of training officers, Bob Voas, the Navy shrink, and Keith Lindell, an Air Force colonel, who marched us around together in those first few months. Voas was the guy who communicated with us the most. He wound up becoming good friends with John Glenn, and later worked in his first political campaign.

Lindell was in Mercury because someone put him there. He had been a pilot at one time, but was flying a desk. He stuck with the program until we transferred down to Houston in 1962.

Our first trip in early May was out to St. Louis, where the McDonnell Aircraft Corporation was building the Mercury capsules. One of the key people in the design and development of Mercury was John Yardley, who later wound up working for NASA.

The idea of Mercury itself took some getting used to. I flew vehicles with wings; this was basically a missile nose cone. In the early 1950s, when the Air Force was looking for the ideal shape for an atomic bomb warhead—the shape that would allow it to arrive at a target from space without burning up—it had turned to the N.A.C.A. Two engineers at the Ames Lab, H. Julian Allen and Al Eggers, had come up with the idea of a blunt cone (as opposed to, say, a needle shape).

Max Faget was a feisty guy who came out of Louisiana and had served in submarines during World War II. He was a science fiction reader, like a lot of engineers at the time, and wasn't afraid to come up with something wild or new. Once he settled on it, that was it, too. He wasn't afraid to argue.

It was Faget who adapted Allen and Eggers' blunt cone for a manned satellite as the quickest, cheapest, and safest way to get a man into orbit and back again safely. It would be small enough and light enough to fit atop the existing Redstone and Atlas missiles (which defined how big it could be, and how heavy). In orbit it would orient itself with a series of small rocket motors, the way the X-15 did. To return to earth it would fire a set of small solid-fuel rockets at its base, which would slow it down enough so that it would reenter the atmosphere. The big blunt base of the cone, covered with an ablative material (something that was designed to burn or boil off, taking some of the heat with it), would take the brunt of the heat.

You couldn't land the damn thing, of course: it came down by parachute. And you couldn't put a big enough parachute in it—those weight problems again—to slow you down enough so you wouldn't break your neck on land, so you had to splash down in water.

These were the factors that determined how we trained. No sooner had we gotten back from the McDonnell plant than we went over to the Navy's amphibious base in Little Creek, just east of Norfolk, to train in underwater egress with Underwater Demolition Unit TWO under Lieutenant Commander Donald Gaither. This was great stuff for someone like Scott Carpenter, who was in the Navy and liked to scuba dive. As a typical farm boy, however, I didn't know how to swim. And I'm not sure Gus was too at home in the water in those days.

After a few days of that, we headed down to Cape Canaveral for our first visit there—except for Wally, who was moving into his house. Our guide was an Air Force colonel, George Knauf. I don't know how enthusiastic he was about having us down there. A couple of weeks later some of the Langley and Lewis engineers went to the Cape to launch a boilerplate Mercury capsule—a metal mockup of the real thing—on top of an Atlas for the first time. This was the Big Joe test. They got stuck in the old Vanguard hangar somewhere with close to zero cooperation and support from the Air Force.

At the end of June we visited the Army's Ballistic Missile Agency at Redstone Arsenal in Huntsville, Alabama. This was the home of Wernher von Braun. Wernher and his team weren't part of NASA yet; they still belonged to the U.S. Army, though it was clear they were anxious to move over. (The Army wasn't eager to lose them.)

Everybody was in a hurry. The Space Task Group was planning the first Redstone launches for early 1960, with orbital Atlas flights to follow in a few months. The whole program was supposed to be completed by the summer of 1961.

After a few of these trips we realized that the only way for the seven of us to be involved in the development of the Mercury program was to assign each of us an area to cover. We tried to match the area to some experience or expertise we already had. Al Shepard, for example, knew Navy aviation and fleet command, so he handled the tracking and recovery teams. Since our pressure suit was also based on a Navy design, Wally Schirra covered that. Scott Carpenter, whose flight test work in the Navy had dealt with communications systems, among other things, covered communications and navigation.

John Glenn dealt with cockpit layout. Gus Grissom covered manual and automatic control systems. Gordo Cooper had the Redstone missile; I had the Atlas.

As time went on, we each found ourselves dealing with these areas separately. We all worked pretty well together, going off to our different specialities and bringing back the data.

For example, I spent a lot of time at the Convair plant in San Diego, where the Atlas was being built. My job was to give them the astronaut's point of view while learning whatever it was they needed us to know, then to communicate that to the rest of the group. (We started calling our get-togethers séances; it was how we made decisions on our own.)

This was my first experience with what we call "integration." You had the Mercury spacecraft, a brand-new vehicle still in development, being built by McDonnell in St. Louis. You had the Atlas, an intercontinental ballistic missile, also still in development, and being modified as a manned space launcher as well, being built by Convair a thousand miles away in San Diego. Eventually they had to be put together and launched into space from a third site, in Florida.

Did the extra weight and different aerodynamics of the Mercury affect the Atlas performance? Did noise and vibration of an Atlas launch do something to the Mercury? I had to monitor both sides of the argument for the other astronauts. Since I figured I'd be sitting up there myself one of these days, I was pretty interested in whether the spacecraft guys were talking to the launch vehicle guys.

It was a lot like being a test pilot for an aircraft company. Walt Williams was a lot of help to us here. He had been made deputy for Mercury operations to Bob Gilruth, by then working with Chuck Mathews, too. Mathews concentrated on helping to set up the Mercury tracking and communications systems while Walt and Chris Kraft concentrated on getting the spacecraft off the ground . . . and learning what to do with it when it was.

Walt knew exactly what a test pilot could contribute to this process, and he made sure we were involved. And listened to.

It helped that we were the guys whose butts were on the line. When we visited Convair, the plant that was building the Atlas, they asked Gus Grissom to give some kind of pep talk. Gus wasn't any more comfortable with this stuff than I was, but he came up with the perfect words. "Do good work," he said. They fell all over themselves applauding, and had signs made up: DO GOOD WORK. Because it's our butts.

The *Life* magazine deal was signed on August 5, 1959, and we started taking heat for it from some of the press. Financial matters aside, it was a great thing for us, no doubt about it, because it kept us from having to answer a million questions about our personal lives. Anything having to do with being an astronaut, Project Mercury, NASA—that was fair game. But our wives and children were off-limits, except to *Life*.

I got to know some of the editors and reporters quite well. Ed Thompson was in charge of the whole thing. Loudon Wainright, John Dillie, and Don

Skanke were the writers I dealt with. I looked on them more as allies than as adversaries.

Our biggest problem, in fact, wasn't with each other, it was with the press. We were all in one office there at Langley, just one big room on the second floor, with desks for the seven of us and our secretary, Nancy Lowe. Every time you'd try to get something done, somebody would be in there with a reporter. It was real aggravating. (The first thing I did when I took over the astronaut office was limit press access to one day a week. Every Friday you had to be ready to be interviewed. The other days you were free.)

Shorty Powers was an Air Force public relations type, a real pain in the ass. Since there was no real reporting structure—all the astronauts worked for Gilruth—Powers took it upon himself to do whatever he thought right. So he and I had a few head knockers later on over my one-day-a-week press idea. He wanted to make it a freedom-of-the-press thing.

The first week of July we were up in Washington, D.C., for some NASA political appearances. I got photographed with Senator Alexander Wiley from my home state of Wisconsin. On Wednesday, July 8, we also had our first group press conference since our selection. (We had testified before Congress in late May. I think that was just to let them get a look at us.)

It was about this time that some of the bloom came off the rose. NASA had just announced it was dropping the Mercury-Jupiter suborbital mission as unneeded. At the same time, the Atlas was facing further delays, pushing the first manned orbital flight into 1961 at the earliest.

We were also having some problems. Some of us were bitching privately about lack of per diem and the fact that we had no planes for proficiency flying.

In the services then, and it's still true today, you need to log a certain amount of flying time per month to earn any extra money. In 1959, for an Air Force captain, it was four hours monthly for $190. I'd been saved from loss of pay by the "three months rule," which had allowed me to spread extra flight time from April over May, June, and July, but that was running out. Here we were supposed to be these hotshot pilots, and we were spending most of our time in airports waiting for our luggage. And losing money at the same time!

Well, some reporter for the Washington *Post* caught Gordo in a corner during this trip, and Gordo being Gordo, the whole thing came out.

It turned into quite a scandal. Some people in Congress thought we were ungrateful; others thought NASA was shortchanging us. And poor

Shorty Powers was having to spend a lot of time denying that "astronaut morale" was low.

Within a couple of weeks we got authorized to use some T-33s—the jet trainer version of the F-80 Shooting Star—assigned to the Air Force headquarters unit at Langley Air Force Base. We used those for a while, Gus and I in particular. We'd go out there on a Saturday and just take off, flying the plane cross-country until our fannies were tired. Catch some sleep . . . fly back the next day. You might spend thirty hours flying in a short time, and by the time you got home you didn't want to look at an airplane for a while. But that was the kind of thing we had to do.

After a while we got access to some F-102s, which were a little better than the T-33s. And finally we were given a couple of two-seaters, F-106s.

Other Voices

KENT SLAYTON

When we were living at Langley, my dad was gone a lot, usually out flying. His schedule would usually bring him back home toward dinnertime, and the flight pattern took him over our house.

So my mom and I would go outside on the evenings when Dad was flying, and watch for him to go over. He would waggle his wings as he passed, then Mom would go in and start dinner, knowing he would be home by the time it was ready.

One day—I don't know what got into him—but Dad came over and instead of just waggling his wings, he lit up his afterburner. Pow! A couple of windows broke and kids started screaming.

I loved it. I was jumping up and down and shouting, "That's my Dad! That's my Dad!" My mom, however, was hauling me inside, yelling, "No, it's not!"

I think he got written up for it, too.

After a few months of this endless publicity, we also started to catch flak from our colleagues, the "real" test pilots. It started out as a joke. Walt Williams introduced a couple of the Mercury guys to Milt Thompson, an X-15 pilot, for the first time, saying, "Here's a *real* test pilot." Well, that was Walt's idea of a joke, but the attitude was quite real.

To be fair, calling the seven of us the best test pilots in the world was stretching things. *Life* magazine had been presenting us like we were God's gift, whether we wanted it or not. It was true that Gordo and Scott certainly didn't have the professional qualifications to compare with Scott Crossfield or Pete Everest or Chuck Yeager or Joe Walker.

But I'd done a little frontline test flying, and so had Wally—not only the Sidewinder weapons tests at China Lake, but when he was selected he was project pilot for the F-4H Phantom. Al Shepard was one of the first Navy pilots to land on a carrier with an angled deck. Gus had done all-weather testing. None of this was as spectacular as flying the X-1, of course, but ninety percent of the test pilots in America never got close to stuff like that, anyway. We were working test pilots who happened to get selected.

The most famous grumbler was Chuck Yeager. He'd had his own book published, *Across the High Frontier*—the one years before *The Right Stuff* or *Yeager!*—so he wasn't shy when it came to publicity. I guess some reporter asked him whether he'd be interested in flying in Project Mercury. And Chuck just laughed and said, "No. It doesn't really require a pilot, and besides, you'd have to sweep the monkey shit off the seat before you could sit down."

I think Chuck was just frustrated. Maybe it was personal jealousy. He wasn't the only guy who didn't get selected because he just didn't fit the requirements for education . . . or even for something like height.

The annual meeting of the Society of Experimental Test Pilots was coming up that October. Scott Crossfield, who had done the first X-15 tests, was supposed to discuss that project. Chris Kraft and I had been asked to deliver papers on Project Mercury, and having had about all this monkey shit talk I could stand, I wrote a paper called "The Space Program: A Pilot's Point of View."

I said that there was some kind of misconception that just about anybody, from a "college-trained chimpanzee to the village idiot," could do as well in space as a trained test pilot. "To send anyone of lesser technical ability would be equivalent to sending a pilot to perform a complex exploratory medical operation. Assuming he didn't kill the patient, he wouldn't know what had happened when it was over," I wrote.

I probably overdid it . . . but it sure made me feel better.

At that same time the Air Force was running a program of its own called Discoverer, launching unmanned research satellites out of a new facility in California called Vandenberg Air Force Base. (Launching from California allowed them to send Discoverers into polar orbit, which meant you could cover every foot of the earth's surface in a twenty-four-hour period, as the earth rotated underneath the spacecraft.

What no one said at the time was that Discoverer was a prototype surveillance satellite developed by Lockheed for the Air Force and the Central Intelligence Agency. Its purpose was to take pictures of military facilities in the Soviet Union, then return that film to earth.

They weren't having a lot of luck with it. Some of the satellites just didn't make it to orbit. Others would tumble out of control once they got there. A couple of them were pointed the wrong way when their retro-rockets fired.

Discoverer VII failed on November 22, 1959, the seventh failure in a row. Some reporters asked me what I thought. Maybe the idea was that since we couldn't even get an unmanned satellite back from orbit, a manned satellite was probably impossible.

I pointed out that things were the other way around. Most of the problems with Discoverer—especially the orientation screwups—could have easily been prevented with a pilot at the controls. That was a typical question, though.

Our training was still incomplete because simulators hadn't been built. For example, we hadn't had any desert survival training because it would be several months before anyone realized that certain situations could land us in the Sahara.

We also had some real beauties waiting for us. Up at the Lewis Center they were developing a trainer called MASTIF—the Multiple Axis Space Test Inertia Facility. This was a cagelike contraption in which you got strapped into a couch with a control stick. The MASTIF would start to spin . . . first in one axis, around and around . . . then another, over and over . . . then in a third, like some carnival ride gone wild. Your job was to use the control stick to stabilize the thing, first damping out the motion in one axis, then the second, then the third. This was supposed to train us to do the same thing in space, if some steering rocket got stuck and threw us into a multiple-axis spin.

I don't know what the odds of that happening were. It never has happened, though Neil Armstrong and Dave Scott got into a bad spin on Gemini 8 some years later. But training on the MASTIF sure wasn't much fun.

Marge and I did manage a vacation to New York that fall. There were family visits, too. But most of the time it was travel and work.

Unmanned Mercury launches actually began a few months after the seven of us reported to Langley. On August 21, 1959, the first Mercury boilerplate was scheduled to be fired out of NASA's Wallops Island range atop a rocket called the Little Joe. The plan was to test the Mercury's launch escape system—a smaller rocket mounted on the nose of the Mercury that would pull it away from an exploding Atlas or Redstone.

The way things went that day was typical of those early tests. Thirty-five minutes *before* the scheduled launch, there was an explosion of some kind at the base of the Little Joe itself. The escape tower rockets fired, pulling the Mercury off the Little Joe to an altitude of maybe eighteen hundred feet. . . . The tower separated, as designed, and its own separation motor fired to get it away from the falling capsule. The small drogue chute came out of the capsule, but the main chute didn't. Wham! Destroyed on impact with the Atlantic.

It turned out the initial explosion had been the Little Joe itself, which ignited prematurely because of some stray electrical signal. The sensors on the Mercury escape tower were triggered by the Little Joe ignition and shut down, and so they fired. It looked bad . . . but it was all part of the process.

The master plan when we first started was to fly seven Mercury-Redstone or Mercury-Jupiter suborbital missions, then go to the Atlas orbital flights. It was assumed we would all fly suborbitals, and the selection for an orbital flight would be based on our performance in the suborbitals.

Assuming the Atlas was ever ready to go. At the time it had about a sixty percent success rate. (If you had told me that by the end of the Mercury program we'd have gotten all our manned Atlas launches off without using the escape system, I'd have said no way.) I spent a lot of time over the next few months sitting in a damn blockhouse at the Cape watching them fail to make launches. It got so we hated to have the Air Force schedule its own Atlas launches before ours; something was bound to go wrong, which would mean ours would have to be pushed back.

This, like the Little Joe tests, was also very necessary. The Atlas failures showed us, for example, that almost any serious problem would give at least a few seconds warning—maybe not enough for a pilot to react, but enough for automatic systems to do the job.

So Convair developed ASIS, the Abort Sensing and Implementation System. This was a series of sensors planted all over the Atlas. They monitored fuel pressure and electrical voltage—all the systems that kept an Atlas flying. At the first sign of trouble, the sensor would send a signal directly to the launch escape system. The rockets would fire, and the capsule would pull away before the inevitable explosion.

On Tuesday, August 25, 1959, I was at the Naval Air Station in Johnsville (now Warminster), Pennsylvania, where they had a centrifuge. (People think we had these things all over, but at the time there were only a couple:

this Navy one, and an Air Force one at Wright-Pat.) Scott Carpenter was there, too, and so were Bob Voas and Bill Douglas.

Other Voices

BILL DOUGLAS

This was one of our first runs on the centrifuge at Johnsville, and the instrumentation was still pretty crude. For example, we used an old brush recorder—tracing pen on paper—for electrocardiograms. It was set so slow that you really couldn't tell much, but we didn't think we needed much.

We put Deke in the centrifuge gondola and before we started up, there was already some trace of sinus arrythmia, which we assumed was his normal condition. (It wasn't uncommon in healthy young men, and it was the kind of thing that often went away with exertion.)

So, figuring the condition would go away, we did the first run—one negative G, meaning Deke went through it hanging in his harness. The second run was for three negative Gs, same position. The third was a tumble run, from three positive Gs to three negative Gs, with the gondola rotating.

The wheel stopped, and when we checked Deke again, we saw that the arrythmia was still present. That's when we thought we would get a clinical EKG, which led us to the conclusion that Deke had an auricular fibrillation—a flutter in his heartbeat.

Some of the other authorities in NASA were afraid we had caused Deke's condition by putting him through the three-G tumble, but he had the symptoms before he even got into the gondola.

Eventually Deke got so where he could recognize the onset. We'd be sitting in meetings and I'd notice him checking his pulse. It was just a fact of life for him: he remained asymptomatic, with no sweating or fainting.

Before I took my first run, Bill was giving me the routine physical when he noticed something funny on my EKG. He thought there was something wrong with the equipment, so he tried it again. Same thing. So we went ahead with the run anyway, which is something you wouldn't be caught dead doing nowadays. We got to three Gs and nothing changed. So they took me down to see some Navy doctors in Philadelphia. They did some tests, and the EKG went back to normal.

None of us was sure what the implications of this problem were. I certainly had no idea how much of a problem it was.

A month later I got sent to Brooks Air Force Base in San Antonio. This was where a couple of Air Force doctors, Charles Kossman and Lawrence Lamb, took a look at me. (I missed out on a Mercury trip to Convair that week. NASA put out a cover story that I was sick with a "viral infection.")

These two guys concluded that while I had this condition—idiopathic atrial fibrillation—it was nothing to take me off flight status. Everyone was optimistic that I was going to get a Mercury flight, though by now it was highly unlikely I would get the first suborbital.

11

FIRST IN SPACE

Although Shorty Powers had announced in November that we had sort of "graduated," we were still covering new ground in training for the better part of 1960. We were out at Holloman in New Mexico for zero-G flights aboard a C-47. At Wright-Patterson for more of the same. At Lewis to fly the MASTIF, another center to use the ALFA Air Lubricated Free Axis trainer. Pressure suit fittings at B. F. Goodrich. (This is where they took that picture of the seven of us in pressure suits, because it was the one and only time we were ever together in those things.) To Johnsville for more centrifuge runs. Down to the Cape to sit in the Mercury spacecraft simulator.

There was also a ton of political or public relations things to do. For example, I wound up being declared an honorary mayor of Sparta, Wisconsin, and got the key to the city. I got an honorary doctorate of science from Carthage College. For a week in July 1960, the seven of us were dumped in the desert a few miles north of Stead Air Base in Reno, Nevada, for survival training. I spent most of my time that year at the Cape, watching Atlases take off and blow up.

The Cape in those days was pretty low rent. At the southern end you had Patrick Air Force Base, at the northern end you had Cape Canaveral—the actual launch center. In between was Cocoa Beach, which was a small town even by the standards of the early 1960s. Of course, with all the increase in NASA construction and activity, it started to grow like mad: pretty soon you had apartments and motels and restaurants all over the place. It was still pretty casual, however. Working down there meant long, tedious days—sitting around waiting for some hold in a countdown to be worked through, that kind of thing. We were always looking for ways to let off steam. Some guys bought hot cars, for example.

There still wasn't much flying, but I managed to squeeze in a little. (There were reports in the local papers one day in April 1960 of "mysterious" sonic booms along the coast, attributed to some "hot pilot" in an F-104 . . .)

We were getting along pretty well as a group, though we had a bunch of technical disagreements. For example, I felt that the Mercury flight control system should be as close to that of an aircraft as possible. You should control your orientation with a stick and pedals. A couple of the other guys wanted to combine all the control functions into a stick. I lost that one: Mercury, Gemini, and Apollo all used that system. (But the shuttle control system is based on an aircraft, so maybe I won in the long run.)

I did push for calling the capsule a "spacecraft." I won on that.

Other Voices

JOHN GLENN

We did a lot of traveling as a group in those days, and at check-in time at the hotel there would usually be a mad scramble—first of all, to be the odd man out, the one with the single room. But also to avoid rooming with Deke. The reason was this: Lots of kids growing up before World War II used to have operations to have their tonsils or adenoids removed, but not little Deke Slayton. I've been in barracks and BOQs for years, but Deke was the worst snorer I had ever heard. He could rattle the pictures off the wall.

Deke was a person of few words, like Gus Grissom. Quiet, not flamboyant. When Deke and Gus flew cross-country, they must have made the least talkative flights ever made by two people anywhere. Even they joked about it. East Coast to West Coast in ten words or less.

But Deke got all the information. He waited, made a solid decision. Then, while the rest of us were still thinking about it, he would say, "Let's get on with it." That was his attitude toward the job. I don't know how many times I heard that. "Let's get on with it."

I got promoted to major while I was training for Mercury, not that it meant anything in NASA. We weren't thinking about our military careers, since it was increasingly obvious we weren't going to be done with the Mercury program by early 1962. That was the limit of our original military details to NASA. So NASA went to the Department of Defense for a series of one-year extensions. We were in this for the duration.

On July 29 we launched Mercury-Atlas 1, the first Atlas rocket to carry a real Mercury spacecraft instead of a boilerplate version. It had rained earlier in the day, so the sky was still overcast when it took off . . . straight up, looking good. At one minute into the flight, at an altitude of thirty-two thousand feet—at Max Q—*wham*! It just blew up.

It looked bad, and now Mercury was looking at a delay of several months while the MA-1 accident was sorted out. For the first time, the press started having second thoughts about how the program was going.

1960 was also an election year, Kennedy versus Nixon. One of the things Kennedy was hammering Nixon and the Republicans about was how they hadn't been aggressive enough in building more missiles than the Russians, or in catching up in the space race. So a lot of people felt that having Kennedy win would be good for NASA.

Other people weren't so sure. Kennedy might decide that since we were hopelessly behind the Russians, NASA wasn't the organization to handle a manned satellite program, and give the whole thing back to the Air Force. That wouldn't have been good for the seven of us.

Some elements in NASA were charging full steam ahead. Back in the early weeks of 1960 Eisenhower had ordered Glennan to accelerate work on von Braun's big new booster rocket, the Saturn. Saturn had originally been conceived in the late 1950s as the carrier for elements of a future military space station as well as a military base on the moon. With the creation of NASA, the Air Force and the Army quit thinking openly about bases on the moon, but von Braun kept developing the Saturn. In November 1959 he and his team were transferred from the Army to NASA, and Saturn came with them.

Gilruth and the Langley guys figured that if you had a Saturn, you could have a bigger manned spacecraft, one that would carry three astronauts. (Three was considered the ideal number for round-the-clock flight operation; each astronaut could perform an eight-hour shift.) A spacecraft that large would certainly be capable of serving as a sort of laboratory in earth orbit . . . and might even be flown around the moon.

Abe Silverstein, the guy who named Mercury, was also the guy who named the three-man spacecraft Apollo.

It was George Low, who had worked with Silverstein at the N.A.C.A. Lewis Center before coming to headquarters, who pushed the idea of a manned lunar landing. He put together a team that considered all the questions: What kind of spacecraft would you need to land on the moon? What was the lunar surface made of? In case anyone decided to go ahead with such a program, Low would have the answers.

For most of 1960, however, this was one step away from daydreaming. Eisenhower had started the Saturn program, but there wasn't much money for it. He refused to authorize any real manned program beyond Mercury.

On Election Day, November 8, we tried another abort test with a Little Joe and a real Mercury spacecraft at Wallops Island. The Little Joe fired and took off . . . sixteen seconds into the flight, at Max Q, the Mercury escape tower was supposed to fire. It did, but so did the tower jettison motor, which was supposed pull the tower off the Mercury after separation from the Little Joe. The whole thing, Little Joe, Mercury, and escape tower, fell into the Atlantic. Strike two.

Kennedy won; two weeks later, on November 21, we got ready to launch the first Mercury-Redstone. It was to be the first time we used the new Mercury Control Center at the Cape, too.

The launch had been postponed a number of times already, something we eventually got quite used to in Mercury. Everyone was there. In the MCC you had Gilruth and his deputy, Chuck Mathews. Walt Williams and Chris Kraft were running the flight direction. Wernher von Braun was in the blockhouse, along with Kurt Debus. (These were the guys from Marshall who had built the Redstone.) And then there were the seven of us.

The countdown reaches zero . . . the Redstone ingites for maybe two seconds, rises a couple of inches, then settles back down on the pad. Then the damned escape tower fires and blasts itself into the sky. Not the Mercury . . . just the tower!

While we're all standing there watching this fiasco, the housing on the nose of the Mercury flies off, spewing radar chaff and kicking out the parachute, just as it's supposed to do on splashdown!

To make matters even worse, about ten seconds later the escape tower comes crashing down down onto the beach a few hundred yards away.

Worst of all, we still had a fully-fueled Redstone sitting there on the pad, and no way to safely unfuel it. And what if the wind started blowing, catching that parachute? It could have pulled the whole thing over.

It was back to the drawing board.

Actually, the MR-1 mess turned out to have a fairly simple solution. There were these two-pronged plugs at the base of the Redstone that were to be rigged to pull out in the proper sequence as the rocket lifted off. Apparently some technician had found it tough to make the short-rigged one fit, so he had lengthened the rigging without telling anybody. The Redstone lifted off . . . and the shorter prong disconnected, but the longer one didn't. The Redstone's circuitry said, wait a minute, something's not right—and shut down the rocket before it got more than a couple of inches off the ground.

Meanwhile, the Mercury spacecraft is up there saying, we've had liftoff, now we've had shutdown. Shutdown was the point where the escape tower would separate from the Mercury, so what the hell—it separates. Then all the usual reentry events happened in sequence.

That's how we learned. In fact, three weeks later we had that same Mercury spacecraft mounted on top of another Redstone. On December 19, MR-1A fired off as designed. The spacecraft sailed up to 160 miles in altitude, reentered, and splashed down safely in the Atlantic.

The question about which of the Mercury astronauts would go into space first wasn't really the issue. We wondered if an *American* would be the first. Even John Kennedy, during his campaign, had said that if anybody went into space during 1960, "his name was likely to be Ivan."

We assumed by this time that the Russians had their own astronauts training even before we did . . . which turned out not to be the case. Their selection process started in late summer of 1959, right about the time I was finding out I had an irregular heartbeat. They recruited operational military pilots, about three thousand of them, and ran them through a set of medical tests that took several months.

They came up with a group of twenty in February 1960. (Why twenty? The way I heard it later, the chief Soviet rocket designer, Sergei Korolev, was asked how many cosmonauts he thought he would need. He said, "How many do the Americans have?" "Seven," was the answer. "Give me three times as many." In fact, they selected twenty-two, but a couple of guys pulled out at the last minute.)

The Russians used quite a different set of requirements for their first cosmonauts. I suppose this reflected the difference in the two programs. We were supposed to assist in or at least monitor the development of our spacecraft. That's why NASA picked test pilots.

The Russians had a different system altogether. There was a cabinet-level organization called the Ministry of General Machine Building that was in charge of missile development. (The Ministry of Medium Machine Building handled nuclear weapons.) Under General Machine Building was a bunch of design bureaus building ICBMs, launch facilities, and tracking networks. The lead design bureau, called OKB #1, was headed by Korolev.

It was Korolev's bureau that built the R-7, the first Russian intercontinental ballistic missile, and the Sputnik satellite and Luna probes.

By 1959 they had started work on a manned satellite they wound up calling Vostok. Korolev's engineers did all the development on the Vostok, which was a fully automated spacecraft. The pilots—who came from the air force—never even got a look at the thing until they'd been in training for six months. So they didn't have test pilots. Most of their early cosmonauts were about twenty-six years old with maybe four hundred hours of flying time.

Following their first three Sputniks, and some Lunas, the Russians had actually been quiet when it came to launching spacecraft. On May 15, 1960, however, they had put up something called the Korabl Sputnik—which meant "Spacecraft Satellite"—which carried a mannequin and was the first flight of the Vostok. The retro-rockets misfired, however, and kicked the Korabl into a higher orbit instead of returning it to earth.

They tried again on August 20, with a Korabl carrying two dogs. This one came back safely a day later. On December 1 they tried a third time, with another pair of dogs. This one went badly wrong: during reentry, the spacecraft burned up.

This was all stuff we heard. We never got any classified briefings. There were rumors going around in the space community that there had also been some damn big rocket explosion in October 1960. (Some reporter linked it to the reported death "in an air crash" of Marshal Mitrofan Nedelin, who was the head of the Soviet Strategic Rocket Forces.)

Combine this with the knowledge that they were putting dogs and dummies in orbit, and you wound up with even weirder rumors, that they had had some cosmonaut killed on the pad, that kind of thing. None of that ever turned out to be true. The October 24, 1960, explosion was an ICBM, which did kill about two hundred people, including Nedelin.

Since we couldn't get much done looking over our shoulders, we just didn't talk or think about it a whole lot.

MR-2 was launched on January 29, 1961. This one carried a chimpanzee named Ham. There were some problems—equipment glitches on the ground delayed the launch by four hours. During the boost, a valve opened by mistake and started bleeding air out of the spacecraft. But Ham's pressure suit protected him.

The Redstone's operation wasn't entirely nominal, either: it climbed too steeply, meaning that the reentry was also steeper than it should have been. Ham wound up pulling fifteen Gs rather than twelve. And on splashdown the spacecraft began to leak and eventually sank.

Ham was one very pissed-off little chimp by the time they got him back to the Cape. But the flight was a qualified success: adjustments could be made to prevent the problems that plagued Ham.

The Atlas problem got solved, too. It had simply blown up on reaching Max Q. Analysis showed that the thin-skinned Atlas just couldn't bear the stress of a heavy payload. Some thicker-skinned Atlases were on the way, but there wasn't time to wait for them if we were going to beat the Russians. NASA engineers put a stainless steel band around the weakest part of the Atlas, and on February 21, 1961, they fired it successfully. A Mercury spacecraft number six rode right through Max Q up to 150 miles, then dropped safely down into the Atlantic eighteen minutes later.

At the end of 1960, Gilruth asked us each for a peer rating, to look at the other six and rank them in order of qualifications for the first manned Mercury-Redstone flight. I don't think he was just handing the selection over to the seven of us. I think he just wanted to know if we agreed with his judgment.

I do know that John Glenn didn't think much of the peer vote process. On some issues, everything from our personal lives to technical matters, there had been splits in the group, usually with John and Scott on one side and the rest of us on the other. John had made himself into the clear front-runner, at least in the public's eyes. He had been enthusiastic about working with everybody, from the doctors to *Life* magazine reporters. He figured he had made all the right moves to win the first Redstone, provided the decision was made from the top. He just figured wrong. Well, it wasn't the last time John ran for something without having the votes he needed.

On January 20, 1961, the day before John F. Kennedy was inaugurated as president, Bob Gilruth called the seven of us into his office at Langley to tell us who would make the first Mercury-Redstone flight. The decision, he said, was based on peer ratings, on evaluations from our training officers, and on his own judgment.

Al Shepard would pilot MR-3. Gus Grissom would pilot MR-4. John Glenn would be backup for both.

I had had that medical problem crop up in August 1959, about four months after we started, so I pretty well concluded I didn't have a chance in the world of getting the first flight. There was too much nervousness about that. Besides, we were sure there wouldn't be such a big deal about that first flight. We guessed wrong. So I felt great for Al. I was also happy that Gus was going to get a flight, too.

So far, so good. Gilruth was the most honest, straightforward guy I ever knew. If he decided Al should fly first, then it was all right with me.

But then the press picked up on the idea that the first American in space had or was about to be chosen. And *everybody* assumed it was John Glenn. I'm sure it annoyed the hell out of him, having people sidle up to him and giving him congratulations.

Since there hadn't been a public announcement, John felt that there was still time to get it changed. I don't know what exactly he did, or who he talked to, but it was clear to the rest of us that he was lobbying for a reversal of Gilruth's decision. There was a lot of backbiting. I didn't pay any attention to it. I went off to the Cape to fly chase on the Mercury-Atlas 2 launch.

But then NASA pulled one of its classic Mickey Mouse routines. On February 20—a month after the actual selection—they announced that Glenn, Grissom, and Shepard had been named the prime *candidates* for the first manned Mercury-Redstone flight. But they didn't say that Shepard was going to make the flight! They gave the names in alphabetical order and said that the decision might not be made until the day of the launch.

As it was, they kept it secret until May 2, when we had the first serious launch attempt, and everybody saw that it was Al Shepard walking up to the Redstone, not John Glenn.

Of course, everybody *inside* the program knew Shepard was going to fly the first one, Gus the second, and Glenn the third. That was the way it was set up to go.

Those three guys kind of became a separate group from the other four of us, because they were already on a flight schedule. I think *Life* magazine got into the act with some horseshit about the Gold Team (Glenn, Grissom, and Shepard) and Red Team (the rest of us). Wally, Scott, Gordo, and I even had to have a press conference at Patrick Air Force Base a couple of days after the announcement to reassure everybody that we weren't depressed or something.

We weren't depressed. We were too busy.

The Kennedy administration took over in Washington. Keith Glennan, being an old Republican, had turned in his resignation right after the election, but the Kennedy people took their own sweet time about finding a new administrator. They got a professor from MIT named Jerome Weisner to put together a report on NASA and Project Mercury that turned out to be pretty negative. Weisner was never a believer in manned space flight, and he wound up as Kennedy's science adviser.

I guess they had a tough time just finding someone who would take the head NASA job. The story that went around was that seventeen people turned it down before James Webb agreed to take it—and even then LBJ, Lyndon Johnson, the new vice president, had to lean on him a little.

Webb was a real operator. He was a country lawyer who had been a Marine fighter pilot back in the 1930s. He had also served in the Truman administration as director of the Bureau of the Budget, and later worked in the oil industry. He was politically connected to Senator Robert Kerr of Oklahoma, who took over as chairman of the Senate Committee on Aeronautical and Space Science when LBJ became vice president. Webb wasn't technically trained, but he knew how to get things done. He turned out to be the best administrator NASA ever had.

After the success of MR-2 with Ham and MA-2, Gilruth and the Space Task Group wanted to move right on to the manned Redstone flight. (On March 9, 1961, the Russians got back in business with their fourth Korabl; this one carried a dog and was recovered safely.)

But Wernher von Braun and a couple of his senior people, Eberhard Rees at Marshall and Kurt Debus at the Cape, were concerned about the Redstone's performance during Ham's flight. They wanted one more test before they put one of the "Gold Team" boys on top of it.

Nobody was happy about the delay, since it would postpone MR-3, the manned launch, for a month. There wasn't even a spacecraft available for the flight—von Braun's guys wound up having to use one of the old Little Joe boilerplates. This added test flight, known as MR-BD (Mercury-Redstone Booster-Development), was launched on March 24. Changes were made in the Redstone itself to improve performance, and they all worked.

The very next day, the Russians launched their second Korabl in three weeks. This one carried another dog, which was also recovered safely on March 26. Anyone who was watching knew they were ready for a manned launch.

So were we. Though, in looking back, we could just as easily have launched Alan Shepard on March 24. Al certainly thought so.

On the morning of April 12, 1961, an R-7 rocket blasted off from the Baikonur Cosmodrome in Kazakhstan carrying another Korabl, now called Vostok, or "East." Inside was a twenty-seven-year-old senior lieutenant named Yuri Gagarin. Vostok and Gagarin made almost one complete orbit of the earth, flying over Japan, the Pacific, South America, and the South Atlantic. Vostok's retro-rockets fired over Africa; the spacecraft descended across Turkey. As the Vostok parachute deployed, Gagarin ejected—by design—and landed in a field near the city of Saratov.

Radio Moscow put out the news while he was still in orbit. The Russians had beaten us again.

We were all down at the Cape getting ready for Al's launch as well as the next unmanned Mercury-Atlas. A reporter got hold of Shorty Powers in the middle of the night and broke the news of Gagarin's flight to him. All that Shorty, the professional public relations guy, could think to do was grumble, "We're still asleep down here." That made for some pretty amusing headlines: RUSSIAN IN ORBIT! . . . AMERICA'S SPACEMEN STILL ASLEEP!

The wheels started to turn at the White House the day of Gagarin's flight. And on April 18 some CIA-backed rebels invaded Cuba at the Bay of Pigs, where they got shot up and arrested by Castro's army. So it was a bad week for Kennedy. He was really looking for something new and exciting. In the next week he and Johnson had James Webb, Robert Seamans, Hugh Dryden, and a bunch of NASA people in to the White House explaining just what the possibilities were that we could beat the Russians at anything in space. The conclusion was that we might beat them to the moon.

The moon was pretty far away to those of us down at the Cape. On April 25, Mercury-Atlas 3 was launched. It had originally been intended as a suborbital test of the abort system, but two days after the Gagarin flight, Gilruth and his people decided to send this spacecraft into orbit.

The Atlas was one of the new "thick-skinned" models. Forty seconds after launch, it failed to roll and pitch to its proper attitude. It was heading straight up, so the range safety officer blew it up. (From the ground the fireball appeared to engulf Gus Grissom, who was flying chase in a 106. It was just the angle we were looking at: Gus was fine.)

The Mercury abort system worked perfectly, so we wound up meeting the test's original objective. For the Atlas, however, it meant more delays.

Mercury-Redstone was the first concern, anyway. We were spending our time rehearsing—*simulating* was the word—Al's flight. Of course, thanks to the NASA Mickey Mouse, nobody was supposed to know whose flight it was, so Al and Gus and John were *all* going through procedures in the simulator in Hangar S.

Al kept things loose by turning himself into José Jimenez. This was a character created by a comedian named Bill Dana. Dana had come up with a whole routine about José as an astronaut. "Astronaut Jimenez, what is the first thing you're going to do once you reach orbit?" "I'm gonna cry a lot." That kind of stuff.

The press descended on the Cape, too. The Russian flight had been totally blacked out: nobody knew what the rocket or the Vostok really looked like. The Russians were lying about the location of the Baikonur Cosmodrome—the launch site was actually near a town called Tyuratam. (Baikonur was the name of a place about two hundred miles away from Tyuratam. The CIA knew where it was, of course. They'd flown U-2 spy planes over the place as early as 1956, before the Russians were even finished building it.) The only Russian cosmonaut anyone was allowed to know about was Gagarin himself; even the people who designed and built the Vostok were a state secret.

America's decision to do the whole thing out in the open was pretty unusual. It was certainly unusual to those of us from military flight test: nobody was covering X-2 flights live on TV. It was a great idea—unless something went wrong.

Everything was ready to go on the morning of May 2 . . . except the weather. Al waited for three hours in his pressure suit as the rain came down. At the point where they were going to have to put him in the spacecraft, they scrubbed the launch.

Since this was the first time the press knew it was going to be AL SHEPARD as America's first man in space, the business of the scrubbed launch didn't seem that important. We were ready again on the fourth, but had to scrub again. Finally we were ready on the fifth.

All seven of us had our roles to play. Al, of course, was the prime pilot. His backup, John, would do the final checkout of the spacecraft before Al got into it. Gordo was the radio go-between—capsule communicator, or capcom—at the blockhouse. Gus joined Gordo. Wally and Scott were up in F-106s, ready to fly chase on the Redstone. I was going to be the capcom at the Mercury Control Center.

At a little after one in the morning, May 5, Bill Douglas woke Al. He had breakfast, then went into suiting. Just before four he climbed into the transfer van and rode out to the pad. By five-thirty he was in the spacecraft; the hatch was bolted on around six.

For this first launch Al carried a personal parachute. I guess the idea was that if there was some failure of the two Mercury recovery parachutes, he might have a chance to blow the hatch and bail out. I don't think it was a real possibility, and since the personal chute made the inside of the spacecraft even more cramped than ever, the idea got dropped.

During this phase Gordo was doing most of the talking to Al, who had decided to name his spacecraft *Freedom*. Since it was Mercury spacecraft number seven, it became *Freedom* 7. A lot of people had the mistaken idea that "7" came from the seven Mercury astronauts. Not at first. It only became true when Gus Grissom named his spacecraft *Liberty Bell* 7.

There were holds in the countdown. Al had been awake since one, so there was some concern in the MCC about his health. Since no one had bothered to think that urinating would be an issue during a fifteen-minute flight, there were no sanitary facilities in Al's suit. He wound up having to turn himself into a "wetback," as he said.

At T-minus fifteen seconds communications shifted from the blockhouse to the MCC, from Gordo to me. At 9:49 the count reached zero; the Redstone lit up. A couple of seconds later, up she went. The vibrations in the spacecraft were pretty severe. Al later said he could barely read the instruments and his voice was shaking, too. "Roger, liftoff, and the clock is started."

"You're on your way, José!" I told him.

Fifteen minutes later he was floating on the water three hundred miles downrange. The flight was, in Shorty Powers's words, "A-okay." (Al never said it.)

Things started changing for NASA after Al's flight. Kennedy and Johnson had bought George Low's idea that the best shot the United States had of beating the Russians in space was a manned landing on the moon. On Thursday, May 25, Kennedy went before Congress to deliver a speech on a bunch of issues. Toward the end of it he said, "I believe this nation should commit itself to achieving the goal, before this decade is out, of a man on the moon and returning him safely to earth." The reaction in Congress was underwhelming. Eventually, however, they went along with massive increases in the NASA budget—just what Eisenhower hadn't wanted.

We were supposed to plan for manned earth orbital Apollo flights beginning in 1965, a manned circumlunar mission in 1967, and a manned lunar landing and return within two years of that. It seems plain and simple in retrospect. Here was a job that needed to be done. I didn't realize I would spend the rest of the decade trying to do just that.

Gus had become my best friend of the seven. Even before we got planes of our own, he and I had managed to borrow a T-33 from time to time. We would go off, spending all weekend flying around the country. Keep going until we got tired, stop and swap seats. Get gas. When your ass got tired, stop and hang it up for the night. We'd go coast to coast, just riding it around, piling up the flight time. We got to know one another pretty well flying, and just doing stuff like hunting.

The Dismal Swamp in southern Virginia is really a swamp—a wilderness with a dike built around it five miles on a side. You walk off that dike, go through an L, and then you're just into the swamp. To hunt down there you need to stand on the dike. We were deer hunting, using my .357 Robinson—Gus was about three hundred yards down from where I was standing, and I heard him shoot. Then he disappeared and I didn't see him for an hour. I figured, well, I better see what's going on. . . . Here he comes struggling back out of the brush, saying, "Hey, I shot a goddamn bear out of a tree and I can't find him."

So we both went back into that thick brush, crawling around on our bellies. I had a compass, but it didn't make any difference: get a hundred yards away from the dike and you didn't know what the hell direction you were going. We couldn't have found our way back.

But by a strange miracle we found that bear lying right under the tree. So I took the guns and Gus grabbed the bear. It must have taken us two hours to get that son of a bitch onto the road. We were seven miles from the car at that point. So we got poles, put the bear on them, and started carrying him back to the car. That's when he got real heavy.

The first reality set in when we took our big trophy down to Gus's garage and hung him up to skin him. Then little Scotty Grissom walked in and said, "Hey, where'd you get that *cat*?"

For Gus's flight, Mercury-Redstone 4, I was the capcom at the blockhouse. Gus had named his spacecraft *Liberty Bell* 7 and even had a crack painted on the side of the spacecraft, just like the crack on the real thing.

Liberty Bell 7, which actually was the eleventh Mercury spacecraft, had some things Al's didn't. For one, Gus had an actual window, so he could see out. For another, the side hatch had been modified so it could be blown off in a hurry, in case the spacecraft started sinking. (This was a response to what had happened to Ham's spacecraft.)

We suffered through one canceled launch on July 19. Two days later, Friday, July 21, Gus took off just the way Al had.

His problems started after he had splashed down. The procedure with Gus was supposed to be the same as it had been with Al. The chopper would hook on to the spacecraft, raise it out of the water. The astronaut would blow the side hatch and climb out into a horse collar that would haul him up to the chopper.

Fine. Gus was floating in the spacecraft as the recovery chopper came overhead. He told them to give him five minutes to mark some instrument readings with a grease pencil, then he'd be ready to proceed. He took off his helmet and gloves and went to work.

Five minutes passed. The chopper had hooked on to the spacecraft, but hadn't started to lift it. Suddenly the hatch blew off. Water started pouring in through the opening. Gus wiggled his way out, into the water.

A Mercury pressure suit was a lot like a balloon. It would float—as long as you didn't have a leak. There was Gus in the water with waves crashing over his head from the prop wash of the chopper. (He had his helmet off, but there was a rubber neck dam—like a turtleneck—that kept water from getting into the suit from the top.) He started feeling water inside his suit, and remembered that there was a valve on the front of it where an oxygen hose fit. That valve was open and letting water in.

A Mercury pressure suit filling with water was going to drag Gus under. He started waving like mad at the guys in the chopper: come and get me! But they thought he was waving to tell them he was fine. And they were worried about pulling up the spacecraft, which was full of water now. Eventually they had to let it sink. Only then did they drop the horse collar for Gus.

Back in the blockhouse I knew nothing of this. I knew Gus had splashed down safely.

Everybody later developed an opinion about what had really happened. There was one group that thought Gus had screwed up . . . that he had blown the hatch early by mistake. This, of course, is my gripe with Tom Wolfe's *The Right Stuff.* He was kind of tough on Gus anyway, and in the matter of the hatch pretty much convicted him.

Other Voices

WALT WILLIAMS

Even by the time of Gus's flight there were still some problems with the way the astronauts were being trained. I felt there was too much emphasis on procedures and not enough on the systems. What happened with Gus is a classic case.

When he splashed down, he started going through the procedures just the way he'd been trained. Helmet off, mark the instrument panel, all of that. One of the steps was to remove the safety pin from the handle that would blow the hatch.

The problem was, the spacecraft was bobbing around on the water. There's no room inside there anyway, and trying to manage a helmet and gloves and marker and safety pin in that environment was a little tricky. And Gus was never comfortable on the water.

If he'd been trained to think of the system as opposed to a set of steps, he would probably have waited until the last possible second to take the safety pin out of the hatch.

I never believed that he panicked and blew the hatch. (Everybody who blew the hatch manually got a bruise, but Gus didn't.) I do believe, however, that it's very possible he bumped the switch by accident with the helmet.

No one should have the idea that Gus was going around being defensive about this hatch thing. I didn't sleep with him, so I have no idea what he may have said to his wife, Betty. But he told me, sure, there was a possibility he had banged the thing by mistake. There wasn't a lot of room inside *Liberty Bell* 7, and it was sloshing around on the water something fierce.

All I know is that when Wally Schirra blew the hatch on his flight, he wound up with a big bruise on his hand. Gus never had one. We got one joke out of it, though. Never again would we launch a manned spacecraft with a crack in it.

Gilruth had canceled MR-6 a couple of weeks before Gus's flight, on the grounds that in spite of the problems we were just about ready to go on to the Atlas. We also looked like wienies with these suborbital flights after the Russians had put Gagarin in orbit.

John was the obvious candidate for MR-5, and for a couple of days after Gus's flight and the lost spacecraft, it looked as though it was going to happen.

None of us really thought a third suborbital flight was necessary. Any time and energy and money spent on them didn't get used on the Atlas, where it was needed most. I was worried that John might not like all of us lobbying for the cancellation of his Redstone flight, but he was right in there with us. Of course, it turned out he had his sights set on that first Atlas orbital mission.

NASA itself was starting to change. The original idea had been that the Space Task Group—Project Mercury—would move from Langley to Beltsville, Maryland, where a new control center was being built for unmanned satellites. This idea got put on hold in early 1960, the reason being it was the wrong time to uproot the whole organization.

By the summer of 1961, with the announcement of Apollo, it was pretty clear that Beltsville—which had become the Goddard Space Center—wasn't going to be big enough to house a program management that large. Keith Glennan had wanted the Space Task Group to move to the Bay Area, where the Ames Center was. But Jim Webb realized that a whole new site was needed.

The director of Ames, John Parsons, headed a search committee to select a site. They looked at locations that were accessible by water and air and were near schools, research facilities, power, and industry. I don't know how many they evaluated—locations in Florida, Massachusetts, Louisiana, California, Texas, even Missouri.

On September 1 the committee announced that the new NASA center would be built in Houston, Texas. Humble Oil donated some ranch land about twenty-five miles southeast of the city on the road to Galveston. The Space Task Group was renamed the Manned Spacecraft Center, and plans were made to start moving from Langley to Houston by the end of 1961 . . . even though the center wouldn't actually be completed for a couple of years. When we got there, we were scattered all over the city in two dozen different locations.

I don't think it was any coincidence that Texas wound up getting MSC, since it was LBJ's home state, and also that of Congressman Olin Teague, head of the House Space Committee. The story is that Webb and Gilruth flew west to look at Houston, and during a bumpy ride Gilruth complained about having to leave Virginia. He loved to sail, for one thing, and there he was on the peninsula, with some of the best sailing in the world.

Webb just looked at him and said, "Bob, what has Harry Byrd ever done for NASA?" Byrd was the senior senator from Virginia.

The Manned Spacecraft Center wasn't the only new thing in NASA. Some land north of Cape Canaveral was taken over for the construction of Saturn launch facilities. An old aircraft plant in Michoud, Louisiana, was turned into the Saturn production facility. More land was acquired in Mississippi as a Saturn rocket test site. An electronics center was set up in Cambridge, Massachusetts.

Contracts for the Saturn V rocket were given to Boeing, North American, and McDonnell Douglas. Von Braun's first Saturn I had a successful flight. And the biggest deal of all, at the time, the contract for the Apollo command and service modules, went to North American as well.

Some of the astronauts were involved in that decision. Twelve aircraft contractors had bid on Apollo, some of them in teams. The two companies that went in solo made it to the final round of selections: Martin and North American.

According to some of the early scoring by the panel judging the proposals, Martin came out ahead of North American. I don't think anyone who knew about aircraft, however, would rate it that way. North American had built the P-51 Mustang, the best prop fighter in the world . . . my old plane, the B-25 . . . the F-80 and F-86 Sabre jet fighters . . . not to mention the B-70 bomber and the X-15. Martin's reputation wasn't nearly as good among people who actually dealt with the hardware, and I think that's what tipped the decision to North American.

It was controversial as hell, however. North American had some pretty arrogant lobbyists working for them, and they weren't shy about taking credit for the contract. There was also the matter of a guy in the vending machine business named Bobby Baker, who happened to be a friend of LBJ's and Robert Kerr's. He had interests in North American.

I don't know the politics involved. I just know that all of us felt North American was a better choice than Martin.

The NASA hierarchy started changing, too. Abe Silverstein, who had been the headquarters guy in charge of manned flight, wanted to go back to the Lewis Center. Webb, Dryden, and Seamans hired Brainerd Holmes from RCA to become the new associate administrator for manned flight.

Gilruth's deputy Charlie Donlan was replaced by Walt Williams. Kenny Kleinknecht became head of the Mercury Project Office.

Webb and Gilruth and the others had realized that some sort of manned program was needed to bridge Mercury and Apollo. During 1961 they had begun to develop what they called an Advanced Mercury. Gilruth announced it during his first official visit to Houston in December 1961 as "Mercury Mark II."

This was going to be a two-man improved Mercury launched by a Titan II rocket. Mark II would be able to maneuver in space, testing rendezvous techniques, for one thing. (No one was quite sure yet just how Apollo was going to get to the moon, but everyone thought rendezvous would be part of it.) Astronauts would be able to stay in space up to two weeks, the proposed length of an Apollo lunar mission. (Mercury missions were limited to three days at most.) It would also be possible to test pressure suits in "extravehicular activity," or space walks—also needed because astronauts would eventually have to leave their spacecraft and walk on the surface of the moon.

Mark II was supposed to commence in March 1963 with an unmanned test flight. An eighteen-orbit manned flight would be next, followed by a series of fourteen-day missions, then rendezvous and docking missions. Ultimately, there were going to be ten or eleven manned Mark II flights, concluding in late 1964.

There were a few goofy ideas in the original Mark II plans. A couple of those fourteen-day missions were to be crewed by one astronaut and one *chimp*. There was even an option for some Mark II circumlunar flights using a Titan-Centaur combination.

Jim Chamberlin became head of the Mercury Mark II project, which got the name Gemini in early 1962.

On August 6, 1961, the Russians sent up Vostok 2 with their second cosmonaut, Captain Gherman Titov. His flight was originally supposed to have been three orbits—the same as our public plan for our first orbital mission. The Soviet Premier, Nikita Khrushchev, apparently pushed Korolev to lengthen the flight to a whole day, just to make us look bad.

He got whatever he wanted. Titov went around the earth seventeen times. There was a TV camera inside Vostok, so you could even see him as he radioed greetings to everybody. We discovered later that he was actually sick for a lot of the flight—probably the first case of space adaptation syndrome—but no one knew that then. There was just this new Russian going around the world every ninety minutes.

One thing Titov's flight did was permanently kill Mercury-Redstone 5. The next flight by an astronaut would be on the Atlas, we hoped by the end of the year.

The Atlas was ready to fly again on September 12, this time on its first orbital flight, a single loop around the earth.

The seven of us fanned out to the various tracking stations. Scott was at the Muchea, Australia, site. Gordo was at Point Arguello, north of Los Angeles. Wally was in Guaymas, Mexico. Al, Gus, and John were at the Cape. I had the toughest assignment, getting stuck in Bermuda for a couple of weeks.

This was the first real test of our worldwide tracking network. There were actually eighteen different tracking sites, including ships, coordinating with the Mercury Control Center at the Cape. Everything went well. Mercury-Atlas 4 lasted one hour and twenty-two minutes; the spacecraft was recovered safely in the Atlantic.

There was another misstep yet to come. George Low and Harry Goett had convinced Silverstein that the tracking network needed a more severe test, so a small satellite was designed and built for launch aboard NASA's new Scout vehicle. For a little program, it had big delays: it was supposed to be launched in early July, but it didn't take off until November 1. And that Mercury-Scout mission lasted about forty seconds. The rocket went out of control almost immediately and had to be destroyed by the range safety officer.

What Gilruth and NASA had planned for the seven of us, I honestly couldn't tell you. We had all assumed we were going to fly Redstones to qualify us for Atlases. Al and Gus could certainly have been selfish and said, "Hey, we think everybody ought to fly a Redstone before they qualify to fly an Atlas." If they'd said that, I'll guarantee you that's what would have happened.

But they were both test pilots and they said, "You don't really need this experience to go fly an Atlas." Late in November there were some leaks in the press that there were three candidates for the first orbital flight: John, Al, and Scott.

In fact, before that Gilruth had already told us: John Glenn was going to make the first Mercury-Atlas flight, with Scott Carpenter as his backup. I was assigned to pilot the second one, with Wally Schirra as mine.

On November 28, 1961, we took the last necessary step in qualifying the Atlas for a flight by an astronaut. Mercury-Atlas 5 was launched carrying a chimp named Enos for a planned three orbits of the earth. I was at the Guaymas tracking station for this one. The chimp couldn't talk to us—he was busy enough pushing levers to get banana pellets—but the engineers had rigged up a recorder to transmit messages from the spacecraft to the ground.

There were some problems with the spacecraft clock, which ran fast. This wasn't just an inconvenience: a few seconds' discrepancy during reentry could mean the spacecraft would splash down miles away from where it was supposed to. The environmental system overheated, too. Worse yet, the automatic control system had a blocked thruster, which allowed the spacecraft to drift out of the proper attitude. At the MCC, Walt Williams and Chris Kraft got to test some of their new rules for flight directors: they had the retro-rockets fired one orbit early. Enos landed safely in the Atlantic, and we looked as though we were ready to go.

The next day the announcement made it official: John Glenn would fly Mercury-Atlas 6 late next month, December 1961. The pilot of Mercury-Atlas 7, scheduled for launch four months later, would be Major Donald K. Slayton.

12

DELTA SEVEN

John Glenn's life became a lot more complicated once the announcement was made. Everywhere he turned, reporters were after him. Scott Carpenter got caught up in some of it. Since Wally and I were only working on the second flight, we managed to slip by unnoticed.

Both MA-6 (John's flight) and MA-7 (mine) were intended to last for three orbits. In the original planning for Mercury, that had been intended as the culmination of the whole program. It was the creation of the Apollo program, then Gemini, not to mention the Russians with their day-long flights, that encouraged us to think about longer flights. (It was also a product of the learning curve. We got more confidence in what the Mercury spacecraft could do.)

But most of the planning for MA-7 depended on what happened with John. And John's flight seemed to take forever to get off the ground.

John wasn't particularly bothered. Nobody inside the program was. This was the first manned Atlas, and given the history of that particular booster, we wanted everything to go right.

When John's assignment as pilot was announced, there was a remote chance that MA-6 would get launched in December 1961. The spacecraft itself, Mercury number thirteen, had been at the Cape since late August. The Atlas booster was also there. According to the original schedule, it should have taken thirteen days from the time number thirteen was mounted atop the Atlas for it to be launched.

It didn't actually get there until January 3. One reason for the delays was the complexity of the Atlas launchpad system; it was much more complicated than the Redstone setup. There was a new "white room" structure at the top of the thing that enclosed the Mercury and the pad crew until about an hour before liftoff. Just getting the procedures on how and when to take that apart took a lot of work. I was part of the spacecraft checkout team for the mission, so I was right in the middle of this.

The first "scheduled" launch date of January 16 got canceled because a problem developed with the Atlas propellant tanks. Things looked like they might go on January 23, and all the press in the world started showing up. . . . Then the weather got bad.

There were six hundred reporters waiting around for four days. Finally, on the twenty-seventh, John suited up and climbed into *Friendship* 7—but the clouds refused to clear. After five hours and no launch, he had to climb right back out again.

The press had been camped out on John's lawn in Arlington, Virginia, all morning. LBJ was around, too. When the scrub was announced, LBJ decided to console John's wife, Annie. The problem was, LBJ wanted to bring about fifty reporters with him, which Annie couldn't handle. And he insisted that Loudon Wainwright, the *Life* reporter with the family that day, get out of the house.

John was in his robe or something on the way to the shower when they brought him the news. He just told his wife he would back her up, whatever she decided. Which turned out to be to keep Wainwright in and LBJ and the reporters out. I understand the vice president was pretty pissed off, and that he wasn't too happy with Jim Webb or Webb's astronauts at that point.

Since the booster had been fueled, it took at least two days to recycle for another attempt. During the unfueling, somebody noticed a leak in one of the Atlas tanks. That gave us a ten-day delay right there. The next attempt was set for February 13. The reporters went home. Even John went back to his house in Arlington. (While he was in the area, he managed to drop in at the White House and brief President Kennedy.)

We didn't launch on the thirteenth, or the fourteenth, or the sixteenth, due to weather. Only on the nineteenth did things start to look good for a launch the following day.

On the morning of the twentieth, Bill Douglas woke up John. I joined them for breakfast, then went out to the pad while John got suited up.

At 9:47, following some more worries about weather and a minor glitch with a bolt on the spacecraft hatch, the Atlas carrying *Friendship* 7 lit up. Al Shepard was the capcom in MCC and I was sitting next to him as John reacted to the Atlas rolling to the proper azimuth . . . then rattled through Max Q. It was a little easier ride for John than it had been for Al or Gus—seven Gs as opposed to eleven.

We got booster cutoff at 2:09 into the mission; the Atlas pressed forward on its single sustainer engine. John mistook some of the smoke from that cutoff for the jettison of the escape tower, which didn't happen until thirty seconds after that. The guidance system began to pitch the vehicle forward so that it would go into orbit horizontally. The Atlas kept wiggling around as the guidance system tried to keep the single sustainer firing through the center of gravity. John later said he felt like he was on the tip of a springboard. As the fuel was expended, gas was pumped into the Atlas propellant tanks to keep them from collapsing. That only made things louder.

At five minutes the single booster shut down, and John was in orbit.

The first few minutes of the flight were full of system checks. John was busy reading a bunch of data down to Al, then to Gus at the Bermuda tracking station.

But then he got to enjoy the sights a little. He saw a dust storm as he flew over Africa. He also proved you could make star sightings from the spacecraft. The city of Perth, Australia, turned on its lights for him, since it was night on that side of the world.

When the sun rose again for *Friendship* 7, John radioed: "I'm in a big mass of some very small particles that are brilliantly lit up, like they're luminescent. . . . There are literally thousands of them!" These became known as John Glenn's "fireflies." They turned out to be frozen particles of gas from the Mercury thrusters which kept traveling right along with the spacecraft.

Before John completed his first orbit, however, there was a problem. One of the telemetry engineers at MCC had a signal in Segment-51, which meant that the landing bag between the Mercury heat shield and the spacecraft itself had somehow inflated. If that was true, there was nothing to hold the heat shield to the spacecraft during reentry. *Friendship* 7 would burn up.

Almost immediately Chris Kraft went to one of his contingency plans: rather than jettisoning the retro-rocket pack, which was held on to the heatshield by three straps, you could leave it through reentry. The straps might hold long enough to keep the shield in place, even if it had come loose somehow.

Max Faget's only worry was that if only two of the three retro-rockets fired, you'd have a little bomb sitting there on the heat shield just waiting to go off. But if only two retro-rockets fired, John wasn't going to be reentering properly, anyway.

There was a lot of debate in the MCC between Kraft, the telemetry officer, and Al about what to do, and what to tell John.

John was dealing with smaller crises. The flight control system was acting up. One time he had no thrust in yaw, the next time he did . . . but the opposite thruster wouldn't work. It made it hard for him to keep the spacecraft in the proper attitude, and cost him fuel.

At two hours, 19 minutes into the mission, MCC via the Indian Ocean tracking ship told John to keep his landing bag switch in the off position. A few minutes later, with John over Australia, Gordo Cooper did the same thing. Gordo went on to ask John if he'd heard any banging noises, which was negative.

Nobody wanted to worry John unnecessarily, especially since there wasn't a damn thing he could do about it. But when John was over the Canton tracking site a few minutes after Muchea, the capcom there spilled the beans about the suspected landing bag problem. "We also have no indication that your landing bag has deployed." Fortunately, John thought MCC was reacting to his fireflies, that the particles might be coming from the landing bag and heat shield.

The debate about what to do continued through the rest of John's second orbit and into his third. On his third pass over Hawaii, as he was setting up for retrofire, the capcom asked him, "We have been reading an indication on the ground of Segment-51, which is landing bag deploy. We suspect this is an erroneous signal." They asked him to put the switch in auto, to see if he got a light in the spacecraft. It was negative, meaning that the spacecraft systems showed the bag to be safely stowed away.

The situation still bothered Williams and Kraft. They still weren't sure what to believe, the ground signal or the spacecraft. Time was running out: *Friendship* 7 was approaching the California coast, time for retrofire.

Since we had also discovered a problem with the clock aboard the spacecraft, Wally—who was capcom at Point Arguello in California—had to count John down to retrofire. Then, just thirty seconds before retrofire, Wally had to tell John to keep the retropack on until he had at least passed over Texas.

John manually adjusted the spacecraft attitude while the flight control system automatically fired the retros, which gave him quite a kick. "It feels like I'm going back to Hawaii." At the Cape they looked at the telemetry to see if that kick did anything to the Segment-51 signal, but it stayed lit.

As *Friendship* 7 began to descend into the atmosphere, passing over New Mexico and across Texas, the capcom at Corpus Christi gave John the news: he would have to leave the retropack on, meaning he also had to override the 0.05-G switch, an automatic program that was supposed to stabilize the spacecraft through reentry, and also hand crank the periscope. All this in about four minutes.

Al Shepard at the Cape was the last to run through these instructions with John before contact with *Friendship* 7 stopped. (The passage of a metal can like Mercury through the upper atmosphere at seventeen thousand miles an hour created a plasma ball that blocked radio signals.) Then we all waited.

Al started calling. "*Friendship* 7, this is Cape flight. How do you read?"

Suddenly we heard John: "Loud and clear. How me?"

A few minutes later he was bobbing in the Atlantic. Postflight analysis showed that the Segment-51 indicator had been erroneous: it was just a loose switch.

What happened after John's splashdown was pretty amazing. It seemed that everybody in the world had followed the flight. Kennedy flew down to Houston to give Bob Gilruth and John the NASA Distinguished Service Medal. On February 26 John was in Washington to address Congress. On March 1 he had a ticker-tape parade in New York City. It was a madhouse.

And I was glad to be out of it. I had other things on my mind, since the next manned flight was going to be Mercury-Atlas 7, spacecraft number eighteen. I already had the name picked out: it was a nice engineering term that described the change in number. Delta, as in delta-vee, change in velocity. My Mercury was going to be *Delta* 7.

One of the things that had started to creep into mission planning for MA-6 and MA-7 was the idea that astronauts should be spacegoing research scientists and physicians. We certainly weren't against this in principle, but Bob Voas, for one, was pushing some idea about having John do a simple test in which he would reach out inside the spacecraft with his eyes open, trying to touch certain switches . . . then *repeat the process with his eyes closed*. Gordo, bless him, was the one who pointed out that this was a pretty stupid thing to be doing inside an experimental vehicle on its first orbital flight.

With the success of MA-6, everybody and his brother came out of the woodwork with some experiment like this. One guy wanted me to release a balloon to measure air drag. Another guy had some ground observations I was supposed to make. One damn thing after another. I had my hands full trying to resist it.

What I didn't know at the time was that some other wheels were turning. The NASA administrator, Jim Webb, had ordered an investigation into my physical condition.

Of course, I was now aware that I had an idiopathic atrial fibrillation. I'd never known what it was before, to tell the truth. I realized that every couple of weeks I would go through a period of a day or two when my pulse would act up. I had no other symptoms. It certainly didn't stop me from working. And I found that if I did some heavy exercise, like running a couple of miles, the thing just went away.

That should have been the end of it. As I said, my only problem to that point was that it kept me from having a shot at being the first American in space.

On February 13, a week before John Glenn's launch, Bill Douglas got a phone call from Colonel George Knauf, one of the Air Force doctors assigned to NASA headquarters. Apparently another Air Force doctor, a Colonel Talbott on the staff of Secretary of Defense Robert McNamara, had been told "by a source higher than the Department of Defense," which usually means the White House, that *John Glenn* had some kind a heart problem. (Talbott had actually gone to Brigadier General Chuck Roadman, the Air Force doctor who was head of NASA's medical staff, who had bounced it down to his deputy, Knauf.)

Douglas denied that Glenn had a heart problem. A few days later, here comes another phone call from Knauf. What about Scott Carpenter, who was then working as John's backup? Bill told him, no, Scott Carpenter doesn't have a heart problem, either. Then Bill pointed out to Knauf that astronaut *Slayton* had a minor heart condition, which everybody knew about. Maybe this was the source of the rumors. That should have been the end of it.

It turned out that Dr. Larry Lamb, one of the Air Force flight surgeons who examined me at Brooks Air Force Base back in October 1959, felt quite strongly that this heart fibrillation should disqualify me from flight. He hadn't said so in 1959, but he said so now. I don't think it was anything personal—this was just his medical opinion. He was pretty much a voice in the wilderness. Bill Douglas, who was also an Air Force doctor, and everybody else had agreed that it should have no effect on my status whatsoever. Everything was noted down in my records, and that was that.

But it turned out that Lamb was also cardiologist to LBJ, who had a whole history of heart problems, including one serious attack back in 1955. Lamb raised hell at a pretty high level—which was how the rumors got started.

So three weeks after John's flight, Jim Webb decided to reopen the investigation into my medical condition. He wanted a panel of civilian doctors to examine me, but General Curtis LeMay said I was an Air Force officer and my ass belonged to them.

I was down at the Cape one night in early March. John's flight was three weeks in the past. Mercury-Atlas 7 was scheduled for launch in May, two months away. Wally Schirra and I were going through the simulator in Hangar S when I got a call: get up to Washington. I didn't know what this was all about, but I got changed, went over to Patrick Air Force Base, and jumped in a T-33.

On Tuesday, March 13, Bill Douglas and I went over to the office of the surgeon general of the Air Force, which was in the Temple Building in downtown Washington. Bill made an opening statement about my history, then they must have had every doctor in the place take a crack at me. It was at least twenty of them.

They concluded that day that I was approved for flight. General LeMay agreed. It wasn't good enough for Jim Webb, however. Secretary of the Air Force Eugene Zuckert suggested that he should have civilian doctors examine me.

So, that Thursday, March 15, I went over to NASA HQ for another board. It was three doctors—Proctor Harvey of Georgetown University, Thomas Mattingley of the Washington Hospital Center, and Eugene Braunwell of the National Institutes of Health.

I was sent into the next room and told to take my shirt off. One by one these guys came in there and poked and prodded me, listening to my heart with their stethoscopes—nothing like a serious medical exam, which was something I knew a lot about by now.

I came back in, and just like that, Hugh Dryden told me I was off the flight. These guys didn't find any medical reason to keep me from flying—what they said was as long as NASA has other pilots without this condition, why not fly them instead?

I hadn't expected anything like this. I was just devastated.

There's a theory that Webb had gotten ticked off at John Glenn, probably the fallout from that business with LBJ and Annie Glenn, and the seven of us in general, because he didn't think he could control us. So, the theory goes, he seized on the issue of my health as a way to send us a message.

I think the decision was political, but not for that reason. NASA knew it would have to publicly disclose my heart condition prior to my flight: there would be medical monitors at tracking stations all over the world who wouldn't know how to react otherwise. Everybody expected this to be a big deal. NASA would be opening itself up to a lot of medical second-guessing. But Gilruth, Williams, and Douglas were willing to take the heat. Gilruth and Williams thought I was the most qualified guy to fly MA-7.

Jim Webb wasn't willing to take the heat on this issue, however. Once he got it into his head that I had a problem that was going to cause newspaper headlines, he wanted me off that flight. It didn't matter that a whole lot of doctors thought I didn't have a problem . . . he was only going to listen to the few who did.

On March 16, NASA actually made me sit through a goddamn press conference about my situation, and it was one of the more annoying experiences I had in the Mercury program. Given what anybody knew about what had led to the decision, it was a joke. At one point a reporter asked me if this problem was brought on by stress, and I tried to make it clear that I hadn't even known about it until someone hooked me up to an EKG machine back in 1959. Then somebody asked what I thought the most stressful part of space flight was. I probably got a bit sarcastic. "The press conference after the flight," I said.

Other Voices

WALT WILLIAMS

Everybody knew there was an outside chance that something would crop up to keep Deke from flying that mission—that medical condition, which we'd known about for a couple of years by that point. To be fair, there were a lot of operational questions you had to deal with. For example, what do you do if he starts fibrillating on the pad? Do you scrub the launch or go ahead? But we were prepared to deal with them because we thought Deke was the best choice to follow John Glenn in Mercury.

But it was awful the way it happened. I don't remember Deke being angry as much as he was hurt—partly for the way it was handled.

I wasn't in the actual meeting where the decision was made, but I think I got the first two phone calls about it. The first was from Marge Slayton, who had apparently heard from Deke. She was just devastated—in tears—that we had done this to him.

The second was from Hugh Dryden, who said, "Slayton's off the flight. Who do we fly instead?" It was a choice between Wally Schirra, who was Deke's backup, and Scott Carpenter, who was John's.

I figured that MA-7 was likely to be more a repeat of John's flight than anything groundbreaking, so why not give it to Scott, since he had already trained for something pretty similar? We were thinking about a seven-orbit flight later in the year, and that would be perfect for Wally. So I was the one who made the decision to replace Deke with Scott.

Wally really didn't like it. I kept telling him to be patient, something better was coming along, but it didn't do much good.

I was pretty angry about the decision, but I pitched in to do what I could to help Scott. So did Wally.

I had worked to keep the experiments to a minimum, since there was a hell of a lot we still hadn't demonstrated with Mercury, such as a reliable flight control system.

Scott had a different perspective. He was always at home with the doctors and scientists—I think he was genuinely curious about the things that interested them. But it bit him in the ass during his flight.

Aurora 7, Scott's Mercury, was launched on May 24, 1962. During his first ninety minutes in orbit he had to race through a bunch of ground observations and other experiments so fast that he kept bumping around inside the spacecraft, activating things he would then have to shut off. A couple of times he accidentally turned on the manual flight control system, which wound up wasting fuel he would need later.

He released the goddamn balloon experiment, and it just sat there outside his window. When the time came to cut it loose, the line connecting the balloon to the spacecraft had wound around the nose.

He wound up using so much fuel orienting himself and correcting his mistakes that he had to spend the last orbit just drifting. This gave him time to investigate John Glenn's fireflies. He discovered he could raise a cloud of them by rapping on the inside of the spacecraft. Fine, but then he fired up the control system and started turning himself around to watch them . . . at a time when he should have been setting up for reentry!

When he did finally remember what he was supposed to be doing, the automatic system wouldn't work—and when he switched to the fly-by-wire system, he forgot to turn off the manual controls, wasting more fuel.

The poor Hawaii capcom tried to get him to stick to the checklist. When the automatic system didn't fire the retros on time, Scott punched them himself—three seconds late, because he hadn't set himself up to do it the right way in the first place. He also had the spacecraft slewed twenty-five degrees to the right.

It was kind of sloppy. *Aurora* 7 survived reentry, but splashed down 250 miles from where it should have. Scott crawled out the nose of the spacecraft, tossed a raft into the water, and waited for the frogmen to get him.

Meanwhile, from what I heard later, things were pretty intense at MCC. Chris Kraft was pretty incensed about the way Scott had handled things, and announced for everybody to hear, "That son of a bitch is never gonna fly for me again."

I was in Australia at the Muchea tracking site. It was a good place to be, all things considered.

13

CHIEF ASTRONAUT

The original decision to ground me only removed me from MA-7. Wally Schirra drew MA-8, now planned as a six-orbit mission later in 1962. Gordo Cooper was assigned as Wally's backup pilot. That still left at least one last Mercury flight, MA-9, which we hoped would go for eighteen orbits, a whole day or more in space, for Gordo or me.

I certainly hadn't given up on flying. I made some changes in my lifestyle, gave up drinking, started working out more regularly—quit doing everything that was fun, I guess.

Bill Douglas worked his tail off to get me a different verdict. He arranged for me to be examined by Dr. Paul Dudley White, the most famous cardiologist in the world. I went up to Boston in June 1962 and let him run me through his tests.

One of the problems, Dr. White said, was that the best way to make a diagnosis on my kind of condition was by autopsy, which fortunately didn't strike him as an option. "Two-thirds of the people with a condition like yours will die young. The other third may never know they've got it and may never be affected." His conclusion was this: "Young man, you're going to live a long time."

I went back to work while we waited for his report, figuring that would be enough to get NASA to clear me for a flight.

Two weeks later he sent in his report, which said, Major Slayton doesn't appear to have a problem . . . nevertheless, if there are astronauts without such a condition, they should be assigned in his place. We'd asked him for his *medical* opinion—was I fit or not?—and he'd given us an *operational* judgment. Sure, but why not fly somebody else?

That screwed me as far as Mercury-Atlas 9 went. Whoever was going to fly it, Gordo or Al or even Gus, it wasn't going to be me.

Of course, the two-man Gemini program was already in development, and I had figured that even if I didn't get MA-9, I'd get my shot in Gemini. But after the Paul Dudley White decision, Bob Gilruth called me into his office. He didn't waste time. "We're gonna have a tough time selling you in Gemini." Meaning I wasn't going to have a chance at a two-man flight, either.

The Air Force got into the act at this point. They had already started to make noises that I no longer met the qualifications for a Class I pilot's license. (Under a Class I you're unlimited; you can fly solo, for example.) Now they were even thinking of disqualifying me as a Class II, which would allow me to fly with another pilot.

At the same time, it turned out NASA needed a senior manager to run the astronaut office, which was about to be expanded for Gemini and Apollo. I was supposed to be helping Walt Williams on some operational problems for Mercury. They had been looking at a couple of military guys when Al Shepard, in particular, decided, hell, if we're going to have a boss, why bring somebody in from the outside and superimpose him on us? He got together with Walt Williams, and with Wally and Gus, and they decided they would rather have me. I talked to Gilruth, and he agreed. I became coordinator of astronaut activities. And my first task was selecting a new group of astronauts.

Complicating things was the fact that a lot of the Space Task Group had already moved to Houston in the early months of 1962, leaving their World War I–era buildings in Virginia to be scattered all over the city of Houston. I wound up working for a while in the very same building on Gulf Highway that my Space Services group later occupied in the early 1990s.

Other Voices

DEE O'HARA

I was an Air Force nurse stationed at Patrick Air Force Base in Florida during the Mercury program, and I sort of grew up with the astronauts. People forget just how special these guys were. Atlas rockets were always blowing up, and they were going to put their butts on the line.

They were all a lot of fun—each of them in his own way. The two who struck me as the most focused were Al Shepard and Deke. Deke had a quiet and gentle sense of humor. Even when there were disagreements, he was always a gentleman. He never did anything malicious.

I've found that health is a matter of pride with pilots. They were used to seeing flight surgeons as their natural enemy. When Deke was grounded, we were all just devastated. I know Bill Douglas tried everything he could to get him reinstated.

I don't believe the decision was politically motivated. Everyone was just afraid—in those days we still didn't know what the stresses were on a man. I think they were just worried that the stress might kill Deke.

He never showed any bitterness that I saw, though he must have felt it. He became more sensitive to other people—not that he wasn't before, but when he was in a position where he had to make decisions that affected the others. He treated them the way he wanted to be treated.

I left the Air Force and moved to the Manned Spacecraft Center in March 1964, and saw how Deke became the astronauts' champion. If you had any criticism of them, you had to go through Deke first. He really protected them.

On the weekend of July Fourth, the seven Mercury astronauts were welcomed at a big Astrodome party. Right around that time Hurricane Carla blew through the area, just to make us feel at home.

A hurricane had blown through the year before, which gave me a great way to look for some property. I flew over the area around the space center site looking for dry land, and found some near a creek on the little town of Friendswood, on the west side of the Gulf highway. (Gus and Wally and the other guys bought on the east side, closer to the center.)

Marge and I hired a guy to build the house, which was going to take a year, and we moved with Kent in a cabana at the Lakewood Yacht Club. It would take a long time before any of us had permanent quarters: when Wally went to the Cape in August for final MA-8 training, he still had the house in Stonybrook, Virginia. After the mission, he went home . . . to a new house in Houston.

Right around this time Congressman Olin Teague of Texas, one of the space program's biggest supporters, got the idea that since there was a West Point for the Army and an Annapolis for the Navy, maybe there should be a space academy for NASA. Teague wasn't the only one who got interested in this, either. I was elected to make the problem go away.

Well, what people don't realize is that the service academies are *undergraduate* institutions. Most of the people in the astronaut office in the 1960s had master's degrees; some of them had Ph.Ds. So you couldn't have had an academy in the accepted sense of the word . . . what you would have had was an institute for the advanced study of manned spaceflight or something like that.

Those who didn't have a master's had something equally important, if not more so, which was test pilot experience. Given the NASA method, which was to include astronauts in the spacecraft development process, a space cadet was about the last thing you needed.

Besides, we had the three service academies to draw from, in addition to about 150 universities. There was no need for a space academy to train half a dozen people every year.

When this was pointed out, Congressman Teague gave up on the idea. It would still come up from time to time, though.

In mid-August the Russians were back in business. On the eleventh they launched Vostok 3 carrying cosmonaut Andrian Nikolayev. Everybody figured this would be a longer flight than Titov's, though nobody could understand why they had waited a year to launch it. The papers were full of rumors that something spectacular was going to happen.

Sure enough, on the very next day, Vostok 4 was launched with cosmonaut Pavel Popovich. Since nobody knew what the hell the Vostok looked like, much less what it was capable of, we were all waiting for Vostok 3 and 4 to have some kind of rendezvous . . . maybe even a docking. If they had, it would have told us the Russians were really ahead of us in their lunar program.

But there was no rendezvous. There were points in the mission where Nikolayev and Popovich could see each other's spacecraft, but only from miles away. They had no ability to maneuver toward each other.

Nevertheless, Nikolayev stayed up for four days—a duration that was beyond the limit for Mercury—and Popovich for three. They were still way ahead of us.

Back in 1961, before Al Shepard ever flew, the armed forces had already been alerted by NASA that more astronauts were to be recruited, so those wheels were already turning. The Air Force and the Navy had done some medical examinations and came up with some candidates, but there were no real inputs—just a list of names. That's where I came in.

I developed a set of criteria that would get me qualified people with minimum fuss. Some requirements changed from the Mercury astronauts to the 1962 group: the Gemini and Apollo spacecraft would be slightly bigger than Mercury, so we raised the height limit to six feet even—or we'd have lost Tom Stafford. We allowed candidates to have a degree in the biological sciences. One thing that got tougher was that we dropped the maximum age from forty to thirty-five; in Mercury we were looking at a program that would conclude in three years. We knew that Apollo would be going until 1970 at the very least. Since I really wanted to be able to interview a guy's boss if I was going to hire him, I insisted on each applicant having a letter of recommendation from his boss.

I figured we'd need about ten new astronauts to make up the Gemini crews and first Apollo crews, knowing full well that another selection would be needed in a year or two.

The new group wasn't going to be restricted to military test pilots, however: civilians were going to be eligible for the first time. We made the formal announcement that we were recruiting in April 1962, and by the deadline of June 1 had received 253 applications. We evaluated the applications, ran a smaller number through a more abbreviated series of medical tests at Brooks Air Force Base in San Antonio, then did interviews for thirty-three—thirteen Navy, six civilians, nine Air Force, and five Marines—at Houston. I chaired the panel interviewing the candidates, with Al Shepard and Warren North.

To be honest, I could probably have just gone back to that group of finalists who went through the Lovelace and Wright-Patterson exams in 1959 and hired another group right there. I'm glad we didn't; that second group of astronauts is probably the best all-around group ever put together. There were nine in all:

Neil A. Armstrong
Major Frank Borman, USAF
Lieutenant Charles Conrad, USN
Lieutenant Commander James A. Lovell, Jr., USN
Captain James A. McDivitt, USAF
Elliot M. See, Jr.
Captain Thomas P. Stafford, USAF
Captain Edward H. White II, USAF
Lieutenant John W. Young, USN

Two of the Navy pilots were from that 1959 selection: Jim Lovell and Pete Conrad. Jim had been dropped for some liver condition while Conrad had managed to show a little too much independence when it came to some of the medical tests. They were both very capable test pilots and I was happy to have them.

The third Navy guy, John Young, had not been eligible for Mercury, since he was going through the Navy test pilot school at Patuxent River at the time. He had gone on to set a couple of world time-to-climb records in jets.

There were four guys from the Air Force. Two of them—Frank Borman and Jim McDivitt—were among the first graduates of the Aerospace Research Pilot School, a postgraduate, space-oriented course at the Air Force Test Pilot School. (Chuck Yeager was the commandant of this school.)

The third air force guy was Tom Stafford, who had been too tall and too green (he was just graduating from test pilot school) for Mercury. Tom had published two textbooks on test flying. There was a fourth, Ed White, who, like Gus Grissom, came from all-weather testing at Wright-Patterson.

Nobody pressured us to hire civilians, though we wound up with two—Neil Armstrong, who was already working for NASA and who had flown the X-15; and Elliot See, from General Electric.

By mid-September Al, Warren North, the doctors, and I had agreed on nine candidates. We passed the list through Gilruth to headquarters, where it was approved for public announcement on September 17. I spent the Friday before, the fourteenth, phoning the candidates to tell them they'd been selected. They were supposed to be in Houston for the public announcement. Then they would have until October 1 to report.

The formal announcement of my new title, coordinator of astronaut activities, was made the same day, even though I'd been doing the job for months by that time.

While I was selecting new astronauts, Wally Schirra and Gordo Cooper were training for the next Mercury flight, Mercury-Atlas 8, scheduled for launch on October 3.

We had all learned a lesson from Scott's flight. You couldn't overload the astronaut with a lot of observations and experiments and expect him to do his best job flying the spacecraft. Wally was pretty ruthless about keeping his flight plan manageable, especially since he was supposed to double the duration of John's and Scott's missions.

On the evening of October 2, Wally and I went out to relax a little, to do some fishing on the beach. There were two things we didn't realize: we were a little too close to Thor-Delta launchpad 23 . . . and there was a Thor-Delta ready to take off with Explorer 14.

One moment we were standing there figuring out how to get a fish out of a net, the next we heard this big roar. We look up . . . and there went the Delta, a couple of hundred yards away! Fortunately, there were no problems with the launch, or we'd have been in big trouble.

* * *

The nine new astronauts attended Wally's launch. I wasn't able to stand around with them, however. As the chief astronaut, I took it upon myself to be the guy who woke Wally up and took him to breakfast with Walt Williams, Bob Gilruth, and Howie Minners, the flight surgeon.

I was in the MCC at the Cape serving as launch capcom as Wally lifted off. Things looked good. At three minutes into the flight I broke in to ask Wally a question: "Hey, Wally, are you a turtle?"

Turtle questions go back to World War II. "What can a man do standing up, a woman do sitting down, and a dog do on three legs?" Answer: "Shake hands."

The proper answer to my question is, "You bet your sweet ass I am"—or the victim has to buy drinks for everybody within earshot.

I thought I had him. But Wally, quick as ever, went to the voice-activated on-board tape recorder. When he came back on the air, all he said was, "Rog." (Walt Williams and I made him play his answer for us on the carrier *Kearsarge*.)

The flight of *Sigma* 7 went very well, a real tribute to both Wally and the flight directors. Where we'd run out of maneuvering fuel both times on three-orbit missions, Wally managed to complete six—with over seventy percent of his fuel still remaining. In fact, he wound up having to jettison most of it for reentry.

In addition to selecting the new astronauts and setting up the flight crew operations area, I was given another job: that of making the crew assignments. The first available mission was MA-9, the eighteen-orbit Mercury flight.

To me the choice was pretty straightforward. Leaving me out of the picture, everybody had flown in Mercury except Gordo Cooper, who had been Wally's backup pilot. The problem was, Gordo had managed to alienate a few higher-ups in NASA, beginning with his public bitch about lack of airplanes back in 1959.

It was true he didn't show the best judgment at times. He took an F-102 into Huntsville airport one time for a meeting at the Marshall Space Flight Center. First of all, the runway there was really too short for an F-102. He shouldn't have tried it in the first place, even if he was a good enough pilot to make it work.

Gordo put in a call to Dobbins Air Force Base near Atlanta, the nearest facility, to send a truck over to refuel him, then went off to his meeting. He came back to find some lieutenant standing there with the truck. The lieutenant asked him if he was really going to try to take off from that runway. Gordo said sure. The lieutenant said he didn't think it was safe, and he wouldn't refuel the plane. He got back in the truck and left.

So what did Gordo do next? *He got in the plane and took off*, even though it had about nine minutes worth of fuel left. He made it to Maxwell Air Force Base near Montgomery, something like two hundred miles away, but he could just as easily have run out of gas.

When we heard about it, the rest of us were pretty horrified.

Incidents like that aside—and that wasn't typical—Gordo was a capable pilot and could do the job, so I recommended him. There was some grumbling out of HQ, so I said, "Either we fly him on MA-9 or we send him back to the Air Force now. It isn't fair to keep this guy hanging around if we're not gonna fly him."

I guess you could say I called their bluff. NASA didn't want to be shipping one of its Mercury heroes back to the military under those circumstances. It wasn't a power play: I knew Gordo would knuckle down and do the job, and he did.

I picked Al Shepard to serve as Gordo's backup, figuring he would not only be able to support Gordo, but also get ready for Mercury-Atlas 10, which we were still thinking about as a possible three-day flight later in 1963.

The Cooper assignment was announced on November 14.

Later that same week I made one of my more notable phone calls . . . to General Curtis LeMay, the Air Force chief of staff.

My physical status had not only screwed up my astronaut career, it was threatening my Air Force career as well. I was an Air Force officer and, as I said, the ultimate judge of my ability to fly airplanes was the service. The Air Force had finally gotten around to telling me I couldn't fly on the grounds that I could not pass a Class II physical.

So I contacted General LeMay to personally ask for a waiver, something that was in his power to grant. But he turned me down. "I've got half a dozen people flying on waivers right now, and I can't have any more."

But he gave me an option. "Here's what I'm gonna do. I'm gonna let you fly for one more year . . . then I'll have to ground your ass."

Well, that didn't make much sense. If it was unsafe for me to fly, it was unsafe now. Was I supposed to be safer in a year?

There wasn't much choice. I took the one-year extension. At least it gave me some breathing room.

14

FLIGHT CREW OPERATIONS

As the nine new guys came aboard in late 1962, I assigned Gus Grissom to supervise them, more or less, since he was already working on Gemini full-time. We also set up an administrative staff and created a mail room to handle the correspondence, because at that time astronaut mail was being handled everywhere, at the Cape, at Marshall, at headquarters. It seemed like a good idea to consolidate the mail in one place. That's also how we handled public appearances.

As we had with Mercury, in addition to classroom training and lectures, we also assigned the new astronauts different areas of specialization. Frank Borman was to work on the Titan II booster and Jim McDivitt handled guidance and navigation, for example.

As the second group moved in, they also got presented with a deal for their personal stories by Field Enterprises. A lawyer named named Henry Batten volunteered to handle that for no fee. This caused some more controversy, like the original *Life* magazine deal, and John Glenn came to the rescue. He was friends with JFK and his brother Bobby by now—they were trying to get him to run for office somewhere, and John wasn't resisting very hard. He explained to JFK why the Field Enterprises deal was not only an okay idea, but a necessary one, and Kennedy signed off on it.

Pretty soon the nine new guys were out buying lots and building homes, most of them a little east of the new Manned Spacecraft Center site, in El Lago. (I was the only one west of the site, five miles away in Friendswood.) Marge got the idea that it might be good if everyone got to know each other better, so she organized the Astronaut Wives Club with Frank Borman's wife, Sue.

I was on the road a lot, going through desert survival training again near

Reno, for example. At this point, less than a year after being taken off MA-7, I still had full confidence that I was going to be medically cleared for Gemini and Apollo. I wanted to keep myself trained and ready.

Gemini ran into delays. What had originally been designed as a slightly improved Mercury became a whole new spacecraft. Just getting it designed and built was taking a lot longer, and costing a lot more money, than anyone had expected.

There were also some dead ends. For example, one of our original goals in Gemini was to test a device called a paraglider. This was a combination inflatable wing and parachute which would have let Gemini glide to a landing at a runway—or on a dry lakebed—rather than be forced to splash down in the ocean. I was always uncomfortable with splashdowns, and not only because I didn't swim. For one thing, they were hellaciously expensive, since you effectively had most of the U.S. Navy on standby for the length of a mission. As we got missions going half a dozen times a year or more, that was a concern.

Also, as we had demonstrated with *Liberty Bell* 7, a spacecraft could sink. It didn't seem like such a great idea to have people returning from missions to the moon and still have to worry about drowning.

Jim Chamberlin really pushed the paraglider, which went by the name of the Rogallo Wing, after its inventor, and North American had the contract to build it. But all through its development in 1962, the program never really came together, and by 1963 it was on the shelf.

Another worry was an escape system. The potential for a violent explosion of the Atlas had led to the design of the Mercury escape tower, which would pull the spacecraft away from the booster in a second or two.

The Titan II used hypergolic fuels which ignited on contact. Even in failure, they tended to burn, not explode. So it was possible to use a modified version of an aircraft ejection seat. Possible. If there was something that didn't go wrong with that development, I don't know what it was. As late as 1964, tests of what should have been a relatively simple idea were still going wrong. During one demonstration, the Gemini hatches over the pilots' seats failed to detach, and the rocket-powered seats bored right through them. John Young was watching this. "A hell of a headache," he said. "But a short one."

There were also problems with the Titan booster itself. Redstone, Atlas, and Titan had all been developed as carriers for nuclear weapons, which were a bit more rugged than astronauts. The boosters vibrated so much during launch that they could not only rattle an astronaut's teeth, they could damage systems in the spacecraft. They also weren't designed for the level of reliability we needed. (I guess if you were firing a hundred nuclear weapons at the Russians, it didn't make much difference if five or ten of them failed on the way.) So getting the Titan II "man-rated"—modified for a smoother ride and greater safety—took more time than we expected.

We also had to buy a whole new type of booster called the Atlas-Agena—a newer-model Atlas carrying a Lockheed upper stage called the Agena, which was to serve as the target in our rendezvous and docking tests. NASA had originally planned to order eleven Atlas-Agenas; in July 1962 the order got cut to eight, but the price almost doubled.

Anyway, by February 1963, Gemini was costing twice the original estimates and was way behind schedule. Jim Chamberlin was removed and Chuck Mathews came in. During the next few months, more realistic schedules came out. Chuck added a second unmanned flight to the schedule, reducing the number of manned missions to ten.

The first unmanned Gemini, GT-1, would simply test the booster and spacecraft combination during launch, getting into orbit with no plan for recovery. That would be in December 1963.

Chuck added a second unmanned mission, GT-2, which would be a suborbital test of the Gemini heat shield. That would take place in July 1964, seven months later. Chuck knew from experience that first flights were followed by months of analysis and changes.

GT-3, the first manned Gemini, an eighteen-orbit mission, was targeted for October 1964. The first seven-day mission, GT-4, would follow in January 1965. GT-5 was supposed to be the first rendezvous and docking mission. GT-6 would be the fourteen-day mission. The rest of the program would consist of three-day-long rendezvous and docking missions, some of them with extra-vehicular-activity space walks.

NASA was doing a bunch of projects other than Mercury, or even Gemini. The biggest of these was Apollo.

At the time the contract to build the Apollo command and service modules (CSMs) was given to North American in November 1961, nobody had a real firm idea just how we were supposed to go to the moon. Were we supposed to load the Apollo on top of one of von Braun's big Saturn boosters and blast the whole thing straight to the surface of the moon? Were we going to launch a manned Apollo on a smaller Saturn and a tank of fuel on another Saturn, and dock them together in earth orbit?

Those were the two major plans—direct ascent and earth orbit rendezvous (EOR). Direct ascent was the favored mode at the Space Task Group (then becoming the Manned Spacecraft Center) in early 1962. Von Braun's team at Marshall favored EOR.

The direct ascent method had some real drawbacks that became pretty apparent. For one thing, a fully fueled Apollo CSM with landing stage would weigh about a hundred thousand pounds, meaning that in order to get it to the moon, you needed a huge booster—the Nova—that dwarfed even von Braun's Saturn V. We were years away from actually building the Saturn V. The earth orbit rendezvous plan required two or more Saturn launches, and then you had to figure out how to hook up the pieces of the spacecraft in orbit. That looked pretty difficult in early 1962.

There was a common problem to both of these methods. Even though the Apollo command module—a little cone about eight feet across—was the only part of the vehicle that would actually return to earth, the lunar-landing spacecraft with its rockets and fuel might wind up being sixty feet tall. So you'd have three guys lying on their backs trying to land something the size of an Atlas missile on the surface of the moon.

A third idea had been cooked up in 1958 by a team of engineers at Chance-Vought Corporation, led by a guy named Tom Dolan, and they had eventually found some NASA people at Langley who were intrigued by it—John Bird and Bill Michaels.

This new method was called *lunar* orbit rendezvous (LOR). They figured that you really didn't need to take a big spacecraft down to the surface of the moon. Why not just park the thing in lunar orbit and have one or two astronauts land in a buggy of some kind?

This had the advantage of the direct ascent method, in that it only used one booster, and a smaller one than the Nova. Using the lunar buggy saved a lot of weight: it didn't have to have a heat shield, for one thing, since it wasn't going back to earth.

Michaels had written a paper on LOR in the spring of 1960. But it hadn't gone anywhere. Another guy at Langley named John Houbolt had come up with the same idea, though, and he wouldn't let it die. All through 1960 he wrote letters and memos pushing for it to be considered. In December he and another engineer, Clint Brown, made a presentation on LOR to Seamans, von Braun, Faget, and Gilruth, and to Glennan, who was still the administrator.

Houbolt didn't make the sale. Everyone pointed out that LOR had the same flaw as the EOR method, because you were counting on being able to rendezous the command module and buggy—worse yet, you would have to do it in lunar orbit instead of in earth orbit.

That's where the idea stayed until Apollo was actually given the goal of landing on the moon by President Kennedy. Houboult really decided to press; he even went outside channels and wrote a letter directly to Bob Seamans.

Seamans had come to NASA from MIT and was more than familiar with the possibilities for successful rendezvous and docking. (About this same time two future astronauts—Dave Scott and Buzz Aldrin—were doing theses on the subject at that school.) He wasn't as afraid of LOR as some of the other heads, and he pushed Brainerd Holmes to give LOR more serious consideration.

George Low was already open to the idea, and said so. And by and by Max Faget and Bob Gilruth saw the advantages, too. Eventually there were two camps: the Manned Spacecraft Center (Gilruth and Faget) for LOR, the Marshall Center (von Braun) for EOR. All through 1962, people were going around about this. They had to make a decision, because you couldn't really build much hardware until you had the method.

Finally, on July 11, 1962, NASA announced that it was taking the lunar orbit rendezvous option. NASA wasn't out of the woods on this decision, however. Jerome Weisner, JFK's science adviser, thought lunar orbit rendezvous was dangerous and stupid, and wasn't shy about saying so. That meant a few more months of back and forth, but by November Grumman Aircraft had won the contract to build the buggy, which was called the Lunar Excursion Module or LEM. (Later they shortened the name, dropping the "Excursion" because they thought it sounded too frivolous. It was always pronounced LEM, however.)

Mercury had originally been designed for just the three-orbit mission. We had stretched it pretty successfully for Wally's flight of six, but getting the vehicle ready for twenty or more orbits meant changing something everywhere. In fact, around the Manned Spacecraft Center Wally's flight was considered to be the last real Mercury—MA-9 got designated MODM, the Manned One-Day Mission. MA-9, Mercury spacecraft number twenty, needed to carry more fuel, more water, more oxygen. It needed to have more electrical power. All of this, of course, translated into more weight—and there was a finite limit on the amount of weight the Atlas could lift.

This concern led to some goofy ideas. One of the heaviest items in the spacecraft was the contour couch we rode in. It had turned out, through five flights, that the G-loads weren't as severe as everyone thought they would be, and that maybe we didn't need the couch at all. Maybe Gordo could just be suspended in a net in the middle of the spacecraft! That one got shot down pretty quickly.

It was quite a job. To modify Wally's spacecraft for its mission, 20 changes were made in its configuration; for Gordo's there were 183. The miracle was, by taking things out and saving weight here and there, Gordon's spacecraft wound up weighing only three pounds more than Wally's did.

The big worry was that since Gordo would be in orbit for over a day, there was no way to keep him in constant contact with MCC. You had loss of signal on every orbit, of course, as the spacecraft flew over different tracking stations, but those usually lasted only a few minutes. Gordo would be out of touch for orbits ten through fourteen. So two additional tracking ships were added to the fleet.

One of the worries was that if anything went wrong during those orbits, not only would Gordo be on his own as far as making decisions, there wouldn't be an easy way to recover him if he had to make an emergency reentry. We managed to set up a prime recovery zone for every orbit, but we really hoped we wouldn't have to use some of them.

Even though we thought we'd learned a lesson about flight planning from Scott's mission, Gordo got loaded up with experiments, too, including that damned balloon and a TV camera. The reasoning was that he would have more time to try them. Gordo would also be the first American astronaut to eat and sleep in space.

Since the mission would last more than a few hours, it was also the first to require round-the-clock flight control operations. Chris Kraft, who had been the only flight director until that time, realized he needed help. He became the red shift director and added John Hodge, one of the Canadian engineers who'd come to NASA in 1959, as the blue shift leader. Chris had his guys trained so well that it didn't matter who was on the console.

There were the usual delays. Gordo's Atlas had to be sent back to the factory in January because of wiring problems.

Gordo himself took this all in stride. He was always a good stick and rudder guy and never looked back. He named his spacecraft *Faith* 7, which gave HQ some nervous moments. (What if something went wrong? They could see the headlines now: NASA LOSES "FAITH!")

Other Voices

WALT WILLIAMS

We were down at the Cape, a couple of days before the MA-9 launch, when somebody started tearing up the place with an F-102. Just buzzing it from one end to the other. Well, there weren't that many F-102s likely to be around there, so I had a pretty good idea who it was—Gordo.

I got on the phone to Deke at the crew quarters. "I think this is our boy."

"I know," he said, then had to shout. "He's going over us right now."

Fun is fun, but this was two days before the launch and didn't show much common sense on Gordo's part. I was very willing to turn the mission over to Shepard at any moment, and I said as much to Deke. "Tell you what," he said. "Let's just sweat him a little."

All right, whatever would get the message across. So Deke was waiting for Gordo when he came in, and he told him Walt Williams was looking for the guy who buzzed the Cape, and that he was going to ground him. He really put the fear of the Lord into Gordo, because Gordo knew I would have put Shepard in his place in a minute.

That night I went out to dinner with the seven of them. Wally was Wally, telling jokes. Everybody was having a good time—but there was Gordo, sitting alone at the table, trying to be invisible, especially when I would start talking about grounding any hot dog pilot who was dumb enough to buzz the place.

Of course, Gordo went on to fly a great mission, especially toward the end, when everything was falling apart. But he needed a lesson.

During a hold in the MA-9 countdown, Gordo was so relaxed he took a little nap. At 8:04 A.M. on May 15, 1963, however, he was on his way.

His earliest problem was the temperature inside the spacecraft. It got hot during Gordo's first orbit, then cooled back down. He tried a meal of roast beef, with some success.

By orbit ten he was supposed to be resting, but he was flying over the Himalayas and some pretty spectacular scenery, so he sat up there clicking away with his camera, and seeing things as small as individual houses—he could see the smoke from their chimneys and track them. It was a long time before anyone believed you could do that, but it was true.

Eventually he powered down the spacecraft and drifted off to sleep. It was supposed to be a full eight hours, but he kept waking up.

During Gordo's second day, as he was back in range of the tracking network, some problems began to crop up. There was a slow oxygen leak somewhere that had to be watched. Carbon dioxide levels were higher than they should have been, too. Either one of those could have been a problem, but Gordo knew how to detect any signs of trouble.

The automatic guidance system failed about this time, which got everyone's attention on the ground. Suddenly it looked like Gordo was going to have to do a manual reentry. We hoped he would have some help from the automatic system, but on the twenty-first orbit, the whole thing dropped out. Gordo commented that "things are beginning to stack up a little bit."

But with John Glenn giving him updates and countdowns, Gordo did a great job firing the retros and keeping the Mercury in the right attitude. He splashed down less than five miles from the carrier USS *Kearsarge.* Thirty-four hours and twenty minutes—a good flight.

Mercury-Atlas 10, the three-day Mercury mission, was still a possibility in early 1963. Even after Gordo's flight we would have two Mercury spacecraft and boosters available. The whole program was gaining confidence in being able to supply an astronaut with enough consumables to keep him functioning for that amount of time.

Jim Webb was totally opposed, however. He said Gemini was already designed to keep astronauts in space for three days, seven days, even up to two weeks, in a more capable spacecraft than Mercury. Why bother trying to prove something once in an obsolete system, then go out and do it all over again? He was also concerned that an accident in MA-10 would delay Gemini and Apollo.

We tried to go around him to JFK, but this time it didn't work. MA-10 died. Gordo had flown the last Mercury.

I got another serious disappointment right about that time. Charles Berry, who had taken over from Bill Douglas as the chief astronaut medical officer, had been working with Bill since the summer of 1962 to get me requalified. After Gordo's flight he took me back to see Paul Dudley White.

Well, this visit was a complete waste of time. White didn't even examine me; he just told me right there in his office that there was no reason for him to change his diagnosis. I was still having fibrillations every few weeks, and nobody knew for sure what that meant—anything up to and including a serious heart disease. He thought it was too risky to subject me to spaceflight, especially since there was some preliminary evidence from Gordo's flight that going into space might be stressful for your heart. (It turned out not to be the case, but they didn't confirm this in time to help me.)

Chuck and I went out to Logan Airport and while we were waiting, we just sat there in a room. I was pretty depressed. Even before we'd gone to see White, we had tried to get Jim Webb to let me undergo some tests, including angiograms. But those were pretty tricky back in those days, and even though I was willing to take the risk, Webb wasn't.

I figured there was no point in trying to get any of these doctors to change their minds: if I was going to fly in space, I was going to have to consider some kind of treatment that would change my condition.

Gordo had hardly finished his ticker-tape parade than the Russians put Vostok 5 into orbit with cosmonaut Valery Bykovsky on June 14.

We had been waiting for the shoe to drop since the twin Vostok flights in August 1962, figuring that if the Russians had been able to put two spacecraft up within a day of each other, it wouldn't be long before they'd be able to dock them together. Once they did that, they'd be well on their way to having the capability to land on the moon. There were rumors in the papers that "something big" was going to happen.

June 15 passed; Bykovsky went around the earth like Titov and Nikolayev and Popovich had, nothing to laugh at, but nothing unusual, either. Then, on June 16, Vostok 6 was launched. The cosmonaut was a woman named Valentina Tereshkova.

It was a hell of an attention getter. She stayed up for three days and made most people forget that Bykovsky was up for five. But there was no rendezvous. We began to think that Vostok couldn't maneuver in orbit, which meant maybe the Russians weren't quite as far ahead of us as we had thought.

One thing Tereshkova's flight did was bring up the whole business of woman astronauts again. It's probably hard for anyone to remember, but in the 1950s there weren't any women doing operational flying for the military. There had been a lot of women who flew support during World War II, ferrying planes from factories to airfields, things like that. They were certainly capable; but none of them had been allowed to fly combat or go onto test pilot school, so none of them were technically eligible to become astronauts.

That hadn't stopped some of them from trying. In February 1960, Jerrie Cobb, a woman pilot, had taken and passed some of the same medical tests we had. With the backing of Jackie Cochran, another famous woman pilot and good friend of Chuck Yeager's, a whole group of women went through the Lovelace Clinic a year later. I think there were even some hearings before a congressional committee—Jane Hart, one of these woman pilots, was the wife of Senator Philip Hart of Michigan. There were supposed to be more tests, but they never happened.

Anyway, it became a pretty big issue again, especially since the Russians played it for pure politics. The fact that Tereshkova was in space meant that Russian women were the equal of men, and Soviet society was more advanced than ours, that kind of thing.

It turned out that Sergei Korolev, the chief designer, had been pushed into this by Nikolai Kamanin, the general in charge of cosmonaut training, who had also sold the idea to Khrushchev. But Korolev was unhappy with Tereshkova's performance. I guess she got sick, like Titov, and couldn't keep up with her experiments—most of which were taking pictures.

There was a whole group of five woman cosmonauts at the time. They were basically sport parachutists, though one of them had been trained to fly. There were tentative plans to fly a couple of them later in the 1960s that never came off. The group was disbanded in 1969, though nothing was ever said about it.

The Russians, for all their superiority, didn't put another woman in space for nineteen years—just as NASA was getting ready to launch the first shuttle missions that had woman astronauts in the crew.

All the while the astronaut selection process continued. We had known going into the 1962 selection that we would need more astronauts at some point in the future, so as the nine new guys completed their initial training, we decided to pick up some more.

The decision was based on planning documents that were starting to arrive from headquarters. Gemini was always intended to consist of ten or eleven manned flights—that was a minimum of twenty seats right there. You could assume that a few guys might fly twice, but there was no way to predict astronaut office attrition. In the summer of 1963 I was still not eligible for flight and Scott Carpenter and John Glenn were out of the running. That left thirteen guys for Gemini.

Then you had Apollo, which in at that point called for the following:

Four manned Apollo earth orbit missions launched with the Saturn I rocket beginning in 1965.

Two to four manned Apollo earth orbit missions on the Saturn IB beginning in 1966

At least six manned earth and lunar orbit missions on the Saturn V beginning in 1967

All this was designed to lead up to a manned lunar landing in 1968–69. It looked like forty-two seats. I didn't necessarily think we'd have that many flights, but I didn't want to be caught short, either, so I estimated we would need a minimum of ten new astronauts in 1963, and would take as many as twenty.

The procedure and qualifications were pretty similar to the recruitment for the 1962 group, though we dropped the test pilot requirement, figuring we had just about drained that pool. Applicants with operational flying backgrounds or advanced degrees in related areas would be accepted.

It was with this selection, however, that I got caught in my first, last, and only political battle over astronaut selection.

The Kennedy administration, particularly the President's brother, Robert, thought there should be a black astronaut. The Navy didn't have anyone remotely qualified, but in the Air Force there was a black bomber pilot,

Captain Edward Dwight, who had applied for Yeager's Aerospace Research Pilot School (ARPS).

The trouble was, his multi-engine background, lack of an engineering degree, and lack of the normal test pilot school caused him to be ranked pretty far down the list of applicants. (The school normally enrolled eight at a time.)

The pressure started with General LeMay, who was ordered by Bobby Kennedy to get Dwight enrolled at ARPS. Yeager resisted—it wasn't racism, it was just that according to the rankings, Dwight had finished in the middle of the pack. Yeager thought it would be reverse racism to enroll Dwight *ahead* of pilots with better qualifications.

They worked out a deal: Dwight would be enrolled, but so would all those pilots on the list ahead of him. That's why that year the ARPS had a class of fourteen rather than eight.

Dwight got through the school and did okay, even though Yeager brought in a tutor for him, all of that. But okay wasn't really enough: had he been white, he wouldn't even have been a serious candidate. Remember, NASA wasn't just looking at the ARPS graduates as potential astronauts: our pool included the Navy and the Marines, civilian pilots and now research scientists, not to mention other Air Force pilots and test pilots, some of whom had really proved themselves in flight test. There were guys like Michael Collins, a good applicant for the 1962 group who had been held back to get another year of experience. Or Dick Gordon, who was one of the Navy's best test pilots.

As I heard it, Dwight himself wasn't particularly driven to become an astronaut: he wanted to move up in the Air Force.

I had already developed a point system that we used in making the final evaluations on astronaut candidates. There were three parts: academic, pilot performance, and character/motivation, ten points for each part, with thirty being the highest possible score. Some of it was cut-and-dried: you got points for a certain amount of flying time and for education. Some of it, by design, was subjective and based on face-to-face interviews. Just based on the flying and technical matters, Dwight finished out of the running. He wound up leaving the Air Force and last I heard was a very successful sculptor up in Colorado.

On October 14, 1963, we announced the third group of astronauts:

Major Edwin E. "Buzz" Aldrin, Jr., USAF
Captain William A. Anders, USAF
Captain Charles A. Bassett II, USAF
Lieutenant Alan L. Bean, USN

Lieutenant Eugene A. Cernan, USN
Lieutenant Roger B. Chaffee, USN
Captain Michael Collins, USAF
R. Walter Cunningham
Captain Donn F. Eisele, USAF
Captain Theodore C. Freeman, USAF
Lieutenant Commander Richard F. Gordon, Jr., USN
Russell L. Schweickart
Captain David R. Scott, USAF
Captain Clifton C. Williams, Jr., USMC

Seven Air Force, four Navy, one Marine and two civilians. Eight of them were test pilots—Bassett, Bean, Collins, Eisele, Freeman, Gordon, Scott, and Williams. The rest were a combination of operational Navy pilots (Cernan and Chaffee), Air Force pilots in engineering jobs (Aldrin and Anders), or former military pilots who were in scientific research (Cunningham and Schweickart).

In a way, the third group was two groups—those with test pilot experience, and the others. In my tentative crew plans carrying through the last Gemini missions into Apollo, I figured it would be the test pilots I would count on for the more immediately difficult work—training as command module pilots for future Apollo missions, where they would, in effect, be solo pilots. The more engineering and research guys would get their chance, too, but on the development end of things. Aldrin had done a Ph.D. dissertation on rendezvous, so he was a natural for support of the Gemini program. Anders had a degree in nuclear engineering; it made sense to assign him to Apollo, where there were sure to be questions about radiation exposure to crews on lunar missions.

Gemini was still struggling to get off the ground at the end of 1963. At the same time there were some high-level musical chairs. Brainerd Holmes, the head of manned space flight, had quit in June in a power struggle with Webb. (Holmes wanted to have a freer hand in running Apollo than Webb wanted to give him.) He would ultimately be replaced by George Mueller, who did, in fact, get the control Holmes had wanted.

Holmes's deputy was a bright young guy from Bell Labs and TRW named Joseph Shea. Mueller wanted Shea to go to Houston and become head of the Apollo Spacecraft Program Office (ASPO), but Shea wouldn't, not as long as Walt Williams was there.

So Williams got moved back to NASA HQ. Shea came to Houston as

ASPO, and George Low replaced Williams as deputy director of the Manned Spacecraft Center. At HQ Mueller brought Major General Samuel Phillips from the Air Force Minuteman program in to be the head of Apollo there . . . and that was the team that got us to the moon.

For most of 1963 the astronaut office continued to report directly to Gilruth as a staff office through Jim Elms, his deputy. One day they asked me if I'd like to take on the additional job of handling everything associated with flight crews as assistant center director. (We had been having some problems with astronauts wanting one thing and other people who were supposed to be supporting them wanting something else, and things were not getting done.)

There was also a mess with aircraft operations, which were reporting to me operationally (keeping track of who was flying what planes where) and to North administratively (keeping track of who was buying fuel and paying salaries). NASA was depending on bailed aircraft—T-33s and F-102s borrowed from the military services—and it was clear that we would have to start buying them.

So in October 1963 I ended up being named assistant director of what was called Flight Crew Operations, which had three units. First was North's division, which became the Flight Crew Support Division. Second was Aircraft Operations Group, headed by Joseph Algranti with Bud Ream as his deputy. Third was the Astronaut Office; I stayed on as head of that because I couldn't find anybody else to run it.

At my new level within the Manned Spacecraft Center were the engineering directorate (Max Faget), life sciences (Charles Berry), and flight operations, (Christopher Kraft). This ignores, of course, a whole gaggle of other functions, such as personnel, administrative support, facilities construction, and contract management, not to mention the Mercury, Gemini and Apollo program offices. And this was just at MSC in Houston.

Max, Chuck, and Chris were the people I dealt with on a day-to-day basis. This is where most of the nitty-gritty decisions got made.

Between the time the fourteen new astronauts were announced and the time they reported on January 1, 1964, the first of a bunch of cutbacks came out of headquarters.

What was happening was that Apollo and Saturn development was taking too long. The Saturn rocket, in particular, was Wernher von Braun's baby, and he had a very methodical approach: First you test the rocket with a working first stage and dummy upper stages. When you're sure that works, you add a working second stage. Then you add a working third stage. Only then would you think about adding a manned spacecraft. This was great if you didn't have a deadline, and we did.

Jim Webb had gotten George Mueller from Space Technology Labs to come in as head of manned spaceflight in October 1963. He wasn't there two weeks when he took a look at von Braun's schedule and said it wasn't going to work. He proposed that we go to "all-up" testing, getting rid of any dead-end equipment that wasn't ultimately going to be used, such as the Saturn I rocket. That eliminated four manned Apollo flights right there.

It clarified my planning, however. By early 1964 I had a pretty good idea that I would have to select and train a minimum of ten Gemini crews and eight Apollo crews *prior* to the lunar landing. So my mission was to create a pool of guys who had the necessary experience in rendezvous and docking, EVA, and long duration before I had to select which three would attempt the first lunar landing.

I made up some guidelines for myself, and this is what they were:

1. Everybody was considered to be qualified and acceptable for any mission when brought into NASA. That is, if I hired the guy and kept him around, he was eligible to fly.

2. But some are more qualified than others for specific seats on specific missions. That is, guys with command or management or test pilot experience were more likely to be handed the more challenging assignments.

3. I would try to match people in a crew based on individual talents and, when possible, personal compatibility. But I didn't give nearly as much weight to the compatibility issue as everybody thought, because it wasn't necessary. Everybody in the astronaut group was talented and motivated, or they wouldn't have gotten there in the first place. They were likely to get along no matter how they were matched up. On the other hand, matching people where possible made life easier all around, and added some fun.

4. I always kept future requirements and training in mind. When I wanted somebody for an assignment in Gemini, I had to think how that would affect Apollo. So I had a long-term plan that was updated regularly.

5. I assumed a ten percent annual attrition rate for all causes—accidents or resignation. (This was a wild-assed guess I made in 1963 that turned out to be pretty close.)

From the Mercury group I had Al Shepard, Gus Grissom, and Wally Schirra. John Glenn was scheduled to retire in January 1964, Scott Carpenter was unacceptable to management. Gordo Cooper was a question mark. Then there were the nine guys from the 1962 group: Armstrong, Borman,

Conrad, Lovell, McDivitt, See, Stafford, White, and Young. The fourteen new guys from the 1963 group were just going into training at that point and wouldn't be eligible for assignment to crews until later in the year.

I figured to handpick the crews for the first four Gemini missions, since they had unique requirements, and treat the last six as more or less identical. In talking things over with some of the guys, I knew, for example, that Wally wasn't too eager to fly a long-duration mission but figured that he'd do a good job on rendezvous. Al had been backup on the last Mercury and was more or less in line for the first Gemini; he was also the most capable pilot we had. Gus had dug in on Gemini systems and would be a good candidate for a long-duration flight; Gordo might work in here, too.

I also wanted to give someone from the 1962 group an early chance at command. So in my first pass at a long-term plan, the early Gemini missions slots were taken like this:

GT-3	(first mission)	Shepard
GT-4	(seven days)	1962 astronaut
GT-5	(rendezvous)	Schirra
GT-6	(fourteen days)	Grissom

Each Gemini crew would actually consist of four guys, a commander and pilot for the prime crew, and a commander and pilot for the backup crew. They would work as a unit with the prime crew commander calling the shots. So as GT-3 was completed, I would have four guys available for reassignment. The logical move was to take the GT-3 backup crew and make that the prime crew for GT-6. (The GT-4 and GT-5 crews would still be awaiting their flights.) I could either use one or two of the original GT-3 guys as backups on GT-6, or work in a couple of new ones. That was where the three-mission rotation began.

The next trick was selecting which of the 1962 guys to assign where. All astronauts are created equal, but some are more equal than others. By this time I had a pretty good idea as to who was more equal in the new group. I also had the judgments of Wally, Al, and Gus to go on.

GT-3, the first manned Gemini flight, scheduled to last eighteen orbits, was going to be Al Shepard's. I assigned Tom Stafford as his pilot. The backup commander, aiming at the fourteen-day flight on GT-6, was Gus Grissom. I assigned Frank Borman as Gus's pilot. Frank was tenacious enough to stick it out.

Wally would have GT-5, the rendezvous mission, with John Young as his pilot.

GT-4 was then intended to be a seven-day flight to further test Gemini's capabilities. We were thinking about EVA and planning some limited rendezvous or station-keeping practice. I thought Jim McDivitt would be a good guy to assign as commander; I assigned Ed White, who had been a classmate of Jim's at Michigan, as his pilot. I wanted Pete Conrad as the backup commander, with Jim Lovell as pilot. (They had gone through the Navy test pilot school together and got along well.)

I announced the first set of crews at one of our Monday morning pilot meetings, and wound up having them changed by outside circumstances within a few weeks.

First of all, the Gemini mission schedule got rearranged. The Atlas-Agena wasn't going to be ready in time for GT-5, so the rendezvous and docking was slipped to GT-6. So my prime and backup crews for GT-3 became Shepard-Stafford and Schirra-Young, with Grissom-Borman now assigned to GT-5.

Then Al Shepard got medically disqualified. He had been having some dizziness problems and back in May 1963 had been tentatively diagnosed with Ménière's disease, an inner ear problem, and had gotten grounded. In August he had come back on flight status long enough to get the GT-3 assignment. Then, in October, the early diagnosis was confirmed, and he was out of consideration.

All right, more changes. I moved Gus Grissom from commander GT-5 to commander GT-3, and brought Gordo Cooper in for GT-5. Figuring that John Young was a better personality match with Gus than Tom Stafford was, I also swapped pilots in the group. (Tom was probably our strongest guy in rendezvous, so it made sense to point him at GT-6, the first rendezvous mission.)

I still wanted Frank Borman on that first long-duration flight, which was now going to be GT-7, so there was another swap—Conrad to GT-5, Borman to backup GT-4. And, finally, here was the initial rotation:

Mission	*Prime*	*Backup*
GT-3	Grissom-Young	Schirra-Stafford
GT-4	McDivitt-White	Borman-Lovell
GT-5	Cooper-Conrad	Armstrong-See
GT-6	Schirra-Stafford	Grissom-Young

In November 1963, one year to the day after my last phone call to General LeMay, I called him again and asked about my flying status. "I've got to ground you," he said.

"Well, then, General," I said, "I'm putting in my resignation."

He thought I'd lost my mind. "You can't resign! You've got nineteen years in!"

"I'm sorry, General, but you haven't given me a choice. If I can't fly airplanes, I can't be an astronaut. And you're telling me I can't fly airplanes as an Air Force officer."

So that was it. I'd done the dumbest thing a career military man could do, resigning from the service with one year to go before I qualified for a pension. (I'd joined up in 1942, of course, but my combined service—active duty, Air Guard and reserve—added up to less than nineteen years at that point.)

Chuck Berry had gone to Bob Gilruth with the idea of creating a special NASA Class III physical qualificiation for me. (Al Shepard had gotten a Class III ticket from the Navy because of his medical condition, but NASA didn't have one.) I would be allowed, as a civilian, to pilot NASA aircraft, provided I had a qualified copilot. It wasn't what I wanted, but it was the best deal I was going to get.

The only qualifier was that Al and I couldn't fly with each other. Two Class IIIs didn't add up to a II. Or, as Chuck put it, "Two half pilots don't make a whole."

So, in November 1963, I resigned my commission as a major in the USAF and became a civilian employee of the National Aeronautics and Space Administration. My title was assistant director MSC for flight crew operations.

Since I was trying to run the astronaut office and this whole directorate with several hundred people, I realized I needed help. (I was having my share of problems. I figured if I only made one mistake a day I was doing good.) So it was good fortune for me and bad luck for Al that he got grounded, because I grabbed him and put him in charge of the Astronaut Office.

The first Titan II modified to launch a Gemini, Gemini Launch Vehicle 1, arrived at the Cape on October 26, 1963. It had been back and forth between the contractor, Martin, and the Air Force a couple of times already. It took until January 31, 1964, to get it erected on the pad at Launch Complex 19. The first Gemini spacecraft, which carried only dummy equipment in the adaptor and instrumention pallets instead of crew couches, was mated to GLV-1, and we were finally ready to go.

On April 8, GT-1 took off from the Cape and went into orbit without a hitch. The primary goal was to verify that the Gemini-Titan stack would fly as predicted and reach the proper orbit. This was important because of plans for rendezvous and docking between Gemini and Agena. We also needed to test tracking and telemetry systems, which we did for three orbits.

GT-1 was Walt Williams's last mission at NASA. He was frustrated at headquarters, so he resigned and joined the Aerospace Corporation out in Los Angeles.

Five days after GT-1, we announced the GT-3 crew of Gus Grissom and John Young, with Wally Schirra and Tom Stafford as backups.

15

DELAYS

Into the late spring and summer of 1964 we got closer to a manned Gemini launch. GT-2, the unmanned suborbital, was still hanging. The ideal training schedule we laid out—which never got used as it was designed, though it served as a model—called for a crew to be ready within twenty-four weeks, roughly six months. Accordingly, on July 27, we announced the GT-4 crew of McDivitt-White, with Borman-Lovell as backups.

What slowed us most was weather. The second Titan launch vehicle for Gemini, GLV-2, was erected on Pad 19 on July 14, 1964. A month later a thunderstorm blew through the Cape; the pad was struck by lightning. (The spacecraft hadn't arrived from McDonnell yet. In fact, it was running late.) Checking that out took a few days.

Then, less than two weeks after the thunderstorm, Hurricane Cleo came close. Just to be safe, the second stage was taken down and stored while the first stage, still on the pad, was battened down. Putting them back together added more time.

On September 8, another hurricane, Dora, was predicted to come right through the Cape. This time both stages were taken off the pad and stored in a hangar. Dora turned away . . . but before the Titan could be put together again, here came Hurricane Ethel. It wasn't until September 14 that GLV-2 was back on the pad. Gemini spacecraft number two arrived for what was supposed to be a month of checkout. (It didn't get mated to the launcher until November 5.) There was no way GT-3, Grissom-Young, was going to be launched before the end of the year.

As the Gemini program geared up, I found myself heading a directorate that numbered three hundred people.

One of my divisions was Aircraft Operations, run by a NASA pilot named Joe Algranti. With two dozen active pilots in the astronaut office who not only had to get themselves cross-country on short notice, but who also needed to keep their flying skills sharp, we needed a regular squadron of planes. We originally had the T-33s and F-102s, plus a Gulfstream NASA had bought for executive flights. Walt Williams decided we needed some high-performance birds of our own. The choice came down to the F-4 Phantom and the T-38 Talon, and we wound up going with the T-38: it was a trainer, and probably a little more suited to our needs than an operational fighter. It was also cheaper.

The Air Force loaned us five T-38s in early 1964 while we waited for our original order of fifteen to be delivered. (Later we added ten more, so we had a fleet of twenty-five T-38s alone.) These were in addition to a KC-135 and a pair of WB-57s.

Another division was Flight Crew Support, which was Warren North's area. This included offices for such areas as project support, crew stations, crew safety, simulators, mission operations, and flight planning.

Finally there was the Astronaut Office, headed by Al Shepard, who had twenty-six guys working for him organized like this:

There was the Gemini branch, with Gus Grissom as chief (he was commander of the first crew). Grissom-Young, Schirra-Stafford, McDivitt-White and Borman-Lovell.

Scott Carpenter was on special assignment as liaison with the U.S. Navy for what became SeaLab.

There was the Apollo branch, with Gordo Cooper as chief. Under him were Conrad, Anders, Cernan, Chaffee, Cunningham, Eisele, Freeman, Gordon, and Schweickart. With the exception of Cooper and Conrad, who were pointed at GT-5, these were the guys I expected to be Apollo pilots, even though some of them wound up moving into Gemini.

Then there was the operations and training branch, headed by Neil Armstrong. Under him were See, Aldrin, Bassett, Bean, Collins, Scott, and Williams. They were the points of contact for boosters, recovery systems, tracking, communications, and simulators. They were also the guys first in line for the next Gemini crews.

Al Shepard had his office right there with the astronauts, eventually located in Building Four with them, handling the day-to-day details and basically serving as a squadron commander. I was with the administrative people over in Building One. We would confer several times a week on everything including crew assignments.

I thought that the twenty-six guys in the Astronaut Office—twenty-eight, if you remember that Al and I still planned to get back on flight status—were enough to provide all the Gemini and Apollo crews through the first landing . . . and then some. Which turned out to be the case. This didn't stop the scientific community, which had started agitating for the selection of more scientists as astronauts.

I didn't have anything against scientists, or doctors, but I wasn't quite sure what I was supposed to do with them on flight crews. People have the mistaken idea that your education or professional training is what you call on during a Gemini or Apollo mission. Bob Gilruth protected me from a lot of this: there was a Navy flight surgeon working at the MSC who thought he should be assigned to the first long-duration Gemini crew. There was even one astronaut who made this mistake—Buzz Aldrin always figured that because he had written a thesis on orbital rendezvous he should be the one flying the first rendezvous mission.

People don't realize it takes two or three people to make sure the spacecraft gets where it's supposed to go, performs all the planned maneuvers, then returns safely. There is no room and no requirement for what would basically be a passenger. When you're sitting in a spacecraft all wired up, with your heartbeat and everything being transmitted to mission control, you sure don't need a doctor sitting next to you. If you have a serious problem, you're coming home! What you need is somebody as qualified as you are to fly the spacecraft.

Buzz's expertise was valuable, no question. Which is why his first technical assignment was working with the team that handled rendezvous procedures. It had nothing to do with his qualifications to fly that particular mission, however.

We'd lowered the requirements for the third group of astronauts to allow people with scientific or medical backgrounds to be considered, providing they met the flying minimums. We got Buzz, who was a Ph.D., and Rusty Schweickart and Walt Cunningham. But it wasn't good enough for the scientific community. The political wheels had started to turn even before the third group was actually selected. Bob Voas, who had been involved with the Mercury selection back in 1959, was basically the point man for a lot of the discussions, which involved Gilruth, Shea, and Low—not me. They eventually got Webb and Mueller to sign off on the idea.

By August 1964, Homer Newell, the NASA associate administrator for space science, and Harry Hess, the head of the Space Science Board of the National Academy of Sciences, had come up with a plan and a set of scientific qualifications. On October 19, there was a public announcement that NASA would select a group of scientist-astronauts in 1965.

Everyone agreed that if you were going to add scientists to flight crews, there were two types you might find use for: geologists for lunar landing missions and physicians for long-duration earth orbit missions. Our preference was to have the National Academy of Sciences come up with maybe twenty candidates in each of those fields.

But they felt this was too exclusionary, that we might also find use for astronomers and physicists and meteorologists. Okay, we said. So the NAS put together its board and started looking for scientist-astronauts. NASA put together a board of its own, to review what the NAS came up with. It included Joe Shea, Chuck Berry, Max Faget, Warren North, Al Shepard, and me. The new addition was Dr. John F. Clark, from Newell's Office of Space Science and Applications.

One of the things that made being an astronaut different from being a scientist was the risk.

Just after eleven on the morning of October 30, 1964, I was home in Friendswood when I got a call from the Ellington tower. A T-38 had crashed.

Other Voices

BUD REAM

With Joe Algranti, I was one of the two NASA pilots in aircraft ops at the air patch—Ellington. We normally didn't work Sundays, though that didn't mean astronauts couldn't get some proficiency flying done.

Ted Freeman had taken off in a T-38 around ten A.M. *for a routine flight, out over the Gulf and back in. He was more familiar with the T-38 than most; before coming to NASA, he had done some of the final test work on the aircraft out at Edwards. As he set up to land back at Ellington on runway four, the tower noticed some other traffic and waved him off.*

So he started climbing again and turning to the east, to go around and come back for another try. As he was in his turn and rolled to the right, a snow goose struck the left side of his canopy, shattering it.

(Hitting a bird is one of the hazards of flying. Some years later John Young and I were in a T-38 at eleven thousand feet when we nailed one.)

Pieces of Plexigas got sucked into both engines. The engines continued to operate for a few seconds . . . long enough for Ted to level out and continue his turn to the point where he was heading northwest.

There was a family out for a Sunday drive, heading south on the Gulf Freeway, and they saw the T-38 going lower and lower. Then there was an explosion.

The explosion was Ted ejecting from the aircraft. Unfortunately, by that time he was very low, and the nose of the aircraft was pointing down, so he ejected forward instead of up, too low for his parachute to deploy fully.

It was only a five-mile drive from my front door to the crash site. One of the NASA doctors got there about the same time I did; we were the first. Not far from the smoking heap of the T-38 we found Ted Freeman, still in his seat, chute unopened, dead.

Ted was buried at Arlington.

A few weeks after Ted's accident I got the thirteen remaining guys in the third group together and gave them a little job. Each guy was to exclude himself and rank the others in the group in the order he thought they should go into space. We had done the same thing in Mercury.

I can't say that I paid a lot of attention to the results. I'd made some early judgments about this group as much as a year ago and had plugged them into the master plan accordingly. It was Bob Gilruth who suggested I might do the peer rating—just as a sort of cover your ass.

It was a good idea. These guys knew each other's abilities better than anyone, and they were pretty objective. The ones who came to the top of the peer ratings were pretty much the ones I had assigned to the later Gemini missions.

On December 9, 1964, we were finally ready for the GT-2 launch. Ignition came at 11:41 A.M. and stopped one second later. A housing on a hydraulic valve in the GLV had blown, and the launcher's malfunction detection system (MDS) had simply shut down the engines.

It was frustrating, but a good lesson. You couldn't have shut down an Atlas like that. It wouldn't have done us much good to have the Titan get off the ground, then blow up, either. So it gave us confidence in the MDS—which we needed a year later.

We recycled and scheduled a new liftoff for January 19, 1965. It went like a dream, first-stage ignition, then second-stage. As the second-stage motors shut down, GT-2 separated, then fired its own orbital maneuvering system motors. The reaction control system—the steering motors—went to work, flipping GT-2 around so it faced backward. The equipment module jettisoned. The four solid retro-rockets fired. Then the retro section was dropped. As planned, GT-2 went through a hotter-than-normal reentry and dropped safely into the Atlantic eighteen minutes after launch.

There were still problems cropping up. One of the new systems in Gemini was a fuel cell that would provide power for the spacecraft for missions lasting up to two weeks. (The limit with battery power was about four days.) That was one reason GT-4, the second manned Gemini, which McDivitt and White were training for, had its planned duration shortened.

At the same time we were trying to lengthen GT-3. Chuck Mathews and Bob Gilruth thought they had to be conservative, that a three-orbit shakedown flight was all they wanted to risk. Gus and John and the rest of us thought a thirty-orbit flight, almost two days, was the next logical step after Gordo's MA-9.

We lost that one, however. Jim Webb had to referee the argument personally, and he came down on Gilruth's side.

Since it now looked as though GT-3 (Grissom-Young) would fly in March, with GT-4 (McDivitt-White) in May, I had to announce the GT-5 crew—it was scheduled for July. On February 8 Gordo Cooper and Pete Conrad were announced as the prime crew, with Neil Armstrong and Elliot See as their backups.

We had started hearing from the Russians again in the last few weeks of 1964. On October 12, they had launched Voskhod—"Sunrise"—a new multiseat spacecraft. It carried a crew of three: a pilot, Vladimir Komarov; a scientist, Konstantin Feoktistov, and a physician, Boris Yegorov. We had very little hard information. For all we knew, this could be the Russian version of Apollo.

But Voskhod only stayed up for one day. While it was in orbit, Nikita Khrushchev got thrown out of office. In fact, I think his last official act as head of the Soviet Union was talking to the cosmonauts while they were in space.

It wasn't until years later that it became clear that Voskhod was just a Vostok with the ejection seat removed and three smaller seats inserted. Komarov, Feoktistov, and Yegorov had been launched without pressure suits (to save space and weight) and with no launch escape system: if that booster had blown up, they'd have blown up with it.

I guess the idea was to score another public relations first—if the Americans are going to put two men in space, we'll put *three* up there. It didn't really mean much, since the flight was too short to get any new medical data in spite of the fact that a doctor was aboard. It also turned out that Feoktistov wasn't a "scientist," he was a spacecraft engineer who had been one of the leading designers of Vostok. At the time, though, it made us feel that we were still falling behind.

* * *

Gus and John were living at the Cape getting ready for the GT-3 launch when Voskhod 2 went into space. This time there were two cosmonauts aboard, both of them pilots, Pavel Belyayev and Alexei Leonov.

As they got to the end of their very first orbit, still over Russia, Leonov climbed out of the Voskhod wearing a pressure suit and floated around for ten minutes. They even showed it on live TV, though the pictures we saw were pretty bad.

I got to know Alexei Leonov pretty well when we were training for Apollo-Soyuz ten years later. He told me he had been assigned to the EVA flight two and a half years before it was launched and had been through some pretty rigorous physical conditioning—strengthening his arms and his grip, doing a lot of diving to test his equilibrium. He needed every ounce of his strength: his EVA suit ballooned so big he barely fit back into the Voskhod 2 air lock. He wound up bleeding some of the air out of the thing so he could squeeze back inside.

Belyayev and Leonov's flight only lasted a day. EVA was obviously the only goal. The Voskhod automatic guidance system failed during the first attempt at reentry on their sixteenth orbit; they had to go around the earth one more time while Belyayev set up for a manual retrofire, which he made successfully.

That extra orbit moved Belyayev and Leonov out of the prime recovery zone in Kazakhstan. They came down in the Ural Mountains in a forest so deep that helicopters couldn't land. (It took a couple of hours for rescue teams just to locate them.)

So the two cosmonauts spent a cold night in the woods before skiers got to them. It took them a whole additional day to get out of there, too.

EVA was one of the goals of the Gemini program. Since we had to learn whether astronauts could function in pressure suits on the surface of the moon, we needed to test the suits and procedures in earth orbit. Our preliminary planning for Gemini had indicated we might try a stand-up EVA of some kind on GT-5. McDivitt and White were pushing for a more ambitious one on GT-4, however, and when the Russians came through with theirs, we decided to move ours forward.

It was pretty clear to us that we were in a race. The Soviet press even said that Leonov's EVA suit was a prototype for a lunar surface pressure suit. Nevertheless, the Russians had beaten us again, this time with the first walk in space.

16

GEMINI

At 4:40 on the morning of Tuesday, March 23, 1965, I knocked on the bedroom doors of Gus and John in Hangar S. I'd done this for Gordo and would ultimately do it for twenty-four American manned space flights. My job was to certify to Bob Gilruth that the crew was ready. The drill was a shower, final medical checks, then breakfast, then suiting, then out to Pad 16, where Wally Schirra and Tom Stafford were setting up switches inside the cockpit of Gemini 3, which Gus had named *Molly Brown*.

(We'd borrowed the pilot's tradition of naming our spacecraft for Mercury, of course, and there was every intention of continuing that in Gemini. Gus was still a little sensitive about what happened to *Liberty Bell* 7, so he hit on the idea of using the name of a big musical that was running on Broadway about that time, *The Unsinkable Molly Brown*.

(When he submitted that, the feeling around MSC was that it was kind of frivolous, so we had to ask him if he had other choices. "Sure," he said. "How about *The Titanic*?" Well, compared to that, *Molly Brown* sounded great.)

There was only one minor hold. The Titan GLV-3 ignited at 9:24 and took off. Gordo Cooper was capcom at the Mission Control Center, and told Gus, "You're on your way, *Molly Brown*!"

Less than six minutes later Gus and John were in orbit, a little higher than planned, since the Titan had overperformed. That was the kind of thing we needed to find out, since future rendezvous depended on getting spacecraft into the right orbits.

As they came across the United States on their first orbit, Gus and John fired the Gemini thrusters for seventy-four seconds to lower the orbit. On the second orbit, over the Indian Ocean, Gus fired not only the aft thrusters, but also the smaller nose thrusters, changing Gemini's orbital inclination by a fraction of a degree.

On the third and final orbit, Gus fired the thrusters again, this time lowering the perigee—that is, changing the shape of the orbit—to less than fifty miles at its closest approach to earth, as opposed to the usual one hundred and twenty. This meant that Gemini could reenter the atmosphere even if there was some kind of retro failure.

Reentry went as planned, though we found out that the Gemini didn't have quite the lift we had hoped. Tracking noted this and Gus was able to change the bank angle of the Gemini slightly. He still landed about fifty miles away from the carrier. Rescue choppers were overhead within seventeen minutes . . . but not quickly enough to keep Gus from losing his lunch in the bobbing spacecraft-turned-boat.

It was a good mission that accomplished its major goal of showing that a manned spacecraft could maneuver. Unfortunately, what it's remembered for is John Young's corned beef sandwich.

Astronaut food had been sort of a joke all through Mercury, with everybody thinking we were eating out of tubes and drinking lukewarm Tang breakfast drink. None of that was exactly true, but there was nothing really exciting about what we had to eat. Gus probably liked the food less than anybody.

So before the launch John Young got Wally to pick up a corned beef sandwich from Wolfie's Deli in Cocoa Beach. With my permission, John stuck it in the pocket of his pressure suit. During the first orbit, when Gus was supposed to try some space food, John said, "Hey, skipper, you want a sandwich?" And held it out to him.

Gus took a couple of bites, then they realized there were crumbs floating all over, so they wrapped the thing up. Of course, there was no way to keep this kind of activity secret, and naturally there were complaints. Eventually they worked their way down to me, and I duly informed all astronauts that they weren't supposed to take unauthorized items, especially food, aboard the spacecraft. I even had to give poor John a formal reprimand . . . not that it affected his career.

As soon as GT-3 splashed down, Wally Schirra and Tom Stafford were free to dive into GT-6 training. They would attempt the first rendezvous and docking with an Agena upper stage sometime in early fall. I wanted a veteran backup crew to help with training, so I recycled Gus and John for that.

The next flight was GT-4, scheduled to double the duration of the longest previous American manned spaceflight. The extra length meant that for the first time the guys were going to have to not only deal with eating and sleeping in orbit, but also with some of the real unpleasantness of spaceflight, such as elimination. We would also try some preliminary rendezvous maneuvers. And GT-4 would be the first mission controlled from the Mission Operations Control Room (MOCR) at the Manned Spacecraft Center. Operations at the Cape would be limited to getting the Gemini-Titan launched; everything that took place after the tower was cleared would be from Houston. There would be round-the-clock operations in three shifts, directed by Chris Kraft, Gene Kranz, and John Hodge.

The big item, as it turned out, was a proposed EVA or space walk by Ed White.

Our original plans had been for Pete Conrad to open the hatch on GT-5 and do a brief stand-up EVA; that was moved forward to GT-4. Even that changed when Alexei Leonov floated around outside Voskhod 2 in March. Less than a week later there was a meeting with Bob Gilruth, George Low, Dick Johnston of Crew Systems, and Warren North of my directorate, about actually letting Ed leave the spacecraft and float, like Leonov, on the end of a tether.

Gilruth told them to go ahead and try to sell the idea to George Mueller at HQ, which they did on April 3. But he wasn't sold. If the plan was to have a stand-up EVA on GT-4, we should stick to the plan. Besides, none of the equipment was proven, from the EVA pressure suit to the tether to a handheld maneuvering unit Ed White was supposed to use.

The crew equipment and systems guys dove into the work while Jim and Ed added it to their training schedule. On May 14 the systems people did a presentation for Bob Seamans, who was won over. He promised to get Webb's and Dryden's approval, but found that Dryden, like Mueller, still wasn't convinced we should go charging off in this direction.

It wasn't until May 25 that everybody signed off on the EVA with the tether. Launch was scheduled for June 3.

Jim McDivitt had wanted to name his Gemini *American Eagle*, but NASA HQ wasn't having any more names. Jim and Ed decided to put the American flag on their pressure suits, which we then did for years.

Al Shepard and I were at the Cape to handle any last-minute items with Jim and Ed. In the new MOCR at Houston were John Young, Gus Grissom, and Wally Schirra. Al served as the launch capcom, or "stoney," while Gus was capcom at MOCR.

GT-4 was launched at 10:16 A.M. There had been a long hold to clear up a problem with the gantry tower (part of it had to be lowered away from the vehicle, and it got stuck for more than an hour).

During the first few minutes in orbit Jim tried to maneuver the Gemini back toward the upper stage of the Titan, which had been equipped with some flashing lights for visibility.

The Gemini had just separated from the Titan and the feeling was that Jim could rotate his spacecraft, fire the thrusters by eyeball, and stay close. But pretty soon after separation the Titan went one way and Gemini went another. Worse yet, the stage was tumbling end over end. Jim tried thrusting toward it and only managed to get farther away. (Basic orbital mechanics. Thrust toward something and you increase your velocity, which actually puts you in a higher orbit . . . where you will be going slower than you were when you fired your thrusters to increase your speed. It's a hard thing to learn, since it's kind of backward from anything you know as a pilot.) Frankly, it wasn't an operation that had been too well thought out, and after Jim had managed to burn a lot of fuel to no particular use, Chris Kraft told him to give it up.

The crew went right into prep for Ed's EVA, and that took longer than anyone expected, too. As GT-4 passed over Hawaii, a little over four hours into the mission, Ed opened the hatch and stood up. He attached a film camera to the outside of the spacecraft, then—tethered and holding the hand-held maneuvering unit—pushed himself out. As he cleared the hatch, he fired the HHMU and floated out past the nose of the Gemini. He flew around out there for a few minutes, until the HHMU was out of gas. Jim reported that he could really feel the spacecraft moving around whenever Ed pulled on the tether.

The EVA had taken place over the United States, and they were about to go out of range of the tracking station, so Gus started telling Jim to get Ed back inside. Ed was trying; it was hard pulling himself around with just the tether. Eventually he got back inside the hatch, then really had to struggle to get it locked down. (That had always been one of our big worries about EVA, other than a suit leak or a broken tether: what if you can't get the hatch closed? It would be a pretty bad day.)

Because Gus had had to repeat himself in telling Ed to get back inside, some reporters got the idea there was some kind of "rapture of the deep" horseshit that made Ed want to stay outside. Actually, it was just a communications problem: Gus could talk to Jim in the Gemini, but not directly to Ed, who was connected through a line in his tether. Ed couldn't hear Gus, and even Jim wasn't hearing everything. There was no other problem.

Pictures of Ed floating outside Gemini became one of the most famous space pictures ever. Ed himself wasn't blind to the public relations possibilities: during the EVA he said something about how he was "feeling red, white, and blue all over." I had the feeling he was setting himself up to be another astronaut-politician.

The rest of the mission went fine, with some minor problems. The biggest one was just figuring out how to deal with all the trash inside the spacecraft.

Jim and Ed splashed down in good enough physical shape that Chuck Berry's people thought we could not only go for a seven-day mission on GT-5, we could go for eight. It would be our first chance to beat the Russians at anything.

While GT-4 was getting ready to go, the scientist-astronaut selection was coming to an end. I had originally hoped the National Academy of Sciences would come up with a list of at least twenty names—ten geologists and ten doctors—figuring we would pick three of each. Any who weren't pilots would go through an Air Force training program to qualify. That way we would wind up with at least one geologist and one doctor for one of the earlier Apollo crews, which was all I planned to use.

Well, the selection committee had broadened the requirements to include physicists and astronomers. And NASA HQ was saying we should have more like ten or fifteen in the group. Okay, fine. But then I wanted at least ten candidates in any field, or a number more like sixty.

NASA had received over thirteen hundred applications for the scientist-astronaut group by the deadline of December 31, 1964. We did a preliminary pass through those, then in February 1965 gave four hundred of them to the National Academy of Sciences for their selection process.

The NAS couldn't find a way to select a big enough pool of qualified people. They submitted sixteen names, total, to us in April 1965. We ran those sixteen through the usual personal and technical interview at MSC. And came up with these six:

Owen K. Garriott
Edward G. Gibson
Duane E. Graveline
Lieutenant Commander Joseph P. Kerwin, USN
F. Curtis Michel
Harrison H. "Jack" Schmitt

Kerwin and Graveline were physicians. Schmitt was the only geologist. The other three were physicists of one kind or another.

As a Navy flight surgeon, Kerwin was a qualified jet pilot. Michel had flown on active duty with the Air Force. The other four were all set to go to flight school at Williams Air Force Base in Arizona beginning in August. They wouldn't be actually beginning astronaut training until the summer of 1966.

That's when the first big problem came up. Since the National Academy of Sciences had handled the initial phases of the selection, none of the candidates had been run through any kind of rigorous background checks.

Duane Graveline was working at the MSC as a flight surgeon in the Life Sciences Division when selected. He had served as an Air Force flight surgeon on active duty until the spring of 1965 and had done some interesting medical work related to the space program—I think he spent an entire week floating submerged in a tank, breathing through a tube, to see how near-weightlessness might affect an astronaut. His qualifications were good on paper.

But it turned out Graveline had a big problem at home. I'm not spilling any secrets here, because it wound up in the newspapers about a month after he'd been announced: his wife filed for divorce, claiming he had an "uncontrollable" temper.

This, of course, was one of the things I had always warned the guys about. It had become obvious back in Mercury that there were lots of opportunities for astronauts to get into trouble, whether it was financial dealings or drinking or chasing women. We had pretty strict rules for financial matters, and the medical rules, if nothing else, encouraged people to moderate their booze intake. Personal matters were trickier. I frankly didn't give a damn what a pilot did in his spare time; how he and his wife got along was their business. If he played around when he was out of town, or even if he didn't, again, it was only my concern if it affected his work.

But one thing I had to tell the guys was this: the program didn't need a scandal. A messy divorce meant a quick ticket back to wherever you came from—not because we were trying to enforce morality, which was impossible, anyway, but because it would detract from the job.

Given the time I spent with the pilots, I had a pretty good idea who was doing what. Some guys were model husbands, some guys played around and were pretty indiscreet, with a few in between. When it affected the way the work was being done, I didn't hesitate to send a message.

Graveline was the first to get it. At that time we didn't need the bad publicity and didn't like what it said about his judgment. He was back in the Life Sciences Division so fast he never even made it to the group photo. (Graveline left NASA a few months later and became a family physician up in Vermont.)

In the meantime we were still taking applications for a new group of pilots to be selected early in 1966.

On July 1 we announced that Frank Borman and Jim Lovell would fly the GT-7 mission, scheduled to last two whole weeks, late in 1965 or early 1966. The backup crew was Ed White and Mike Collins.

Other Voices

KENT SLAYTON

Dad started taking me hunting when I was about six years old, though I really didn't do much until a few years after that. What we would often do is leave Houston on a Friday night and drive three to four hours to the hill country near San Antonio or Austin. Usually Dad would have arranged to do some hunting on the King Ranch property or something like that. We would spend the night in a bunkhouse, hunt all day Saturday, then drive back to Houston on Sunday. Sometimes Wally or Gus would come along.

We took family fishing trips to places like Wyoming. One trip in particular was to some property that backed right up against Yellowstone Park. Dad was pretty concerned about what I would do if a grizzly came through the camp. He wanted me to make sure I always had an escape route planned.

I can remember how excited I was when I shot my first deer. We brought it home and I was learning how to skin it in the garage when Mom appeared in the doorway. Now, Mom was not a pouter or anything like that, but when she saw what I had done, she got tears in her eyes and said, "I can't believe you did that." And she turned around and walked away.

Even though I had an administrative job and was still restricted to flying with a copilot, I kept up with astronaut training. Shortly after we announced the scientist-astronaut group, I joined some of the 1963 guys in the Apollo branch, Gordon, Cunningham, Eisele, Schweickart, and Anders, on a geological trip to Oregon. One night while we were hanging out in the motel, I got them together and told them about their first crew assignments. They already knew Mike Collins would be backup pilot on GT-7. Dave Scott would be prime crew pilot on GT-8, with Neil Armstrong; Dick Gordon would be backup, with Pete Conrad.

For once the technical team got ahead of the astronaut crew. McDonnell was going to have GT-5 ready for launch around August 1, less than two months after GT-4. But Gordo and Pete Conrad had only been training for GT-5 since February, meaning they were going to have about five months to get ready for the longest space flight anyone had ever tried. They were killing themselves, working sixteen-hour days and weekends, and we realized they just needed more time. I got on a plane and went to see George Mueller to ask him for help, and he delayed the launch by two weeks.

GT-5 was going to be the first Gemini to be powered by fuel cells rather than batteries alone. Given the weight and size constraints, the batteries we could carry on a manned spacecraft in those days would only support a mission about four days long. To go to the moon and back, you needed a minimum of eight days; a landing mission could take even longer. So proving the fuel cells was pretty important, though it wasn't the kind of thing that made for exciting coverage in the press.

One of the other experiments planned was a test of the Radar Evaluation Pod, a transponder and flashing light unit that would be deployed from the GT-5 adaptor at some point in the mission. Gordo and Pete would then try to rendezvous with it.

The real challenge was going to be crew comfort and plain old housekeeping. Jim and Ed had had enough problems stowing all the garbage on their four-day flight. They basically had to sit there inside their heavy G4C EVA pressure suits for the whole time. Jim, who was a little taller than Ed, kept scraping his helmet on the hatch above his head. He finally took the helmet off, which was just the kind of thing that made everybody nervous. We had confidence in the spacecraft environmental systems, but we always liked that extra backup of having the guys in their suits in case there was some kind of sudden pressure loss. There was no way they could climb back into things in a hurry.

Gordo and Pete, who were upset when they lost their EVA to GT-4, figured that since they didn't need EVA suits, they should be able to use a lighter pressure suit we were developing called the G5C, which had a hood instead of a helmet and could be taken off. Gordo and Elliot See had even done a simulation without suits at all back at McDonnell in June. But nobody was ready to commit to flying a crew in a partial pressure suit—much less long johns underwear. So Gordo and Pete were stuck with the G4Cs for the whole eight days.

Gordo and Pete were still trying to find a way to individualize their spacecraft. Gus and John had managed to name GT-3; Jim and Ed had worn the American flag on GT-4. Pete hit on the idea of a patch, the kind of thing every Navy air squadron has.

It showed a Conestoga wagon with the slogan, "Eight Days or Bust." When Jim Webb saw that, however, he had a fit. He didn't want the motto, for one thing, and decided upon a whole set of guidelines for what he called "Cooper patches" that each crew commander could design and wear.

Other Voices

JAMES WEBB

MEMORANDUM FOR: Mr. Donald K. Slayton, MSC, Houston, Texas August 14, 1965

. . . I must say I have some concern about the fact that . . . the most urgent and important factors affecting the Gemini program seem to get involved in a morale matter such as this and at the last moment. I believe it is your responsibility to avoid this in the future.

When we are dealing with matters which affect the way elements of these programs are viewed in many different countries by many different nationalities, we cannot leave to the crew the decision with respect to these matters no matter how strongly they feel that they would like to have some element of individuality. In this case, both Dr. Gilruth and I have a very strong concern about the "8 days or bust" motto. I wish it could be omitted. If the flight does not go 8 days, there are many who are going to say that it was "busted." Further, whether we get the 8 days or not, the way the language will be translated in certain countries will not be to the benefit of the United States.

I was back at the Cape to wake up the crew on the morning of August 21. We had already been through one scrubbed launch on the nineteenth, caused by some worry about the fuel cells.

Gordo and Pete took off in GT-5 right on schedule at nine A.M. Before they were half an hour into the flight, a heater in one of the fuel cells failed, and pressure began dropping. It was a slow drop, however, something to keep an eye on.

The first milestone was to test the Radar Evaluation Pod. On the second orbit Gordo yawed GT-5 ninety degrees and released the REP, which Pete Conrad called the "Little Rascal." The onboard Gemini radar picked her up. Before the crew could do any maneuvering, however, the fuel cells started dropping below minimums. There was no power to spare for chasing around; in fact, the whole mission was close to being aborted after less than one day.

Kraft and his people decided to wait it out. GT-5's reentry batteries would provide power for at least thirteen hours, in case the fuel cells failed altogether. By the sixth orbit, however, the pressure level was so low that Gordo and Pete actually had to start packing for reentry.

Then things stabilized and actually began to creep back up. GT-5 got a go for at least one day, and at the end of that Chris Kraft was confident we'd get a whole eight days.

Buzz Aldrin, who was working in the back room, even suggested a way to salvage the Little Rascal experiment. The pod itself was out of range, but Gordo and Pete were able to use their orbit attitude and maneuvering system engines to make a rendezvous with a phantom "pod," a point in space. It worked pretty well, and also gave the crew something productive to do.

On day four they did some observations of events on the ground—Gordo had said he could see the smoke from individual chimneys during his Mercury flight. So we set up a couple of Minuteman launches out of Vandenberg, which were easily visible. At Holloman Air Force Base in New Mexico they did a test of a rocket sled, which the guys saw, too.

On day five Gordo and Pete passed Bykovsky's record of five days in space. Not only was this the first time we'd passed the Russians, we kept on going and never looked back.

The last four days were pretty tough on the crew. There was junk all over the cabin—food wrappers, cameras, helmets, you name it. Every time they needed to get something out, they had to repack. It made the crew kind of testy, Gordo in particular. When Chuck Berry asked him if he was getting any exercise, he said, "I hold Pete's hand once in a while. I use a cleansing towel. Then a couple of days we chewed gum." Gordo and Pete were supposed to be doing isometrics with a bungee cord.

Given the physical situation aboard the spacecraft, the attitude was understandable. But it didn't help Gordo's reputation any.

On the last day of the mission, August 29, Gordo and Pete had a conversation with Scott Carpenter, who was two hundred feet down in the Pacific off La Jolla, California, inside SeaLab II. (Scott was still an astronaut, but he had been given leave to return to the Navy for this project.)

There was some last-minute worry about a hurricane in the landing zone, so we targeted for reentry one revolution earlier than planned. Some computer problems and navigation mistakes—not the crew's—forced GT-5 to come down a hundred miles from the target point.

Nevertheless, an hour and half after splashdown, Gordo and Pete were walking on the deck of the carrier USS *Lake Champlain*.

Gordo and Pete's world publicity tour following splashdown and debriefing took them to Athens, Greece, for an international space congress, and I went along.

What was interesting about it was that the Voskhod 2 cosmonauts, Pavel Belyayev and Alexei Leonov, were also there. This wasn't the first time

American and Soviet cosmonauts had gotten together: Gherman Titov met up with John Glenn during a visit to the United States back in 1962, and Jim McDivitt and Ed White had shaken hands with Belyayev and Leonov at the Paris Air Show earlier that summer.

We managed to get together in a relatively small group, three American astronauts and two Russians, plus a couple of translators, and had a couple or three drinks. I looked right at Alexei, who was in full colonel's uniform, and proposed a toast to the two of us sharing a drink in orbit someday. I don't think I'd have bet any money at any odds that we would do just that, but damned if we didn't.

In the first three Gemini flights we'd accomplished a lot—proving that the vehicle would fly and maneuver, getting some preliminary EVA work, raising our time in space to eight days, the minimum required for a flight to the moon and back.

But the real work was about to start. GT-6 was going to be part of a dual countdown with the first Gemini Atlas-Agena D. The Agena would launch . . . one orbit later GT-6 would launch. Then the Gemini would maneuver to a rendezvous and docking.

It was a very complicated flight, since at the time there were a lot of people who didn't think rendezvous was practical. That's why I had pointed Wally Schirra and Tom Stafford at this from the beginning.

As we analyzed the results of GT-5, I was able to go ahead with crews for GT-8 on September 20: the commander would be Neil Armstrong and the pilot would be Dave Scott, the first of the 1963 group to be assigned to a prime crew. Backups would be Pete Conrad and Dick Gordon.

It started to look like Gemini would be flying additional manned missions. The Air Force had adopted it as the carrier vehicle for the Manned Orbiting Laboratory, a polar-orbiting spy satellite that they hoped to launch in 1968. MOL was given the go-ahead by LBJ while Gordo and Pete were in orbit.

The driver in the GT-6 planning, aside from practice in rendezvous, was the Agena, the Atlas upper stage we were adapting as the Gemini-Agena Target Vehicle (GATV). We had originally wanted to use the Agena-B, but adding docking radar, a forward docking adaptor, improved stabilization, and an engine restart capability forced NASA to choose the more advanced D model back in June 1962. There were the usual delays and dead ends; by April 1964 we had accepted the first GATV. Rather than launch it, we used it for ground tests and saved NASA the cost of an Atlas booster.

Come May 1965 and we still couldn't approve GATV-5001. Fortunately, we had already received the second Agena, and went ahead with plans to get that ready as GT-6's target.

We tried to keep the mission simple. GT-6 was the last of the battery-powered Geminis, so the duration was going to be two days, max. Wally gave a thumbs-down on the idea of an EVA. He and Tom wanted to relight the Agena once they docked with it, but we didn't think the Agena would be quite ready for that.

We not only had to accomplish rendezvous, if we were going to set the stage for Apollo, we had to accomplish it in different modes. There were three of them: tangential, coelliptic, and first-apogee.

In the tangential mode the Agena would be in a circular orbit, the Gemini in an elliptical one, with the Gemini apogee (highest point in the orbit) equal to the Agena's altitude. During several revolutions, with several minor burns from the propulsion system, Gemini would loop closer and closer to the Agena, ultimately arriving at the same point in space the same time as Agena.

The coelliptical mode started out with Agena in a circular orbit, but Gemini would be placed in a smaller circular orbit. As the two spacecraft revolved around the earth, Gemini would sort of lap the Agena. At the right moment, Gemini would fire thrusters and pop up to rendezvous and dock.

Then there was the first-apogee mode, which was basically a quicker version of the tangential mode. It had the disadvantage of burning a lot of fuel.

What was chosen for the first Gemini-Agena was the coelliptical mode, which would have Wally and Tom docking with the Agena over the Indian Ocean a little less than six hours into the mission. They would stay linked for the next seven hours, then separate. The whole mission was only going to last forty-eight hours, but it was a busy forty-eight hours.

On the morning of Monday, October 25, 1965, Wally and Tom got into GT-6 on Pad 19 while our first Atlas-Agena was being counted down on Pad 14, a few miles south. A few seconds after ten A.M., the Atlas lifted off. Five minutes in, the Agena separated, firing its own motor.

Suddenly there was no more telemetry. The Canary Island tracking station had no Agena. We waited until forty-eight minutes after launch, when the Carnavon tracking station in Australia should have picked it up. "No joy," they said. The Agena had probably exploded at the time of ignition.

No target, no docking. Wally and Tom had to climb out of their Gemini and come back down.

* * *

For a day or so we thought about recycling Agena 5001, the ground test bird that hadn't ever come up to specs. Even then, we still had to figure out what had happened to 5002. By this time Air Force was launching at least one Atlas-Agena a month out at its West Coast site at Vandenberg Air Force Base, California. Most of these were photo reconnaissance satellites that flew in polar orbit. So there was a hell of a lot of operational experience we could draw on with that system, but a failure like this one inevitably meant some kind of delay.

It was during a conversation at the Cape between John Yardley and Walter Burke of McDonnell Douglas, and Frank Borman and Jim Lovell, that the idea of maybe docking GT-6 with GT-7 came up. (The Martin Company had proposed something like this earlier in 1965, so there were even some paper studies on it.) Frank was dead-set against the idea of turning GT-7 into a docking target, however; the vehicle hadn't been designed for it.

There were also some pretty serious logistical problems. We had just managed to show that we could get Gemini-Titans ready for launch every two months. To put GT-7 and GT-6 in space at the same time meant we would have to launch them within *two weeks*.

Here were some of the problems: stacking the Titan—that is, putting one stage on top of the other on the pad and testing and integrating all the systems—took several weeks. The question was, once you had the Titan stacked, could you just take it apart and put it back together without starting from scratch? (Somebody had the idea we might lift the Titan off Pad 19 and move to another one where it could be stored vertically. That would be one way to cut down the time between launches.)

GT-6 was unable to sustain a mission longer than four days; it was designed for a rendezvous and docking mission. There probably wouldn't be another Agena available until early 1966. The GT-7 spacecraft was configured for a two-week mission and could be launched as soon as it was ready. So the first question was, why not simply put GT-7 on top of the GT-6 Titan?

Well, GT-7 was just a little too heavy. So the launcher for GT-6 was going to have to be taken off the pad. The Martin teams thought they could make it work, since they had already stacked the thing once.

There were communications and telemetry issues. The tracking network was set up to handle, at most, one manned vehicle and one unmanned vehicle—not two that were manned. Chris Kraft's people, especially John Hodge, figured that during rendezvous they could treat GT-7 like a Mercury, where the telemetry came to mission control by teletype, letting the active rendezvous craft, GT-6, have the real-time channels that were available.

(It's kind of funny to realize how big an issue this was at the time. Same thing with the very tiny computer memory we had available onboard the spacecraft and even at the MOCR.)

There were a number of quick meetings on this, not only at the Cape, but also in Houston and Washington. I'd flown back to Houston Monday afternoon. On Tuesday at four in the afternoon I was in a meeting in Bob Gilruth's office with Chuck Mathews, the Gemini program director, George Low, Yardley, Burke, and Kraft. We ran through the issues; nobody had any show-stoppers. Certainly not from flight crew operations. Wally and Tom were already trained for rendezvous. Serving as relatively passive targets wouldn't put any undue strain on Frank and Jim.

So we agreed to launch GT-7 in December, followed by GT-6 in ten days. They would rendezvous and not dock.

We even got Jim Webb's approval a day later. He liked the idea of four American astronauts in space at the same time. It was another thing we could beat the Russians at. And that night, LBJ announced the new plan.

It was maybe sixty hours from the Agena failure to White House approval. That was how things got done in those days.

I had pointed Frank Borman at one of the Gemini long-duration missions from the very beginning, because of his tenacity. The original planning had him flying with Gus Grissom on the first seven-day mission; the rearranged schedule gave him command of the two-week one. Jim Lovell, the pilot, was a very capable guy you could have assigned to any mission. He got along with everybody, so that was a big help, because basically we were asking these two guys to spend two weeks in a space equivalent to the front seat of a Volkswagen.

Training for the mission concentrated on a lot of the housekeeping. There were also about twenty medical experiments. I didn't like loading up the guys with a lot of this stuff, but except for a few hours of proximity operations with GT-6, they were going to have a lot of time.

One new item for GT-7 was the G5C pressure suit. It was lighter than the earlier models and designed strictly for protection: you didn't want to do an EVA in it. For a while we were talking about having Jim Lovell and Tom Stafford do EVAs during the rendezvous, swapping seats. But it made GT-7 more complicated than we wanted, and would have screwed up a lot of the medical data on Jim. Instead of a heavy helmet, G5C had a hood that could be unzipped and opened. (Frank and Jim wore pilot-style helmets underneath, to protect their ears from noise during launch.) You could also take it apart and get out of it in tight quarters.

Things were pressing ahead at such a pace that on November 8 we went ahead and announced the GT-9 crew: Elliot See as commander with Charlie Bassett as pilot. They would fly a three-day mission with rendezvous and docking, and perform the first test of a Buck Rogers EVA backpack called the Astronaut Maneuvering Unit.

The backup crew would be Tom Stafford and Gene Cernan. Eliot and Charlie were already at work; Tom and Gene were still busy with GT-6.

On Saturday, December 4, 1965, GT-7 went into orbit as planned. During the first orbit Frank kept station with the Titan upper stage, and had a lot more luck with it than Jim McDivitt had on GT-4. Then he and Jim Lovell settled down to work.

The real work was out at Pad 19. Normally there's a big relief after a vehicle is launched, but they had to get another one stacked out there in nine days. The whole vehicle—Titan first stage, Titan second stage, and what we were now calling GT-6A—were erected the very next day.

In orbit, Jim Lovell had shucked his pressure suit and was living in relative comfort. Frank was still wearing his, and he didn't like it much. There were some problems keeping it cool inside GT-7. But flight rules dictated that one guy should be in a suit at all times. Frank privately asked Gene Cernan, who was capcom, to ask me about it. All I could do was get Jim to put his back on, so Frank could have a turn in his long johns.

Preparations at the Cape were going great, aiming at a launch on Sunday, December 12. I sent Gordo down there to help out Wally and Tom while I stayed in Houston. Frank and Jim had fired their thrusters to put GT-7 in a close to circular orbit. A few seconds after 9:54 A.M., as GT-7 passed over the Cape, the GT-6A countdown reached zero.

I was sitting in the firing room watching what happened. The Titan's motors ignited for about a second, then shut down. I waited for Wally to pull the D-ring inside Gemini, firing the ejection seats to get him and Tom out of there.

But nothing happened. The Titan didn't blow. Wally and Tom sat tight. Chris Kraft told him, "No liftoff, no liftoff." Wally started reporting that fuel pressure in the Titan was lowering.

We moved the erector back up to the Titan and got the crew out of there as the ground team safed the launcher. They assumed they would have another chance within maybe forty-eight hours.

By the next morning analysis of telemetry from Titan showed that a dust cap had been left in place in the tail of the vehicle, blocking flow to a gas generator. The Titan had sensed that the gas flow wasn't right and had shut down as designed.

Wally said later that he never felt the vehicle lift, so he felt confident in waiting it out. None of us had too much confidence in those ejection seats anyway, especially at what was practically zero altitude. It was a situation in which by pulling the ring you risked possible death to escape certain death. Wally did the right thing.

And three days later, on December 15, GLV-6 fired as designed and finally put Wally and Tom in orbit.

Frank and Jim were feeling a little "crummy," they said, by now, day eleven of their mission. They weren't sleeping well—Frank in particular—but they were hanging on. They had to put their suits on in preparation for the rendezvous.

GT-6A wound up in an orbit slightly lower than we wanted, which gave us a chance to test some of our training right away. Wally made the first OAMS burn at the end of the first revolution to correct that.

The second burn made forty-some minutes later, added velocity to GT-6A, which now began to "chase" GT-7. There were three other burns, one of them lasting just a second, to put Wally and Tom in a circular orbit close to Frank and Jim. At a distance of 268 miles, GT-6A got a radar lock on GT-7. There was another burn, then it was time to close in.

Some rendezvous lights had been added to the exterior of GT-7, as an aid to Wally and Tom. They saw one at a distance of about thirty miles. At this point both vehicles were over the Indian Ocean, in darkness.

At five hours, fifty minutes into the flight of GT-6A, a mile from GT-7, both vehicles moved into sunlight. Wally let his vehicle coast up to GT-7 until they were just a few yards apart. Wally and Tom could see Frank and Jim in their windows—they'd grown beards. The two Annapolis grads in GT-6A stuck a "Beat Army" poster in their window, to tease Borman, who was from West Point.

There was pandemonium down in MOCR. Everybody was clapping and lighting cigars—well, making that first rendezvous was a pretty big deal. We weren't going to the moon without it.

They stayed together on station for a few hours, then Wally maneuvered away. He and Tom landed safely the next day.

Poor Frank and Jim still had three more days to go. After the fun of the rendezvous—the first space visitors—things really got tedious for the crew. Some of the spacecraft systems were starting to act up, too.

But they stuck it out. They came down on December 18. Frank and

Jim were tired, but they were able to walk unassisted on the deck of the carrier.

At the end of 1965 the Apollo program was gearing up. Joe Shea and George Low thought it was likely that we'd have the first manned earth orbital mission ready for launch before the end of 1966, so it was time to name the crews.

I went back to my two-year-old schedule, somewhat modified. My first choice to command the first Apollo would have been Al Shepard, but he had had more medical problems in the summer of 1965 and was about to be ruled out of consideration. Gus Grissom was going to be coming off the backup assignment to GT-6A and so was a pretty natural choice for commander of the first mission.

Since we weren't going to fly a lunar module with the first Apollo, the crew didn't need a whole lot of experience and it would be a good place to try out some of the guys who, frankly, I thought were weaker. My original rotation had Donn Eisele and Roger Chaffee as the senior pilot and pilot working for Gus.

The backup crew was another matter, since I wanted to aim them at the first lunar module test flight down the line. (This crew would also be a potential lunar landing crew.) I picked Jim McDivitt, who had actually been working Apollo since September, as commander. The command module pilot needed to have flight experience and, preferably, rendezvous experience. Dave Scott, who would be coming off GT-8 in a couple of months, was the first choice. The lunar module pilot would be Rusty Schweickart.

The first thing that changed was that Donn Eisele messed up his shoulder during a zero-G flight in a KC-135 just before Christmas. Nothing too serious, but he was off flying status for a couple of months. He was going to be behind the training curve right from the start, so I simply swapped him with Ed White, who I'd originally had down as senior pilot of the next crew, Apollo 2.

I brought Gus, Donn, and Roger into the office and told them they were the first Apollo crew. The public announcement was still a few months off.

17

OPERATIONAL

Looking back, 1966 was NASA's best year. We were in the middle of Gemini, Apollo was building up. Unmanned programs like Ranger, Surveyor, and Lunar Orbiter—all of them designed to support Apollo—were on track. A new program to follow Apollo—Apollo Applications—was being laid out, to perform lunar landings and extended stays on the surface of the moon, in addition to operations with orbital workshops in earth and lunar orbit. The agency would never have as much support as it had in 1966. Or as clear a goal.

The first real Apollo flight—unmanned—took place on Saturday, February 26, 1966. Called AS-201 (George Mueller had come up with a system that called Saturn IB flights the 200 series with Saturn V the 500 series), this was a suborbital flight of a Block I Apollo command and service module, spacecraft 009. AS-201 took off from Pad 34 and went into orbit. Twenty minutes into the mission, the service propulsion system (SPS) was ordered to fire, to send the command module back into the atmosphere at a higher speed than we ever expected to encounter in a mission, in order to test the heat shield.

The SPS didn't work perfectly—there was a pressure loss in one of the engine chambers—but we still got the test. Thirty-eight minutes after launch, spacecraft 009 was floating in the Atlantic.

AS-202, scheduled for launch in early summer 1966, was supposed to be a repeat of 201, carrying a more capable Block I Apollo. AS-203, also in the summer, would not carry an Apollo CSM, but would test the S-IVB upper stage—the same one we would use on the Saturn V to send Apollo out of earth orbit and on its way to the moon. If all these went well, AS-204 would be the first manned Apollo, Apollo 1.

Other Voices

KENT SLAYTON

One of my most vivid memories is of waking up in the morning and feeling the house shaking. It was my dad, who always got up early, working out with his Exer-Genie. This was a kind of pulley device you could attach to doorframes. He must have put in half an hour a day with that thing, not only every day at home, but even when we went on trips. I would sit there and eat my breakfast before school, and talk to Dad as he worked out. He was religious about it. He ran, too.

He worked long hours, but whenever he was in town, Mom would wait dinner until Dad got home. So most nights we didn't eat until eight-thirty or nine.

The year 1966 started out looking good for me, too. I was still able to fly the T-38s, though never as much as I would have liked, thanks to the second-pilot rule. On the other hand, while I spent a lot of time away from home, it wasn't as much as any of the guys actively training for flights. Marge was busy acting as unofficial chairwoman of the unofficial Astronaut Wives' Club. Kent was coming up on his ninth birthday, an age when he could really start to enjoy hunting and fishing.

I still wanted to move back into active training, but that wasn't going to happen until I got a bunch of doctors to change their minds. Meanwhile, I had to select a new group of pilot astronauts.

And there were a gaggle of Gemini missions coming up, one every two months. Neil Armstrong and Dave Scott to try the first rendezvous and docking on GT-8 in March. GT-9, Elliot See and Charlie Bassett, in May. GT-10, John Young and Mike Collins, in July. GT-11, Pete Conrad and Dick Gordon, in September. GT-12 would wind up the program in November with Tom Stafford and Gene Cernan.

On the Apollo program, Gus Grissom, Ed White, and Roger Chaffee were already spending a lot of their time at North American Aviation, in Downey, California. With them were two of their three backups, Jim McDivitt and Rusty Schweickart. (Dave Scott was still getting ready for GT-8.)

I had put another crew in training for the second manned Apollo without making a public announcement. Wally Schirra would be the commander, with Donn Eisele and Walt Cunningham as the senior pilot and pilot. It wasn't a crew I planned to use on lunar landing missions. Wally was making noises about retiring, and I figured to move Donn and Walt over to Apollo Applications. The mission wouldn't break any new ground. As the second and last flight of a Block I command module it wouldn't even be equipped for rendezvous and docking with a lunar module. It was on the schedule to give us a chance to pick up whatever might get missed on the first manned Apollo flight.

The backups were Frank Borman, Charlie Bassett, and Bill Anders, with Charlie still tied up with Gemini until May. I thought they would be a good lunar landing crew.

At 7:35 on a Monday morning, February 28, 1966, the two GT-9 crews took off from Ellington on their way to McDonnell in St. Louis. They were supposed to do rendezvous training on the Gemini simulator there. The prime crew, Elliot See and Charlie Bassett, were in the lead T-38, with their backups, Tom Stafford and Gene Cernan, in the second one.

The McDonnell plant in St. Louis was located right next to Lambert Field, the big airport. The sky was rainy as Elliot and Charlie broke through the clouds. Elliot was flying and must have realized he was coming in too low and slow to reach the runway. He hit the afterburner in order to go around and try again, but turned right into a building near the airport. The T-38 hit the roof and exploded, killing Elliot and Charlie, and injuring fourteen people on the ground—none of those people seriously, fortunately.

The building they hit was McDonnell 101—their GT-9 spacecraft was inside it. Tom and Gene landed safely a few minutes later.

The investigation concluded that the crash was pilot error, but I didn't need a board to tell me that. Of all the guys in the second group of astronauts, Elliot was the only one I had any doubts about. I had flown with him—so had other guys—and the conclusion was just that he wasn't aggressive enough. Too old-womanish. I mean, he flew too slow—a fatal problem in a plane like the T-38, which will stall easily if you get below about 270 knots.

He wasn't even in the best physical shape. I hadn't been able to assign him to GT-8 as pilot (with Neil Armstrong) because I didn't think he was up to handling an EVA. I made him commander of GT-9 and teamed him up with Charlie Bassett—who was strong enough to carry the two of them. Now they were both dead. I had let myself get sentimental about Elliot, wanting to get him a flight. It was a bum decision.

I also had a lot of plans for Charlie Bassett—after GT-9 he would have moved on to command module pilot for Frank Borman's Apollo crew. Elliot was going to be backup commander for GT-12.

The accident wound up having a lot to do with who wound up landing on the moon, because with the loss of the GT-9 prime crew, Stafford and

Cernan got the flight. All the backups changed, and Jim Lovell and Buzz Aldrin wound up being pointed at GT-12. Without flying GT-12, it was very unlikely that Buzz would have been in any position to be lunar module pilot on the first lunar landing attempt.

John Young and Mike Collins were already training for GT-10. I made Al Bean and C. C. Williams their backups. (They had been scheduled as backup pilots for GT-11 and GT-12.) A lot of people think I underestimated Al Bean, giving him a dead-end Gemini job, then moving him on to Apollo Applications at the end of 1966. But Al was just a victim of a numbers game: I had so many seats for the early Apollo missions, and also needed him on Gemini. I would only point to the fact that he was the first guy from his group assigned as a crew commander. I was confident he could do the job if anything happened to John Young. And when I made him head of the Apollo Applications branch it was because I trusted him to get it in shape.

I put Bill Anders into the Gemini rotation as a backup. He would still be available for reassignment to an Apollo crew. The GT-12 backup crew was a real dead end; there was no point in training somebody new. I figured I could rotate Gene Cernan back to this and still move him on to Apollo as originally planned. Gordo Cooper, who was basically marking time, stepped in as backup commander.

I needed an experienced command module pilot for Frank Borman's crew to replace Charlie Bassett, so that became Tom Stafford.

This is how it wound up looking:

Mission	*Prime Crew*	*Backup*
GT-9	Stafford-Cernan	Lovell-Aldrin
GT-10	Young-Collins	Bean-Williams
GT-11	Conrad-Gordon	Armstrong-Anders
GT-12	Lovell-Aldrin	Cooper-Cernan
AS-204	Grissom-White-Chaffee	McDivitt-Scott-Schweickart
AS-205	Schirra-Eisele-Cunningham	Borman-Stafford-Collins

Neil Armstrong and Dave Scott still had to get ready for GT-8. (Neil was pretty shaken by Elliot's death; they had worked together very closely as the GT-5 backup crew.) The Air Force, Lockheed, and NASA investigation into the failure of Atlas-Agena 5002 had found a probable cause: an electrical failure stopped the flow of oxidizer to the Agena's engine while also allowing unburned fuel to build up in it. The whole thing had just exploded.

So repairs to the Agena electrical system were in the works. NASA and

McDonnell also took steps to back up the Agena by building an Augmented Target Docking Adaptor (ATDA)—they literally took the docking adaptor that was mounted on the front of the Agena and made it into a small vehicle of its own, an Agena without engines and fuel, something an Atlas could simply put in orbit to serve as a passive docking target. The ATDA was ready very quickly, by early February.

In addition to rendezvous and docking with either Agena 5003 or the ATDA, Neil and Dave's biggest goal was to demonstrate a full-blown EVA.

Ed White's EVA on GT-4 had lasted twenty minutes: Dave Scott's would last more than two hours. Dave would have better oxygen and propulsion systems. In fact, to get himself around, he would use a maneuvering gun connected to a backpack called the ESP, the Extravehicular Support Package.

The EVA would take place a day after docking with the Agena. Dave would open the hatch, then Neil would back Gemini away from the Agena. Dave would crawl around to the adaptor section at the rear of the spacecraft and strap on the ESP (there was no room inside the cabin for it). He would then fly over to the Agena. Neil would then back the Gemini farther away from the Agena, hauling Dave with him . . . then move back and redock.

Dave would be tethered to the Gemini the whole time, however. We weren't ready to have astronauts go flying off into space. In fact, one of the reasons for the redocking was to test how an immobilized EVA astronaut might be rescued.

The second Agena, GATV-5003, lifted off on schedule at 10:00 A.M., March 16, from Pad 14. Neil and Dave were already inside GT-8 on Pad 19. Five minutes into the flight, where 5002 had blown up, we still had telemetry. Two minutes after that, 5003 was safely in orbit. Neil and Dave lifted off at 11:41.

We were still in the coelliptic rendezvous mode. Agena was in a circular orbit, GT-8 was in an elliptical orbit fifteen hundred miles behind and below it. Several burns reshaped its orbit as it caught up. Six hours into the mission, over the Pacific, Neil reported that he was on station with the Agena.

They had to sit there, a few feet from the Agena, for the better part of half an hour. We wanted the docking to take place where we had the best tracking and telemetry. Keith Kundell, the capcom on the *Rose Knot Victor,* a tracking ship in the South Atlantic, passed up the word to go for docking. Neil slowly edged Gemini forward just as they were passing into darkness. There was one big unknown—some people were worried that two spacecraft might carry an electrical charge which would arc between them on contact.

(That was one of the reasons we hadn't even tried to add an adaptor to GT-7 so it could dock with GT-6.) But there was no arc. "Flight, we're docked," Neil radioed.

The first task then was to check out the command links between Gemini and Agena. (The idea was Neil and Dave should be able to fly the Gemini-Agena combination as if it were one big spacecraft.) Within a minute Neil had yawed the whole combination. Things were looking good.

Dave sent some more commands to the Agena—this consisted of punching certain numbers into the onboard computer—and didn't get the response he wanted. He and Neil figured the problem was with the Agena's control system, and they turned it off.

Then they noticed that the Gemini-Agena combination had yawed out of alignment by thirty degrees. They were literally bending in two. Since the spacecraft were now in darkness, it was hard to see that by looking out the window: it was only something that showed up on the instruments.

Over the next few minutes the Gemini-Agena yawed and rolled a bit. Then it settled down. Neil and Dave kept trying to turn off the Agena control system, figuring that was still causing the problems.

Then, eight minutes after Dave had first noticed the big problem, the whole thing started rolling pretty seriously, literally beginning to spin. The vehicles were also yawing—moving from side to side. Neil and Dave were worried that the docking adaptor was going to break.

They managed to get things to slow down enough so they could undock. The moment they backed away from the Agena, however, Gemini starting rolling and tumbling, just like that damned MASTIF trainer. Nothing Neil could do with the hand controller seemed to make any difference. Pretty soon they were doing a 360 every second. Their vision was getting blurred and they were worried that pieces of Gemini were going to start flying off. "We have serious problems here . . . we're tumbling end over end," Dave radioed. It was hard to hear anything from Gemini; the antenna was pointing all over the sky.

They shut off the OAMS system, which not only fueled the thrusters that allowed them to change orbits, but also the tiny thrusters all around the spacecraft that controlled its attitude. They used the reaction control thrusters on the nose of the Gemini to stop the roll and yaw motion. Those thrusters were only supposed to be used during reentry. It took about thirty seconds, but eventually they got it stopped.

Most of this happened while GT-8 was between tracking stations. As they came into view over the *Coastal Sentry Quebec*, the controllers there got their first indication of trouble: the Gemini and Agena were separated when they should have been docked. And it was hard to get a fix on Gemini; no wonder, since it was rolling over and over.

Once the spacecraft was stable, there was only one option. Since the crew had had to go to the reentry thrusters, they would have to come back early. As the controllers in Houston set up for reentry at the first opportunity, Neil and Dave checked out the OAMS system, hoping to have some more control. Sure enough, one of the thrusters started them rolling again. The problem wasn't the individual thrusters; you could turn any one of the sixteen of them off. It was the control system itself, something electrical.

The emergency caught most NASA and contractor management in the air, flying from the Cape back to Houston or Washington. It was a tribute to Chris Kraft's system, however, that the whole problem was handled. Chris had moved up to head flight operations, leaving Gene Kranz heading one shift and John Hodge the other. (Hodge was training Cliff Charlesworth to become a third flight director.) Those teams had built up a lot of confidence based on the first five Gemini missions. They were ready for anything. There was even a shift change in MOCR right in the middle of the "emergency."

GT-8 fired its retros on the next orbit, over the Congo, and came down in one of the emergency recovery zones in the western Pacific, six hundred miles south of Japan. Neil and Dave floated alone in the ocean for about forty minutes until rescue planes reached them. They dropped drivers and a flotation collar, which was then attached to GT-8. The guys stayed in the spacecraft until the destroyer *Leonard F. Mason* pulled alongside two hours after that.

Meanwhile, I tracked down Wally Schirra and Frank Borman in Yokohama, Japan, where they were on a publicity tour with their wives after the December Gemini missions. Wally met the crew on Okinawa, then took them on to Honolulu, where Frank had gone on ahead to get the medical teams ready.

The mission was pretty disappointing, especially to Dave, who lost his EVA. But we had proved that docking was possible. And Agena 5003 got put through its paces—it was controllable from the ground, so we fired its engine ten different times over the next couple of days, eventually parking the thing in a 250-mile orbit, where some other crew might use it later.

The medical exams and interviews for the fifth group of astronauts took place in March. We wound up with thirty-five finalists, and I recruited some of the GT-10 crew—Mike Collins (Air Force), John Young (Navy) and C. C. Williams (Marines)—to help out in the selection. I locked them away in the Rice Hotel with Al Shepard, Max Faget, Warren North, and a couple of other board members.

I had mixed feelings about this selection. I knew I had enough guys in the first three groups to get us through the first moon landing, and then some. But beginning in 1964 there had been serious discussions about what we might do in manned space beyond Apollo. A program of additional lunar landings and earth orbital missions called AES, the Apollo Extension System, was developed, though the name was changed to Apollo Applications Program in September 1965.

The preliminary schedules coming out of George Mueller's office were impressive. They called for flights to three different Saturn IB "wet" workshops. (These would be S-IVB upper stages that would be launched as fueled stages, then modified for habitation after reaching orbit); three different Saturn V "dry" workshops; and four flights of an astronomical experiment package called the Apollo Telescope Mount. Each workshop mission alone would require visits by two or three astronaut crews, and all this was scheduled for operations beginning in April 1968.

At a minimum—we weren't even talking about more lunar landings, or lunar orbit missions—I was looking at eighteen additional Apollo flights, or forty-eight seats. Even if everybody flew twice, that was still twenty-four astronauts.

So when the selection panel asked me how many guys I wanted to hire out of the thirty-five finalists, I said, "As many qualified guys as you find." They came up with nineteen:

Vance D. Brand
Lieutenant John S. Bull, USN
Major Gerald P. Carr, USMC
Captain Charles M. Duke, Jr., USAF
Captain Joe H. Engle, USAF
Lieutenant Commander Ronald E. Evans, Jr., USN
Major Edward G. Givens, Jr., USAF
Fred W. Haise, Jr.
Major James B. Irwin, USAF
Don L. Lind
Captain Jack R. Lousma, USMC
Lieutenant Bruce McCandless II, USN
Lieutenant Thomas K. "Ken" Mattingly II, USN
Lieutenant Commander Edgar D. Mitchell, USN
Major William R. Pogue, USAF
Captain Stuart A. Roosa, USAF

John L. Swigert, Jr.
Lieutenant Commander Paul J. Weitz, USN
Captain Alfred M. Worden, USAF

Their selection was announced on April 4, 1966.

The group was about half split between those with test pilot (or at least test pilot school) experience and those who were operational pilots. One of the guys with an operational background, Don Lind, was actually a NASA scientist and could just as easily have been in the scientist-astronaut group.

Some of the others were already doing space-related work. Joe Engle had already won Air Force astronaut wings for flying the X-15 over fifty miles altitude. Ed Givens was already working at the MSC on the Astronaut Maneuvering Unit program. Ed Mitchell had worked as a Navy representative on the Manned Orbiting Lab program. Duke, Engle, Haise, Mattingly, Roosa, and Worden had come out of the Aerospace Research Pilot School at Edwards.

Their addition brought the number of active astronauts to fifty-seven. I was ready to add myself to the list, and I knew Al Shepard didn't think of himself as retired, either. Since the three non-flying scientist-astronauts—Jack Schmitt, Owen Garriott, and Ed Gibson—were ready to finish their pilot training and return to the MSC, it was easy to treat the nineteen new pilots and five scientist-astronauts as one big group, which I did.

We weren't through selecting. George Mueller told me that most AAP missions—all earth-orbiting AAP missions––were going to have scientist-astronauts in the crews. He suggested I start thinking of a second scientist-astronaut group numbering fifteen to twenty guys, for selection in 1967.

I needed help in the directorate. I had recruited Tom McElmurry to be the flight crew operations representative at NASA headquarters in Washington in 1964. He was a very qualified guy who had set up the Aerospace Research Pilot School, then gone through the first class with Jim McDivitt and Frank Borman. Since none of the managerial changes had panned out, Tom was dying to get out of there. So I brought him to MSC and made him my deputy for Apollo Applications. That way I didn't have to think much about it.

Down at the Cape, even while Gemini was still going full bore, other things were starting to happen. There had already been the one Apollo flight, AS-201. North of Pad 34, on Merrit Island, Rocco Petrone and his crew were finishing the Vehicle Assembly Building (VAB), a monster cube fifty stories tall designed to stack von Braun's Saturn V.

On May 25, five years to the day from JFK's speech to Congress about going to the moon, the forty-five-story doors of the VAB opened and out rolled a mock-up of the Saturn V on its transport, which was a hell of an achievement just by itself. (Imagine a tractor trailer the size of a football field which carries a weight of six million pounds a distance of three miles . . . then crawls uphill! And doesn't tip more than an inch in any direction.)

As the Saturn V mock-up rolled out to Pad 39A for fit tests, two Saturn IB launchers were undergoing checks nearby. On Pad 37 was AS-203, scheduled to be the test flight of the S-IVB upper stage; on Pad 34 was AS-202, another test of the Block I CSM, spacecraft 011.

GT-9 had become Tom Stafford and Gene Cernan's flight after the accident that killed Elliot See and Charlie Bassett.

Their major goal was to perform a rendezvous and docking with an Agena. This time we would try to more closely simulate a lunar orbit rendezvous. Plans called for the lunar module to rendezvous with the Apollo command module three orbits after lifting off from the moon, so we were going to try what they called an M=3 rendezvous. There were also a couple of other rendezvous maneuvers, such as testing the ability of the CSM to "rescue" a lunar module stranded in a lower orbit.

Gene Cernan was going to do an EVA even more ambitious than the one Dave Scott was supposed to do. Instead of a handheld maneuvering unit, Gene was going to strap himself into an honest-to-God backpack called the Astronaut Maneuvering Unit (AMU). This was something the Air Force had developed for the Manned Orbiting Laboratory. I guess they thought they might have the chance to inspect somebody else's satellites. The project officer on the AMU, Major Ed Givens, was one of the new astronauts we'd just selected.

On Tuesday, May 17, Tom and Gene were strapped into GT-9 as the countdown for their Atlas-Agena—GATV 5004—reached zero. Two minutes after liftoff, however, one of the Atlas engines swiveled "hard over," throwing the whole vehicle off course. The Air Force flight controllers had to shut down the Atlas sustainer engine; the Agena wouldn't be able to fire, either. The Agena separated from the Atlas, but both of them fell into the Atlantic.

A quick investigation showed what the problem was with the Atlas. Unfortunately, GATV-5005, the next Agena, was already targeted for the GT-10 mission.

Luckily we'd ordered the ATDA, the Augmented Target Docking Adaptor, earlier in the year. We had it ready to go, and it was launched two weeks later, on June 1. Tom and Gene waited in GT-9A, as we now called it.

Two problems came up right away. Telemetry from the ATDA couldn't confirm that its shroud had separated. This was the conical nose of the ATDA: it was supposed to break in two and fall off after reaching orbit, giving Gemini access to the docking collar.

We had already decided we would go ahead with GT-9A whether the ATDA made it or not. We could always learn something from phantom rendezvous, and Gene's EVA didn't depend on the docking.

But then we had a computer problem aboard GT-9A. We couldn't get it resolved in time to hit the launch window (the Gemini had to be launched in a period of a few minutes or it would have no chance to rendezvous with the target), so Tom and Gene had to crawl out of the spacecraft for a second time.

We were able to recycle in two days, though, and on June 3 they were off. On the first orbit they were firing the OAMS thrusters, starting the chase to the ATDA. Right on schedule, they were closing in during the third orbit. They saw the ATDA's flashing light, which suggested that the docking adaptor had separated—the light wouldn't have deployed otherwise.

As they closed in and hit sunlight, they could see that the news wasn't good. The shroud had split, letting the light deploy. But the shroud hadn't fallen off: it was still attached to the front end of the ATDA, looking like "an angry alligator," Tom radioed.

There was a lot of activity in the next couple of hours. At MOCR we sent commands up to the ATDA, figuring that might help. All it did was close the jaws of the alligator a little more.

Jim Lovell and Buzz Aldrin, the backup crew, had flown back to Houston by then. They wound up in a strategy meeting that night with Gilruth and Kraft. I guess Buzz piped up with the idea that Gene could simply float over to the ATDA and cut the strap. Up at McDonnell, Jim McDivitt and Dave Scott were looking into just that possibility. They found it would work, but that the straps could snap around and maybe cut Gene's suit.

Gilruth and Kraft were just aghast. Then the meeting plowed on, as if nothing had happened.

A week after GT-9A ended, Gilruth called me into his office to discuss Buzz's pending assignment to GT-12. He wondered if it was such a good idea. Maybe somebody else should have that flight. Since I hadn't been at the meeting, I went looking for Buzz. I got his explanation, told him to wait in my office, then went back to face Gilruth. I literally spent a couple of hours convincing him that Buzz was still a good choice. (In fact, he did the longest and most successful EVA in the Gemini program.) Then I went back to the office to release Buzz from house arrest.

Looking back, of course, what Buzz suggested wasn't that outrageous. We did things that were at least as tricky on Skylab and the shuttle; so did the Russians on some Mir EVAs. But this was early. We only had twenty minutes of EVA experience with Ed White.

On day three of the GT-9A mission it became pretty clear we had a long way to go in understanding EVA. The major goal was to get Gene hooked up to the AMU, so he wouldn't be depending on a tether for his communications and oxygen. (He was still going to have a safety tether connected, however.) That was how an astronaut would have to operate on the lunar surface.

Wearing a chest pack that controlled the flow of oxygen and cooling fluid through the tether, Gene was able to move around the spacecraft pretty easily, setting up cameras and mirrors. The tether made things difficult for Tom inside the Gemini: just as Jim McDivitt found out on GT-4, the spacecraft was getting whipped around by Gene's movements.

As Gene moved back to the adaptor and tried to put on the AMU, he found that the few restraints and footholds built in were totally inadequate. He couldn't position himself to even pull down the arms of the AMU. He started getting hot inside the pressure suit, so he decided to take a breather for a moment while waiting for Gemini to pass into night.

Once it was dark and he was cooler, he tried again. Things weren't much better: he couldn't even keep himself in one position. Eventually he got the armrests down and even hooked himself up to the AMU, but the communications degraded at that point. Worst of all, the visor on Gene's helmet had started to fog over. He tried to clear it up by resting and cooling off, but it didn't work. Finally Tom told him to give it up and come back in.

Tom and Gene splashed down safely on June 6. And we clearly had to rethink EVA for GT-10.

On July 5, 1966, AS-203 was launched—before AS-202. The S-IVB successfully started a number of times, then the whole thing was broken up deliberately as a stress test.

Less than two months later, on August 25, AS-202 got off the ground. This was an actual CSM—spacecraft 011—that was placed in a shallow trajectory that took it all the way from the Cape to near Wake Island in the Pacific, where it reentered and was recovered. We were still learning about the lifting capabilities of the Apollo command module, however; spacecraft 011 landed 160 miles off course.

These two flights, together with AS-201, cleared the way for the first manned Apollo, AS-204. Gus Grissom, Ed White, and Roger Chaffee were spending their time at North American Aviation in Downey, California, getting spacecraft 012 ready for deliver to the Cape. It was far enough behind schedule, however, that we had given up the idea of launching it before the end of the year.

The crew for GT-10 was John Young and Mike Collins. They were both real hard workers who got along well with people. But where John was usually uncomfortable in public, Mike was very smooth and articulate. He'd come from a military family and gone to West Point. After leaving NASA he wound up at the State Department also and wrote a terrific book about his astronaut career.

As always, the goal of these later Gemini missions was rendezvous and docking. We had yet to succeed in docking a Gemini to an Agena, and maneuvering the two of them. For Gemini 10, we got pretty ambitious. We not only planned to launch a new Agena, GATV 5005, for one rendezvous and docking, we planned to have the crew hook up with the Agena left parked in orbit by Neil and Dave, GATV 5003. If the first docking went well, John and Mike would use its engine to boost the whole combination into a higher orbit than ever before, four hundred miles. (We weren't just looking to set some kind of altitude record—if we were going to dock with the GT-8 Agena, orbital mechanics pretty much dictated that we get to that orbit.)

Given what we'd seen with Gene's EVA problems, we adjusted the planning for a couple of EVAs by Mike, including one where he would go over to the GT-8 Agena and pick up an experiment package that was mounted on it. The one thing going for him in some pretty complex training was that his specialty before being assigned to a crew was the G4C pressure suit. He was more familiar with that than anybody.

Given the Agena problems, we even had a whole backup mission: if GATV 5005 failed, GT-10 would become GT-10A, and we would launch them to rendezvous directly with GATV 5003.

Fortunately, none of that was needed. Atlas-Agena 5005 went into orbit on the afternoon of July 18. John and Mike were launched at 5:20 in the afternoon.

The first docking was made less than six hours into the flight, even though John wound up using more fuel than he wanted to get there. (Not his fault; there had been a mistake in loading the initial guidance program into the Gemini computer, so the spacecraft was never quite where the instruments said it was.)

John and Mike put the Gemini-Agena combination through some maneuvers, to make sure we weren't going to have a replay of the GT-8 situation. There were no problems, and after they'd been docked for an hour and a half, they got the go-ahead to light up the Agena's motor.

Since the Gemini and Agena were docked nose to nose, John and Mike were facing backward and had a good view of the engine firing. I guess it was pretty spectacular, even though it only lasted eighteen seconds. That was enough to put them in an orbit whose apogee was 475 miles high.

On day two, while GT-10 was still docked with the Agena, Mike started his first EVA. It was going great until both he and John started getting some serious eye irritation. It turned out lithium hydroxide—a substance normally used to scrub carbon dioxide out of exhaled air—was being fed directly into their suits by mistake. They terminated the EVA, then switched off a compressor that was the cause of the problem.

The next day they fired up Agena 5005 again, to reshape their orbit for rendezvous with 5003, then undocked. The two spacecraft had been together for thirty-eight hours by that time, which was a nice confidence builder for the future.

John and Mike closed in on Agena 5003 very smoothly, which was another accomplishment, since that spacecraft was essentially dead: it wasn't giving off any radar signal and its lights weren't working. They really just had to eyeball their way in.

Once they were station keeping, Mike opened his hatch and went outside. Using his own handheld unit, he got himself over to the Agena and literally grabbed hold of the docking cone. It hadn't really been designed for that, so he kept slipping off. He improvised, however, grabbing hold of a bunch of wires hanging out of the Agena and pulling himself back to the micrometeorite package he was seeking. He got it, and after a few minor misadventures, that was the EVA.

Given what we'd accomplished through GT-10, there was some discussion of stopping the program at that point. But we'd already bought the hardware and we really didn't have anything close to an operational system when it came to Gemini-Agena, for example. EVAs were still a big problem. And there were other rendezvous modes to test.

The other factor was that Apollo wasn't ready to go yet and would be lucky to get launched in the first quarter of 1967.

So even though we were given a deadline of January 31, 1967, to complete the program, GT-11 and GT-12 remained on the schedule.

Naturally, everybody came out of the woodwork with some "dream" ideas for the next mission. One of them was to hook up the Gemini and Agena with a tether and let them rotate around up there, simulating artifical gravity. The other was to use the Agena to boost Gemini to an altitude of eight hundred miles. (This was left over from some earlier idea of using a Centaur upper stage to send a Gemini completely around the moon—the LEO, Large Earth Orbit mission. But we'd only have tried that if we'd thought the Russians were going to get there first. And they hadn't had a manned flight in eighteen months at that point, something nobody understood.)

The crew for GT-11 was Pete Conrad and Dick Gordon, two Navy test pilots who had flown together at Patuxent River. Like McDivitt and White, who were friends before coming to work at NASA, they were a natural pair. They were both not only good pilots, but a good time.

Pete had really wanted to try the LEO, the loop around the moon, but got overruled. He was stuck with aiming for an altitude record, and, more importantly, accomplishing an M = 1 rendezvous—reaching the Agena before the start of the second revolution. That was how things were supposed to go in an Apollo lunar landing mission.

The Agena target for GT-11, number 5006, got off the pad in good order on September 12. With the M = 1 goal, the window for launching GT-11 was even more constrained than usual, just two seconds long. We made it; twenty-three minutes into the mission, Pete and Dick were making their first OAMS burn.

A bunch of other burns followed in short order. At one hour and twenty-five minutes into the mission, Pete radioed: "Mr. Kraft? Would you believe M equals one?" A few minutes later they were docked. I think a few people were surprised.

The next day was Dick Gordon's EVA. The flight plan allowed four hours for the crew to get ready, which turned out to be a mistake: they were ready in less than an hour, then found themselves having to sit around in their suits, all buttoned up with nowhere to go. They couldn't just start early—tracking and lighting would be wrong. So they just waited.

When they finally got the word to go ahead, Dick suddenly had a problem with his sun visor. Struggling with it overheated his suit, and he was pretty hot and sweaty when the hatch was finally opened.

Eventually he made his way over to the Agena, which was docked nose to nose. Dick was supposed to hook up the tether for the artificial gravity experiment. That's when he discovered he had nothing to hold on to. He wound up wrapping his legs around the docking cone. Pete told him he looked like a cowboy riding a bronc.

Once Dick managed to attach the tether, he was supposed to crawl back to the Gemini adaptor, pick up another handheld maneuvering unit, and try that out. But when he stopped at the cabin, Pete saw how tired and sweaty he was, and called off the rest of the EVA. It wound up lasting thirty-three minutes instead of the planned two hours.

The next day they fired up the Agena and boosted themselves up to 850 miles, where the earth finally began to look round. Two revolutions later, they fired 5006 again, to lower the orbit. Then Dick did a two-hour stand-up EVA, which went so smoothly that during a nightside pass, when there was nothing programmed, Pete took a little nap.

The artificial gravity experiment was next. Pete undocked and backed Gemini away from the Agena. The plan was to have them separate the whole hundred-foot length of the tether, but the damn thing got stuck in its deployment bag. When Pete fired the thrusters to loosen it, it came out, but it wasn't straight. The tether never did straighten out: it had its own loop and, according to Pete, looked like a jumping rope between the two vehicles.

Eventually they were rotating around their common center of mass, about six rotations a minute. That was as fast as she got. Some fraction of gravity was "created," not enough for the crew to feel, then the tether was cut loose. Agena 5006 went drifting away.

The next day Pete and Dick did another rendezvous with the Agena, which also went well. They came back on an automatic reentry, which put them closer to the carrier than any other manned flight.

So Gemini was down to the last mission. Even at that point you could say it had done everything it was supposed to—demonstrate rendezvous and docking. Perform long-duration. Even prove out the EVA pressure suits, though it was clear that we still hadn't accomplished a truly effective EVA.

Buzz Aldrin, who was going to do EVA on GT-12; his backup, Gene Cernan; and the guys from Crew Systems and Flight Support really tackled this. The basic problem everyone seemed to have was lack of effective restraint: there was no adequate place to wedge yourself to do even basic things like pulling down the arms of the AMU.

One item that did get dropped from the flight plan, just a few days after the end of GT-11, was a proposed test of the AMU. I don't think any of us believed we had enough experience with EVA to let an astronaut go jetting off at the end of a tether just yet.

The idea behind GT-12's EVA program was to make EVA itself the experiment—not a method to do some other job, like hooking up a tether.

In addition to the handrails mounted on the adaptor section behind the crew cabin—they'd been standard since GT-8—we added other rails and handholds all over the spacecraft. We even put some on the Agena.

There were two "workstations," one on the nose, the other one built into the rear adaptor. The one in the adaptor came complete with "slippers" that Buzz could lock his feet into.

Getting a docking target took a bit of juggling. The original contract had called for six GATVs—we had flown numbers 5002 through 5006. So we refurbished the first one, bringing it up to spec as GATV 5001R.

We didn't even have an Atlas rocket programmed for the flight at one point. (The ATDA launch had used up the Atlas that would have carried Agena 5001.) We finally "borrowed" one from the unmanned Lunar Orbiter program.

As Jim Lovell, the GT-12 commander, and Buzz walked out to the launchpad on Friday, November 11, they wore signs on the back of their suits. Jim's said THE and Buzz's said END.

Once they and the Agena were in orbit, things went pretty smoothly. They had closed in on the Agena to a range of about seventy-five miles when the rendezvous radar just quit.

It was pretty funny that this happened to Buzz, Dr. Rendezvous. He hauled out his sextants and charts, and with input from MCC managed to get Jim to the right spot in the sky. They docked safely.

Some problems crept up with the Agena, so Glynn Lunney, the flight director, canceled any plans to fire it up and boost the spacecraft to a higher orbit.

Over the next three days Buzz did three EVAs, starting with a two-hour stand-up that went very well. The second EVA was the ambitious one, with Buzz—using his handholds and railings—crawling from the Gemini to the Agena and back. He also did some tests back in the Gemini adaptor, all without getting tired or overheated during another two-hour session. There was a third stand-up EVA that lasted less than an hour.

GT-12 landed on November 15.

Changes were on the way in Apollo. Nobody was knocked out by North American's performance as a spacecraft builder. I kept hearing complaints from the crews working there that a lot of the workers were more interested in goofing off than in doing their jobs. It was hard to get ideas across, not just with North American, but also in the Apollo program office. Things were way behind schedule.

That was just at the astronaut level. I didn't know much about it at the time, but General Sam Phillips, the Apollo program manager, had already put North American on notice once that things needed to get better, not just with the Apollo command and service module, but also with the S-II stage for the Saturn V, which NAA was also building. George Jeffs was brought in to help out Dale Myers on the North American side, and there was some improvement.

By August, NAA had delivered eight Apollo boilerplates, which were launched on Saturn I and Little Joe II rockets. There had also been the three Block I Apollo command and service modules for unmanned flights. On the nineteenth, spacecraft 012, the first one intended for manned flight, went up for its contract acceptance readiness review. Joe Shea, the head of the Apollo program office at MSC, was there. So was the Apollo 1 crew, Gus, Ed, and Roger, in addition to Chris Kraft and Max Faget.

There was still a long list of problems to be fixed in the spacecraft, but it was conditionally accepted.

As the meeting ended, Gus presented Joe Shea with a picture of his crew bowed in prayer in front of an Apollo. "It isn't that we don't trust you, Joe, but this time we've decided to go over your head."

The second Block I Apollo, 014, was even further behind schedule. Mission planning for Apollo 2 was getting all gummed up, to the point where the crew—Wally Schirra, Donn Eisele, and Walt Cunningham—rebelled. They sent around a memo with a bunch of changes they wanted.

Wally's heart was never in Apollo 2, and I can't say that I blamed him. He wanted a flight that was going to be a challenge, not a repeat of something Gus should have already done.

Other people were wondering the same thing: why do we need Apollo 2, anyway? When it was originally scheduled, it still wasn't clear when the Block II CSM would be ready. Now we had a pretty good idea, so what did we need a second Block I flight for? We wouldn't learn much that would apply to a Block II flight.

Wally's bitching may have been the straw that broke the camel's back. Right after GT-12 splashed down, George Mueller canceled Apollo 2 as an unnecessary duplication of Apollo I.

This was part of a new mission schedule developed primarily by Owen Maynard, laying out the series of steps it would take to get to a manned lunar landing and return. There were seven missions, A through G:

A was an unmanned Apollo CSM flight, already accomplished.

B was an unmanned lunar module flight—really behind schedule at this time.

C was a manned CSM flight, the one Gus and his crew would fly.

D was the first manned lunar module flight.
E was a high earth orbit manned lunar module flight.
F was a lunar orbit simulation of a landing, with lunar module.
G was the landing itself.

We came up with some variations on this as time went by, but these were the basic steps.

Canceling the second, redundant C mission left Wally, Donn, and Walt without a flight. It wasn't just a matter of giving them the next mission on the schedule, either. That was intended to be the D mission, a dual flight of the command and service module and lunar module in earth orbit. I had pointed Jim McDivitt at that flight for over a year and a half. Aside from the fact that Jim was more familiar with the lunar module than anybody else, I had a rule that the command module pilot on flights with a lunar module should have flown before, preferably on a rendezvous flight. Donn hadn't; Dave Scott had.

To me the choice was clear—move the McDivitt-Scott-Schweickart crew from backup Apollo 1 (Apollo 204) to prime crew for the new Apollo 2 (Apollo 205/208).

It wasn't clear to Wally, however. He didn't think he should be backing up anybody ever again. He had done that for Gus once before already, on GT-3. Here it was, almost three years later, and he was right back where he started.

We had a pretty good argument about it, with me trying to convince him to put his hurt feelings aside and do the job. Wally finally gave in.

I had some more juggling to do. I wanted Frank Borman to command the first Saturn V flight, scheduled to do the E mission, Apollo 3. Since he'd be flying earlier in the sequence now, it wasn't absolutely necessary for him to have two experienced guys with him. So I moved Tom Stafford off Frank's crew and made him backup commander for Apollo 2, with John Young as command module pilot and Gene Cernan as lunar module pilot. They would back up the McDivitt crew while looking ahead to either a high earth orbit or lunar orbit test of the lunar module on Apollo 5. Along with the McDivitt and Borman crews, they would be candidates for the first lunar landing.

Bill Anders replaced Tom on Apollo 3, which meant that Mike Collins had to move to command module pilot (he was flight experienced). For the Apollo 3 backups, I got Pete Conrad, Dick Gordon, and C. C. Williams. They were pointed at a lunar orbit test of the lunar module on Apollo 6, or one of the later lunar landings.

It looked like this:

Mission	*Type*	*Crews*
Apollo 1 (204)	C	prime Grissom-White-Chaffee
		backup Schirra-Eisele-Cunningham
Apollo 2 (205)	D	prime McDivitt-Scott-Schweickart
		backup Stafford-Young-Cernan
Apollo 3 (503)	E	prime Borman-Collins-Anders
		backup Conrad-Gordon-Williams

Jim McDivitt had gotten convinced that with Apollo spread all over the country, meetings that required a representative from a flight crew were being missed. So at this time I also increased the size of the crew unit from six (three prime and three backups) to nine by creating a support crew for each mission. These would be newer guys who would help out wherever the flight commander needed them.

All of the support crew guys came from the 1966 group of pilots, who had by then broken down into command and service module (CSM) and lunar module (LM) specialists. For Apollo 1 it was three command module guys, Ron Evans, Ed Givens, and Jack Swigert. For Apollo 2 it was Al Worden (CSM) with Ed Mitchell and Fred Haise (LM). Apollo 3 was Ken Mattingly (CSM), John Bull (LM), and Gerry Carr (LM).

On December 22, we announced the switch. Gemini ended. Apollo was ready to go.

18

FIRE

Gus Grissom was probably my best friend of all the astronauts, but I don't think it really colored my opinion of him. Along with Borman and McDivitt, he was one of the guys Bob Gilruth and Chris Kraft liked, too. He could be pretty gruff sometimes and he was never happy dealing with the press, but he was a good pilot and a good engineer, and he was willing to get in there and do whatever job was necessary.

He liked to let off steam, no question. He and Gordo got involved in a lot of extracurricular activities, like racing cars. But Gus always made sure he did what he was supposed to do. Gordo wasn't quite as faithful about it.

When Gus came off his Mercury flight, it was obvious we weren't going to have enough flights to come back around to him, so we put him working with the Gemini program manager, Jim Chamberlin, to get Gemini going. He probably had more to do with designing the system than anybody—he and Chamberlin. Everybody else was still off worrying about Mercury, but Gus had shifted over.

Mercury was basically an unmanned system. It was designed to operate unmanned and that's how it operated for three flights, at least, before we put anybody in it. That was Chuck Yeager's complaint, which he wasn't at all shy about stating: what do you need a pilot for?

But Gemini would not fly without a guy at the controls. It was a manned system, from scratch. It was laid out the way a pilot likes to have the thing laid out, the controls and displays. Gus was the guy who did all that. In fact, Tom Stafford, who was about six feet tall and weighed two hundred pounds, later complained that the Gemini cockpit had been sized for Gus—who was about five-six, one-sixty.

Gus and Jim McDivitt were the first two guys available to move from

Gemini to Apollo, which is why I assigned them as prime and backup for the first flight. Gus would keep an eye on the command and service module while Jim would start following development of the lunar module.

People think that being an astronaut is all flying missions, but most of the job was just tedious work—for Gus in particular, who was going to make the first manned flight in two different vehicles, it meant spending weeks away from home at McDonnell in St. Louis, or at North American in Downey. The job was basically monitoring the way the vehicle was being put together—testing things when they were ready, changing them if they weren't. It was tedious as hell, most of the time, since it meant sitting in a spacecraft for hours at a time or just going to endless meetings. That was how a crew learned the spacecraft, and that was how the spacecraft got built.

Apollo was a big frustration for Gus. The major problem was that North American Aviation had been involved in the X-15 rocket plane, but not in manned spaceflight. So there was some mutual education to be done between the company and NASA.

The design of the Apollo command module kept changing, too, to the point where we were looking at those two versions, Block I and Block II. There were hundreds of differences between the two, the major one being that Block I vehicles didn't have the docking tunnel that would allow you to dock with a lunar module.

With Gemini, if Gus didn't like something, he had been able to go straight to Jim Chamberlin or even to old man McDonnell himself. That just wasn't possible with Apollo. It was too damned big: you couldn't have an astronaut ramming some little change through without fitting it into the overall system. At least that's what we thought in 1966.

It wasn't just problems with the spacecraft. Everything was late, especially the simulator down at the Cape. The computer programs never quite worked right, even the physical design of the thing didn't match spacecraft 012. It would, but changes in the flight vehicle hadn't caught up with the simulator. Gus was so annoyed that he hung a lemon on it.

One other thing came up about this time: Gus started talking about going back to the Air Force. The Vietnam War was going on, of course, but there were other factors.

When the Mercury astronauts came in, before we ever really started, one of the first decisions we faced was whether we should bail out and resign from the military. (NASA once had the idea we should all become NASA employees.)

Nobody in his right mind was going to throw away whatever he had invested in a military career to join NASA on something like Mercury, which could just as easily have been canceled. So we all opted to maintain our status as military officers.

But then we brought the next group on, including Borman and McDivitt, and it became pretty obvious to them that the military guys in the program weren't getting promoted with their contemporaries. They were worried that they might be screwing themselves back at the home ranch by being involved with this NASA.

For example, Grissom, McDivitt, and White were majors at the time they flew the first two Geminis. Gus and Jim were getting pissed off because they didn't see themselves getting ahead. They were sitting there as majors while their contemporaries were lieutenant colonels.

But right after GT-4, with Ed White's spacewalk, McDivitt and White got spot promotions to lieutenant colonel from LBJ. Fortunately, LBJ made it retroactive, and included Gus and John Young, who went from lieutenant commander to commander in the Navy.

This started a tradition of promoting every military astronaut one grade after his first flight. So instead of having guys lagging behind their contemporaries, for a while we had them all jumped ahead: Frank Borman was a full bird colonel at the age of thirty-seven; so was Dave Scott. Pete Conrad was a Navy captain at the fairly young age of thirty-nine.

It wasn't as bad as the Russians, who promoted Yuri Gagarin from senior lieutenant to full colonel—skipping the rank of captain altogether—in the space of two and a half years. I hear some of the regular Soviet air force guys who ran the cosmonaut program got pretty annoyed at that, since getting to full colonel usually took twenty years.

This automatic promotion stayed in effect until 1985. Since then your first flight has to be interplanetary—that is, you have to go to the moon or Mars—in order to qualify.

Anyway, the original NASA-Department of Defense plan had been that officers would have a three-year tour, then cycle back to their parent service. The tours got extended in blocks, and after you got to the nine-year block it was pretty obvious there wasn't a lot of sense in going back to the service. The goal in being a pilot is to have your own squadron, which you aren't going to get after nine years as an astronaut because you haven't had the necessary qualifying jobs. For the Navy guys, they weren't going to get the sea duty they needed. (If you wanted to command an aircraft carrier, for example, you had to have had your own squadron, and your own ship.)

The first guy who actually left NASA to go back to the service was Buzz Aldrin, who became commandant of the Air Force Test Pilot School at Edwards in 1970, without any great success. I don't believe that encouraged anybody else to try it.

In 1966 Gus was looking at his friends who were serving in Vietnam, and he thought he should do his part. He was angry about the situation over there. But those same friends, guys he'd flown with in Korea, told him, "You don't want to be here. It's a bad idea." So he stayed with Apollo.

On the morning of Friday, January 27, 1967, I was at the Cape in the crew quarters. Several other guys were there, too: Gus and his crew, Wally and his, plus the support crew—Ron Evans, Ed Givens, and Bill Pogue. And Stu Roosa, who was going to be the launch control capcom.

It was just more convenient than staying in some hotel down in Cocoa Beach, since we always had to be at work early in the mornings. We had our own cook, an ex-Marine named Lew Hartzell, so we had dinner and sat catching up on checklists . . . the forty-nine things that didn't get done during the day.

Gus's crew was scheduled to lift off from Pad 34 on February 21 for their fourteen-day mission. Their Apollo was mounted on its Saturn IB launcher on the pad. The day before, the backup crew of Wally, Donn, and Walt had donned pressure suits and performed what we called a "plugs in" countdown test, where we rehearsed the whole business of getting the crew suited, to the pad, into the spacecraft, and hooked up, right to the point of launch. In a "plugs in" version, the vehicle is still dependent on power and support from the ground.

Gus, Ed, and Roger were scheduled to do a "plugs out" test that Friday. We had breakfast, just like launch day. Then they got suited up and headed for the pad.

There had been a lot of problems with Gus's spacecraft. Rocco Petrone, the launch director, and Joe Shea, the Apollo program manager at Houston, had just had a big head butt over schedules and hardware. As far as the astronauts were concerned, the big bitch—we had a lot of big bitches on that spacecraft—was that communications were pretty lousy. So Gus and I talked on Thursday about having me get in the cabin with the crew for the test so I could listen in.

But talking it over at breakfast, we realized it didn't make a lot of sense: what I was going to hear in some jury-rigged headset wasn't going to be the same as what the crew was hearing on its system, anyway. I would also have to stand up for several hours, or crawl under one of the couches, since there wasn't much room in the spacecraft. So I would stay in the blockhouse, as usual.

I rode out to the pad and went up to the white room, to the spacecraft itself with the crew. On launch days I normally dropped off at the firing room, but on this day I went up there with them, until they were strapped in.

I had no official function in the test. Back in the blockhouse I was just piped in to the capsule communicator, Stu Roosa. Mission control back in Houston was in the loop, too.

We started running way behind on the test. The crew didn't actually get into the vehicle until after one in the afternoon. By then Wally and his guys had said good-bye; they were going to fly home to Houston for the weekend.

The communication problem cropped up again. Gus complained, "How the hell can we get to the moon if we can't talk between two buildings?" There were other nagging problems. By six-thirty things were awfully quiet, nothing much going on.

Then Roger announced, "Fire in the spacecraft." That was the first clue that they were having a problem.

I was sitting at the console next to Stu Roosa talking to Rocco Petrone, and all we could do was listen. We had a TV camera trained on the hatch window, which normally showed little more than a dark circle. That circle lit up and went almost white.

Mostly it was Roger we were hearing. "We're burning up! Get us out!" Then he screamed, and that was it. We later figured out it took eighteen seconds.

For a while we were optimistic we could go in and get the crew out of there. I don't know how long it took to figure out that that was probably not too likely. Apparently it was just mass confusion up there at the white room level, where the spacecraft was on the gantry. Smoke, heat, and toxic fumes all over the place. There was some worry that the launch escape rocket mounted on the nose of the Apollo could be set off the by the heat. (That rocket was actually more powerful than the Redstone that launched Al and Gus.) So the pad crew just initially evacuated the area. Then they tried to get back in there and get the crew out. But by the time they got the hatch open, five minutes had elapsed. It was obviously too late.

The hatch itself wasn't something you could open quickly. It was actually three different hatches, one of which—the inside shell—was pretty damned heavy and had to be pulled back into the cockpit. The command module pilot—Ed White—had to insert a ratchet handle and crank it open in six different places. Then the inside shell had to be pulled inside the command module. The process just couldn't be done in less than a couple of minutes under the best circumstances.

I went looking for Fred Kelly, the flight surgeon, and told him he'd better get up there. "You know what I'll find," he said.

We were the first guys from the blockhouse to reach the pad.

I had lost guys before. Ted Freeman. Elliot See. Charlie Bassett. During the war I'd flown on people's wings when they got zapped with flak. Someone's death is never easy, but it was probably easier to take in combat. You're all busier than hell anyway—everybody's deeply involved. You know it's deadly.

At the Cape we thought we were in a relatively benign environment. So it was a hell of a surprise.

I didn't stay at the white room very long, because there wasn't any reason to. The crew was still in the cockpit, so we had to get them out of there, but it was going to take some time.

Houston knew something had gone wrong, since they had been piped in. I went back to the crew quarters and talked on the phone to John Hodge, the flight director, and to my assistant, Don Gregory. George Low and Joe Shea were in the MOCR when the accident happened. Joe had just flown to Houston from the Cape: like me, he had thought about going through that test inside the spacecraft.

We delegated some guys to go talk to the wives. Mike Collins went to Martha Chaffee and Pete Conrad to Pat White. Chuck Berry and Marge went to tell Betty Grissom. Wally, Walt, and Donn were just flying in to Ellington about this time. Bud Ream, one of our pilots at the air patch, broke the news to them.

I called Bob Gilruth, who was up in Washington. In fact, in one of those weird coincidences, Gilruth was at a big dinner at the International Club on 19th Street, a few blocks from the White House. With him were all the top NASA guys—Jim Webb, George Mueller, Sam Phillips, plus Wernher von Braun and Kurt Debus. The head of North American, Lee Atwood, was there. So were Vice President Humphrey and Congressman Teague. I wound up talking to a lot of people.

It wasn't until after one in the morning that we got the crew removed from the cabin. They were taken to a makeshift morgue in the Biomedical Operational Support Unit, an Air Force clinic about a mile from Pad 34. An autopsy was performed in the morning, which showed that Gus, Ed, and Roger had been asphyxiated—killed by carbon monoxide. Their burns were probably survivable.

I think I finally got some sleep around four in the morning. When I got up, I flew back to Houston with Stu Roosa.

It was a bad day. Worst I ever had.

* * *

One thing that would probably have been different if Gus had lived: the first guy to walk on the moon would have been Gus Grissom, not Neil Armstrong.

Nothing against Neil. He did the job, but even on the day he was assigned as commander of Apollo 11, there was no guarantee that mission was going to be first to land. At that time we still had to fly Apollo 9 and Apollo 10; if there had been problems, things would have been different.

I realize that people still think that putting these guys on missions was due to some magic formula or public relations—especially since Neil was a civilian—but it's just not the case. My "master plan" from 1964 was still in place when the landing occurred. It had been changed, of course, by accidents and a lot of other things.

But Bob Gilruth and headquarters and I agreed on one thing, prior to the Apollo fire: if possible, one of the Mercury astronauts would have the first chance at being first on the moon.

And at that time Gus was the one guy from the original seven who had the experience to press on through to the landing. John Glenn had retired years back. Scott Carpenter was spending his time on a Navy project. We had Gordo Cooper, who wasn't even in line for an Apollo crew at the time. Wally Schirra kind of mentally decided he was tired of working that hard anyway. It was pretty obvious before he flew Apollo 7.

In fairness to Wally, it's pretty hard to go out there and cycle from prime to backup to prime, spending eighteen hours a day, day after day, month after month. That begins to wear people down. Some people, like John Young, didn't care: they were going to hang in there for whatever flights they could get. Guys like Wally weren't that dedicated to begin with. He decided he'd had enough of this job. "I've been devoured by this business," was the quote.

That's why Frank Borman didn't go on to more flights. He just elected to stop; it was his personal decision. He told me before he flew Apollo 8 that he wasn't going to fly anymore.

The funerals were held on Tuesday, January 31. I attended the ceremonies for Gus and Roger at Arlington; Ed White was buried at West Point.

Not long after this, the widows took me aside. They had a presentation for *me*.

A bit of explanation is in order. When astronauts made their first flights, they were presented with gold astronaut wings. Of course, at that time I didn't have gold wings and damned little prospect of ever having them.

Gus, Ed, and Roger knew this, and had gotten together to have a special set of gold wings made, complete with a diamond. They had planned to present it to me after the Apollo 1 splashdown. Betty, Pat, and Martha gave them to me instead.

I've worn them ever since . . . except for one week in July 1969, when Neil Armstrong carried them to and from the moon.

One of the immediate jobs back in Houston was pulling together a board of investigation. Bob Seamans had actually started this the night of the fire; it was a routine thing in aircraft accidents, but this was quite a bit bigger. There would be somebody from the Bureau of Mines, a chemist, an Air Force inspector general, plus guys like Max Faget and others from NASA. As chair they got Floyd Thompson, the head of Langley.

We needed a representative from the astronaut office. The most senior guy was Wally Schirra, who had Bob Gilruth's support, but a tedious job like this wasn't really his style. Joe Shea wanted Jim McDivitt or Frank Borman, but I didn't want to pull McDivitt off the lunar module, so we turned to Frank Borman, who had been off on a family vacation at the time of the accident. I called him back to Houston, then got him to the Cape.

Spacecraft 014, the one originally intended for Wally's Apollo 2 flight, was flown from Downey to the Cape. It was taken apart in tandem with 012, the burned vehicle.

Spacecraft 017, which was scheduled to fly unmanned on the first Saturn V later in the year, also got a good inspection, which turned up nothing but problems. This was bad for all the parties involved: NASA had accepted 017 as it was. It was already in the process of being mated to the Saturn V. But they started pulling panels off it and finding sloppy wiring and counting up something like fourteen hundred discrepancies—part of which, North American claimed, was due to NASA having changed 017's flight from a manned to unmanned one fairly late in the game. A lot of wiring had had to be added at the last minute to make the thing fly unmanned. In fairness, it turned out that only about two of the fourteen hundred discrepancies were really important.

But I was surprised at some of the sloppy workmanship.

It was just a feeding frenzy. Everybody was going around pointing fingers at everybody else. The strain was incredible. John Yardley, one of the guys who had made Mercury and Gemini work, had come to work for NASA, and he wound up with a nervous breakdown.

Then there was Joe Shea, the Apollo program manager in Houston. He was a brilliant guy, and on his way to being famous. (*Time* magazine had planned a cover story on him. It was supposed to run the week Apollo 1 flew.) He was the same age as the astronauts and used to play handball with some of the guys. He had come out of a systems engineering background, however, and didn't quite have the perspective on flight testing that some of us had. Gus Grissom had known the risks; he was the guy who said, "If there's ever a fatal accident in the program, it's likely to be me."

Joe took the fire very personally. It wasn't that he collapsed; just the opposite. He went on overdrive, as if he could personally redesign and rebuild that spacecraft so it would never fail again.

That was impossible, of course. These things were horrendously complex; fixing them would take time. And even when you thought you were done, there was sure to be something else to come along and bite you in the ass.

Joe made a speech in Houston during that first bad week after the fire, and it worried some of us. I even called him at home to tell him the fire was nobody's fault—we were all in this thing together.

He just went on overdrive. We made the mistake of letting him be part of the investigating board, and he just continued to drive himself into the ground. Eventually they got him to see a shrink—who decided that Joe was doing just fine, that his reaction was perfectly normal.

Well, maybe it was. But Gilruth got worried about him, and eventually he and Mueller cooked up a plan to move Joe to headquarters in Washington. I think they also wanted him out of the way while the congressional investigation took place: everybody knew that, fair or not, somebody was going to get blamed by Congress.

In March they moved Joe and got George Low to come in as ASPO. Joe figured out pretty quickly that he was on the shelf in Washington; in July he quit and went to work for Raytheon.

While the investigation ground ahead, we had to plan for the resumption of flights. George Mueller wanted to skip the C mission—the manned CSM in earth orbit, with no LM—entirely, and just go on to the D mission when we were ready to fly, with a manned CSM and a manned lunar module.

Bob Gilruth got in the way of this one. For one thing, the Apollo CSM was a sufficiently complex piece of machinery that it needed a shakedown flight of its own. Why try to test two manned vehicles for the first time at the same time? We thought a CSM-only flight was the way to go before the fire, and nothing we were going to learn was likely to change that.

He also thought it would be kind of an insult to the memory of the crew to treat their flight as if it hadn't really mattered.

I'm not sure that would have mattered much to Mueller. The real problem was that we weren't going to have a lunar module ready to fly manned any time soon. The CSM would be ready months earlier.

So he finally gave in. The first manned Apollo mission, once we got flying again, would be a shorter version of what we'd planned for 204: an eleven-day earth orbit test of the CSM.

Once that decision was finally made, I was able to get Wally, Donn, and Walt together at the Cape in early March. I told them they would have the flight, with Tom Stafford's crew as backups. There would be no formal announcement until the investigation was completed.

Even before Joe Shea was moved out of harm's way, Congress had started in on NASA. Olin Teague was the chairman of the House Space Subcommittee, and he was willing to let the NASA Review Board finish its work. But the Senate wasn't as patient. Clinton Anderson, the chairman of the Space Committee, was under pressure, and called a hearing on February 27. There was Jim Webb getting sandbagged by Senator Walter Mondale, who wanted to know about something called the Phillips Report.

It turned out that Sam Phillips had done some troubleshooting as far back as 1965 on Apollo. There had been problems between North American and NASA from the very beginning, with North American basically acting as though they were the experts on spaceflight while NASA was just a bunch of government people who really didn't know what they were doing. NASA, meanwhile, drove the contractors crazy with unrelenting schedule and money pressure, and thousands of last-minute changes in design. It wasn't a happy relationship, and so General Phillips put together a Tiger Team that inspected the program. What he found made him pretty unhappy; he wrote a memo to Lee Atwood that really laid down the law, and also sent an "eyes only" letter to George Mueller in which he said he had "lost confidence" in North American.

A lot of people at North American felt this was unfair, that Phillips's team had concentrated on a paper trail rather than technical accomplishments. Some people changed jobs on both sides, and things had been better during 1966. But the memos existed, and in light of the fire and some complaints by North American employees about shoddy workmanship, it looked like a cover-up, especially when Jim Webb said that he'd never heard of the thing. And Senator Mondale was sitting there reading from his very own leaked copy right there in the committee room.

Webb went back to headquarters and found the Phillips memo. Some of it was classified, so he couldn't just release it to the public or even the committee itself. So what he tried to do, then, was leak it to Mondale. That blew up in his face, too. Finally he and a couple of other guys just went over to the Senate and asked Mondale what they could do to prove they weren't hiding anything. The way I heard it, Mondale wasn't particularly interested in their opinions. He was no friend of NASA, and so he was going to get all the political mileage out of this that he could.

The review board published its report on April 9, 1967. It ran to 2,375

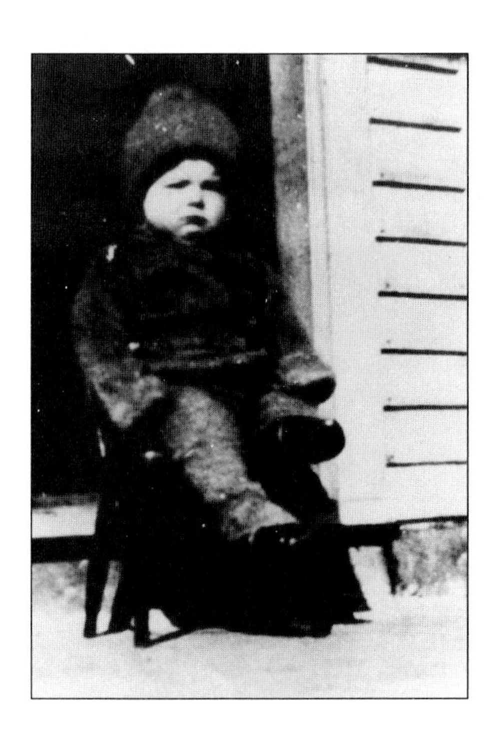

Don Slayton, age two, and
not too happy about it.

Howard and me,
all dressed up
for church.

Half the student body
of Leon Township
elementary school, 1938.
I'm in the front row, third
from the left.

High school graduation, 1942.

My official U.S. Army photo.

Air cadet, goggles and all, Vernon, Texas.

Home on leave with Dad, who still won't let me drive that car.

Back in the States, ready to head off to Japan.

On Okinawa, August 1945, with my A-26.

Wild times on the G.I. Bill in Minnesota.

Climbing into the cockpit, 1959.

The seven Mercury astronauts. (BACK ROW) Al, Gus, and Gordo. (FRONT ROW) Wally, me, John, and Scott.

The original capcom.

Handball with Gus, my best friend.

Egress training in Chesapeake Bay, with a guy who can't swim.

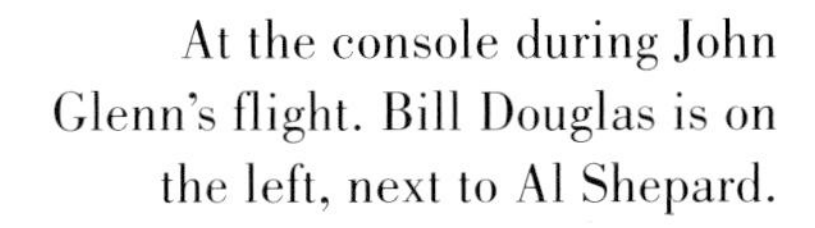

At the console during John Glenn's flight. Bill Douglas is on the left, next to Al Shepard.

July 1962, the Mercury astronauts are welcomed to Houston.

September 1962, my first astronaut selection. (BACK ROW) Elliot See, ~~Jim~~ McDivitt, Jim Lovell, Ed White, Tom Stafford. (FRONT ROW) Pete Conrad, Frank Borman, Neil Armstrong, and John Young.

October 1963, the third group of astronauts. (BACK ROW) Mike Collins, Walt Cunningham, Donn Eisele, Ted Freeman, Dick Gordon, Rusty Schweickart, Dave Scott, and C. C. Williams. (FRONT ROW) Buzz Aldrin, Bill Anders, Charlie Bassett, Al Bean, Gene Cernan, and Roger Chaffee.

Official NASA portrait, 1964. Still hoping to get cleared any day.

June 1965, the first scientist-astronauts: Owen Garriott, Curt Michael, Jack Schmitt, Duane Graveline, Ed Gibson, and Joe Kerwin.

The Gemini 5 patch—minus the "Eight Days or Bust!" motto.

The crew that died in the Apollo fire, Ed, Gus, and Roger. The worst day.

September 1967, the "Excess Eleven." Standing are Joe Allen, Karl Henize, Tony England, Don Holmquest, Story Musgrave, Bill Lenoir, and Brian O'Leary. Sitting are Phil Chapman, Bob Parker, Bill Thorton, and Tony Llewellyn.

Jim Lovell, Bill Anders, and Frank Borman, the Apollo 8 crew. They could have had the first landing.

The Apollo Eleven crew:
Neil, Mike, and Buzz.

Jim Webb, the best
administrator
NASA ever had.

Bob Gilruth, my boss
for thirteen years.

Walt
Williams,
head of
flight
operations
for
Mercury.

George Low, the Apollo program manager after the fire.

In mission control during Apollo Thirteen. I'm showing Chris Kraft and Bob Gilruth (FAR RIGHT) what the crew will build to scrub carbon dioxide out of their air.

The Thirteen crew back home: Fred Haise, Jim Lovell, President Nixon, and Jack Swigert.

Marge, Kent, and me, family portrait in 1974.

The Apollo-Soyuz patch.

Official crew portrait for ASTP. Tom Stafford and Alexei Leonov are standing. I'm sitting in the front with Vance Brand and Valerei Kubasov.

Visit to Baikonur Cosmodrome, April 1975. I'm to the left of Vance, who is standing next to the Sputnik memorial.

Inside the command module. About as roomy as it looks.

Apollo launch, July 15, 1975.

Soyuz seen from Apollo.

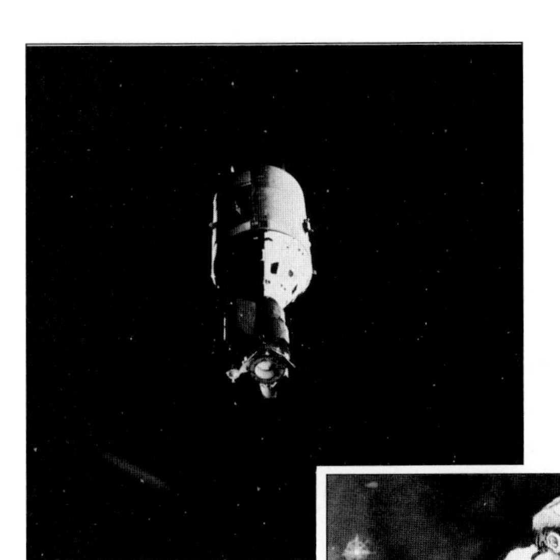

Apollo seen from Soyuz.

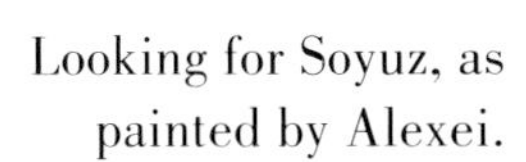

Looking for Soyuz, as painted by Alexei.

A toast in orbit, with some of Alexei's "vodka."

White House visit with ASTP crew and President Ford.

March 1979. Columbia makes a stop at Kelly AFB, Texas, on its way to the Cape, and I get interviewed.

Turkey shoot with Bobbie.

Wedding day, 1983.

Conestoga launch, September 1982.

Commercial launch salesman.

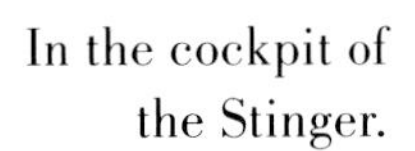

In the cockpit of the Stinger.

Stinger in flight.

pages and a hefty fifteen pounds, and stated that the board had been unable to fix a cause for the accident. There was a probable cause—some unexplained arc on a wire inside the cabin that ignited the hundred-percent oxygen atmosphere. Factors leading to the probable cause were laid to "deficiencies in command module design, workmanship and quality control." For example, there was uncertified and untested equipment in the Apollo cabin including flammable materials. A lot of safety tests simply hadn't been conducted and there wasn't adequate fire safety equipment at the pad itself. One of the biggest problems was that nobody considered a "plugs out" test on a launchpad to be "hazardous," and acted accordingly. It was about as scathing a document as you'd ever see from a government agency toward itself.

The hearings before the House that followed a couple of days later didn't make things better. Floyd Thompson and his board members were the first on the hot seat. Over the next month there were four other sessions with testimony from Mueller, Phillips, Seamans, and Berry, in addition to Lee Atwood and Dale Myers of North American. I was present for some of this. Our guys went up there and took their lumps; Frank Borman, in particular, was very straightforward. But it was still tough to watch. Even congressmen who were sympathetic to the space program found things to criticize. They even rehashed the whole Apollo contract selection business.

One of the most pathetic moments was when John McCarthy of North American raised the possibility that Gus himself might have inadvertently started the fire by kicking a loose wire inside the Apollo cabin. (To be fair to McCarthy, he was only trying to answer a question from Congressman John Davis of Georgia.)

Given Gus's history with the Mercury hatch, of course, this just played right into the image of Gus as some sort of screwup. It really pissed me off, and I wasn't the only one, especially because there were no grounds for the story—it was pure speculation, not to mention physically impossible—and because Gus wasn't around to defend himself the way he was on the *Liberty Bell* incident.

Even a couple of the congressmen thought this was going too far. James Fulton of Pennsylvania, in particular, stood up for Gus. But it was a two-day wonder for the newspapers. One of several.

The fire and the investigation, and the redesign of the Apollo spacecraft, made it possible for us to reach the moon within President Kennedy's deadline. If it hadn't happened, I don't know. We would have made it, but we would have had other problems.

19

APOLLO

Right in the middle of the Apollo hearings, on April 23, 1967, the Russians got back in the manned space business. They hadn't flown in over two years, which nobody could understand. Here they'd started fast right out of the gate with Sputnik, then Gagarin and Tereshkova, then three guys in one vehicle, then Leonov's EVA . . . then nothing. Some people were beginning to say there wasn't really a race to the moon, and on the evidence you had to admit the possibility.

What had hurt them was the unexpected death of Sergei Korolev back in January 1966. He was their von Braun and Gilruth and Webb all rolled into one—not only a great engineer but a hell of a politician. He was good at zigging this way and zagging the other to get what he wanted, which was to go to the moon.

Vasily Mishin, who took over as chief of the Korolev design bureau, was less of a politican, but probably more committed to a manned lunar landing. He helped kill the Voskhod program after only two manned flights—up to seven had been planned, including one by a crew of two women. The ministers in charge of everything decided Voskhod was a diversion of resources from the Soyuz program, which was the thing that was supposed to get them to the moon. Given what we know about Voskhod, it was the right decision: Voskhod was nothing more than a slightly modified Vostok. It couldn't maneuver, so you couldn't do any of the rendezvous and docking tests you needed.

Soyuz, or "Union," consisted of three modules—a spherical orbital module, where a crew of from one to three cosmonauts would spend most of their working time; a bell-shaped command module, where the crew sat during launch and reentry; and a cylindrical service module, which contained maneuvering engines.

Soyuz was about halfway between a Gemini and an Apollo in capability. It had more living space than Gemini, though its guidance system was much cruder, and could match some Gemini's maneuverability. It wasn't much of a match for Apollo, however. For example, while there was a docking collar on the front of the Soyuz orbit module, there was no tunnel! There was no way to get from one goddamn vehicle to the other without going EVA. That should give you some idea how desperate they were to get to the moon, because that was the same kind of quick and dirty plan we'd rejected years back.

Still, this was the vehicle the Russians were going to take to the moon—or, at least, around it. There was a version of the Soyuz that operated without the orbit module and was capable of being launched on a Proton rocket—a launcher the size of the Saturn IB—navigating around the moon, and surviving a high-speed reentry. The lunar Soyuz had an unmanned model called the L-1, which got tested under the cover name of Zond, and a manned model, the L-2. If this first manned test flight went well, a Soviet manned circumlunar flight couldn't be too far off. They might have been able to fly it as early as November 1967 . . . which was, coincidentally, the fiftieth anniversary of the Bolshevik Revolution.

Anyway, by April 1967 they had flown several unmanned versions of Soyuz and were ready for a big manned test. (There had been rumors in the papers of a big spectacular of some kind.) It was launched on April 23. There was just a single pilot aboard, Vladimir Komarov, the same guy who had commanded the first three-man mission back in 1964.

Those rumors had said that another manned vehicle was going to be launched the next day, carrying as many as five cosmonauts. (The first Soyuz was even called Soyuz 1, which sort of implied there was a Soyuz 2 sitting around somewhere.) And, in fact, Soyuz 2 was sitting there ready to go on April 24. It would carry three cosmonauts—Valery Bykovsky (the pilot of Vostok 5), Yevgeny Khrunov, and Alexei Yeliseyev. Soyuz 2 would be a rendezvous and docking target for Soyuz 1; once Komarov had linked up, Khrunov and Yeliseyev were supposed to do an EVA from 2 to 1, then return to earth with Komarov. It was a pretty damned ambitious flight plan for a first manned mission. First rendezvous, first docking, first effective EVA.

Things went wrong right away; one of the two solar panels on Soyuz 1 never deployed, cutting the power available in half. There were apparently guidance problems, and even weather problems at Baikonur, so the launch of Soyuz 2 was canceled.

Komarov set himself up for a retrofire burn on the sixteenth revolution, but couldn't get oriented properly. I guess there was quite a lot of scrambling around at the Soviet flight control center, which was in Yevpatoriya in the

Crimea. They tried a manual reentry on the seventeenth revolution . . . again, some problem forced a wave-off.

They finally got aligned and fired the retros on the eighteenth revolution. By now, of course, Komarov wasn't headed for the primary or even the secondary landing zone. But Russia is a big country; he was actually primed to come down near the town of Orel, where Yuri Gagarin had landed.

But something went wrong during the entry phase. The main Soyuz parachute streamed out, but never deployed. Komarov went to the backup parachute, but it got tangled with the main one. Soyuz 1 crashed in a field and basically blew up. (Like Voskhod, in addition to parachutes, Soyuz used solid-fuel rockets that fired about six feet off the ground to soften the landing, which was not survivable even with parachutes. It was those solid-fuel rockets that caused the explosion.)

I guess it was chaos, just like what happened to us with the Apollo fire. Since Komarov had come down in a backup zone, it took a while to reach the site with rescue helicopters. All telemetry between the spacecraft and mission control had ceased on impact, but nobody knew quite what the problem was. And all this was being conducted on open channels.

The first rescue pilot took a look at the situation and saw right away that it was a fatal accident. But he also knew he was on an open loop with Yevpatoriya and the Ministry of Defense satellite control center in Moscow. So all he said was, "The cosmonaut is going to need emergency medical treatment outside the spacecraft," at which point the lines were cut by somebody in the rescue units.

So here you had these rescue guys landing at a crash site, knowing there wasn't much they could be doing. Kamanin and Gagarin were on a plane headed for Orel, and because of the communications blackout, they didn't know that Komarov had been killed until a couple of hours after the accident, not until Kamanin and Gagarin landed at Orel. I guess it was Kamanin who had to get on the phone and break the news to Dmitri Ustinov, who was the minister of defense.

We knew none of this at the time, however. All the Soviet press had done was report the launch, then note a few hours later that the flight was "proceeding normally." Then they announced that Komarov was dead.

We thought it would be polite to send representatives to Komarov's funeral, and had Frank Borman and Gordo Cooper all ready to go. But the Soviets sent back word that the funeral was a "private matter," so no American astronauts attended.

Komarov was buried in the Kremlin Wall and the Russians were right back where we were.

* * *

Getting through the board and the congressional hearings was only part of the battle. Now we had to redesign the Apollo and get her flying again. It had been pretty obvious right away that the pure oxygen system was the culprit. No matter what had sparked the fire, it was worse because the cabin contained pure oxygen at a pressure of sixteen pounds per square inch—more than sea level and three times as high as we usually had in the spacecraft. (The higher pressure was used in the countdown to test the command module structure.)

There had been a lot of discussion early in the program about using mixed gas, the way the Russians did. But it would have complicated the environmental control system and made a spacecraft intolerably heavy. So we wound up in Mercury with a one hundred percent oxygen atmosphere at five pounds per square inch, a partial pressure equivalent to what you'd find at twenty-seven thousand feet.

The other complicator was that the suits were pressurized to 3.5 pounds per square inch. To walk on the moon you needed to get out of the spacecraft—do an EVA—and with a fourteen psi, mixed-gas system you'd have had to prebreathe for hours, lowering the pressure and getting the nitrogen out of your system so you didn't get the bends. And, of course, if there was an emergency and you had to use the suit, you'd really have been in trouble.

So there were a lot of reasons why we had a pure oxygen system. It was also simpler structurally. If you went to mixed gas, then you had to have a structure that could handle fifteen pounds per square inch.

I heard that Gus Grissom told a friend of his in the X-20 program, a pilot named Hank Gordon, that they shouldn't let themselves get talked into a pure oxygen system, and the X-20 did go to mixed gas. But the X-20 was designed for short missions with no EVA. And I don't recollect anybody, including Gus, sitting around being concerned about that. We'd been through the whole damn Gemini program with the same thing. It wasn't like we were saying, hey, this is something we've got to fix. Nobody that I know was smart enough to have predicted that problem. It would have been fixed.

None of us had any idea that it would be eighteen months—until October 1968—before we would get Apollo flying again. There was too much day-to-day work to concentrate on. There was an exhaustive plan—the first thing was to find out what caused the fire. To this day I don't think anyone knows the answer—some kind of ignition source near Gus's feet inside the cockpit.

One of the biggest jobs was getting rid of all the flammable materials.

We did a lot of basic materials research, developing some new materials. We made a thing called a "Velcro map." We used to stick Velcro everyplace because it was so convenient. But it was flammable, so we realized we were going to have to limit its use. We actually developed a great, detailed map figuring out just exactly where all the pieces of Velcro were in the cabin . . . being sure to have the proper spacing between them, those kind of things. It was a fairly major effort and that's just one simple example.

The hatch was the other big problem. The hatch on Apollo 204 couldn't be opened in less than a couple of minutes. The new hatch opened in about five seconds. In fact, it had been on the boards along with a bunch of other improvements that were to have been incorporated into the Block II spacecraft.

We shouldn't have been flying the Block I. The lunar module wasn't ready; nothing was ready. We were going too fast.

My job was to get training back on track so we would be ready when the spacecraft was. The sequence of missions was still roughly the same, which still—on paper—could lead to a lunar landing on the fifth or sixth manned Apollo. If we got flying again in the spring of 1968, we would make the deadline.

As far as the schedule went—the challenge of meeting JFK's goal of a landing on the moon and return before the end of 1970—the wild card was the redesign and reconstruction of the Apollo command module. No one could really say how North American would respond. Frank Borman's Tiger Team, along with other groups headed by people like Kenny Kleinknecht, were working hard. I felt confident we'd be back in the air in mid-1968.

But that was just with the command module. Grumman's lunar module was way behind. Grumman was doing a good job with an incredibly complex piece of machinery; it was just taking them a long time to get it built.

Before the fire I had six crews in training. That was now five, and the head of one of those (Borman) was occupied with the investigation. So the choice for the first postfire manned Apollo crew was really between the 204 backup crew (Schirra-Eisele-Cunningham) and the 205 prime crew (McDivitt-Scott-Schweickart). McDivitt's crew had been pointed at the first lunar module mission since the beginning, and there was no reason to change that now. And though Wally had bitched about being assigned as Gus's backup, I knew how capable he was. Making that first flight after the fire was just the sort of challenge he would respond to.

So in early March, six weeks after the fire, I sat down with Wally, Donn, and Walt in the crew quarters at the Cape and told them they would have the next flight. Tom Stafford, John Young, and Gene Cernan would be their backups. Ron Evans, Jack Swigert, and Ed Givens would continue as the support crew.

(A few months later, on June 6, Ed Givens was driving back to Houston from a party with a couple of Air Force buddies, Frank Dellatora and Bill Hall. He must have taken a turn too fast, because he just ran off the road near Pearland, about ten miles west of the Manned Spacecraft Center, and rolled his Volkswagen. The two passengers were injured; Ed was killed. It was really a shame. Ed would have been one of the first guys from his group, the April 1966 astronauts, to be on a crew as a command module pilot. Bill Pogue replaced him on the support crew.)

The assignment to the C mission Apollo left McDivitt-Scott-Schweickart as the crew for D, along with Borman-Collins-Anders for E. I kept Conrad-Gordon-Williams as a crew and had them backing up McDivitt's guys. I put together a new crew of Neil Armstrong, Jim Lovell, and Buzz Aldrin and assigned them as backups to Borman. Eighteen guys, prime and backup crews for the first three missions. These were the guys who were going to get us to the moon and make the first landing, though not necessarily in those crews or that order.

The mood in America was different in early 1967 than it had been six years before. Although LBJ was probably a bigger supporter of the space program than JFK, he had other problems on his mind, notably the involvement in Vietnam. NASA was having to fight for money, and there just wasn't as much. I caught some of that in the spring of 1967.

George Mueller and headquarters had plans for manned space activity beyond Apollo, in a program called Apollo Applications. One phase was going to see the flight of a series of manned orbital workshops, with crews going aboard them for stays of a month or longer. (The first workshops were going to be S-IVB upper stages that a crew would outfit as a crude workshop. A later version would be outfitted on the ground and launched by a Saturn V.)

Another element would involve a series of more ambitious manned lunar landings, eventually involving roving vehicles on the moon and unmanned shelters giving a crew the ability to stay on the surface for two weeks. There were plans for earth resources Apollo missions and for lunar orbit mapping and survey missions. By late 1966 it laid out as an average of six manned Apollo missions every year, beginning in late 1968 and continuing into the mid-1970s.

I had already added those nineteen pilots to the astronaut office in April 1966. By the spring of 1967 they had just completed their preliminary training and were moving into technical assignments. Now we could add more scientist-astronauts.

We didn't want to make the mistake we had made with the first group, where the Academy of Sciences basically sandbagged us. This time we got involved from the start. We announced that we were accepting applications, then asked the Academy of Sciences to do a rating of the applicants based on their scientific credentials. NASA would handle the rest. We hoped to select ten to twenty new scientist-astronauts, and expected them to be module pilots (as we were calling them then) on AAP missions beginning in 1970 or 1971.

No sooner had the fiscal year 1968 budget gone into Congress in the spring of 1967 than the cuts started coming. The pressure on the NASA budget was made worse by the fire; money was needed for changes in the Apollo spacecraft. The Saturn IB and Saturn V production lines were to be closed down. There would be fifteen of each, no more—unless some later administration freed up the money. The first thing this killed was the open-ended series of manned lunar missions. Eventually a plan for ten Apollo landings would emerge, but they would use the existing Saturn Vs, if and when they became available. They would not be part of AAP.

Then the earth orbital program got whacked, too. The initial launch was put back from 1968 to 1970—a smart move in any case, because I thought we'd have our hands full accomplishing the lunar landing.

The "wet" workshop was cut—another good idea, in retrospect—in favor of the "dry" Saturn V workshop. But because there were only fifteen Saturn Vs in the pipeline, that meant there would be only two workshops rather than four or five. Meaning there would be a maximum of six manned Apollo Applications missions as opposed to a dozen or more.

It was clear to me that we weren't going to need this second group of scientist-astronauts. The five we had were going to be all we had room for. Nevertheless, the selection process had already started. And it took until the summer of 1967 before it was clear just what these cuts would be, and what that would mean.

On August 11, 1967, we announced the selection:

Joseph P. Allen
Philip K. Chapman
Anthony W. England
Karl G. Henize
Donald L. Holmquest
William B. Lenoir

John Anthony Llewellyn
F. Story Musgrave
Brian T. O'Leary
Robert A. R. Parker
William E. Thornton

It was a real mixed group in some ways. Two of the new guys were technically over the age limit of thirty-five—Thornton was thirty-eight and Henize was forty-one. Both of them had written to me early in the selection process to raise the question. Gilruth and Kraft and I had told them the age limit would be waived for especially qualified individuals, which they were. At the other end of the scale you had Tony England, who was twenty-five years old and just finishing up his Ph.D. Same thing with Brian O'Leary, who had just gotten his doctorate in astronomy.

Two of the new scientists were naturalized American citizens—Llewellyn had been born in Wales, Chapman in Australia.

The one thing they all had in common was that none of them were pilots . . . not even privately licensed.

I was in the uncomfortable position, however, of sitting down with the eleven of them their first day of work, September 18, 1967, and telling them they just weren't needed. I wanted to give them a chance to get out. I guess I didn't expect anybody to walk out on the first day, and I hated like hell to do it to them, but I had to be honest about it.

All of them stuck around for at least a while. O'Leary and Llewellyn had problems in flight school and dropped out within the year. But Joe Allen and Bill Lenoir hung in there and wound up on the first operational shuttle flights. Henize, England, Parker, and Musgrave eventually got missions. Hell, Musgrave went on to fly four different shuttle missions and, last time I looked, was scheduled for a fifth.

Even though I had no chance at the first lunar landing, I still wasn't ready to give up the idea of making a flight.

One thing that had been made clear, however, was that as long as I still had my heart condition, no one was going to take the risk of clearing me to fly. (I was still having fibrillations, of course. They normally lasted two or three days. I could tell by taking my pulse, and eventually got so attuned to it I could just tell. The symptoms always disappeared within an hour or so, however, if I went out and got some exercise, just as Martin Caidin wrote in his novel, *Marooned*.)

So I had to do something about the condition itself.

Other Voices

DR. CHARLES BERRY

Deke and I started out as institutional adversaries when I took over from Bill Douglas in 1962. (I had worked with Bill from the very beginning. In fact, I was one of the military doctors present during the very centrifuge run at Johnsville during which Deke's atrial fibrillation was discovered.) I would go so far to say we stayed institutional adversaries; he represented the flight crew position and I represented the medical community.

But though we had some tremendous battles, we got along very well on a personal level. We were able to do that because we fought in search of answers to problems, *not to score bureaucratic points. He also knew that, like Bill Douglas, I was sympathetic to his situation and open to anything that would give him some hope of getting a flight.*

Deke had heard about a drug called quinadine that would control fibrillation. So in April 1967 we ran him through a day of tests at a diagnostic clinic, to be sure the treatment wouldn't be worse than the problem. And for the next couple of months he was on quinadine. Sure enough, it converted the fibrillation to a normal rhythm.

The problem was, he had to start out taking it four times a day at pretty regular hours—he even had to wake up to do it. Later it tapered off to three times a day, but it was a strict regime, and he admitted he just started getting sloppy. And by July he was fibrillating again.

In May 1968 we tried again, with quinaglut, another version of the antifibrillation drug. He took it quite faithfully into the summer of 1969. But right about that time he heard that anybody on a quinadine/quinaglut regimen was automatically ineligible *for FAA certification as a pilot. So for Deke's purposes, the treatment was worse than the disease. I'm not sure this was true, but Deke believed it was. And he went back to fibrillating every week to ten days.*

The period between the fire and the resumption of flights was a strange one. Here we were in a race to the moon, and nothing seemed to be happening. Of course, there was a tremendous amount of work going on down at the Cape, where Rocco Petrone's people were getting the pads and the Saturns ready. There was a constant stream of people from Houston to Downey, from Houston to Long Island.

I got the guys into helicopter training, since we all concluded that was the best preparation for flying a lunar landing. So things were going on; it just didn't look that way.

At one point Marge and a couple of the other wives organized a party to be held at the King's Inn, and it was a pretty big blowout. Especially since they kind of underestimated the amount of booze a bunch of astronauts could drink. I wound up laying out a couple of hundred bucks to cover the bar tab.

On October 5, 1967, we lost another astronaut.

Other Voices

BUD REAM

C. C. Williams had been down at the Cape and was flying back to Ellington in a brand-new T-38. I don't think it even had fifty hours on it.

Because this new plane had a broken transponder, it was limited to a ceiling of twenty-four thousand feet during its flight. As C.C. approached Tallahassee, he was supposed to make a turn to the west. No one knows exactly what caused this, though I personally think it was simply C.C. rolling his plane slightly to check his position—"Yeah, there's Tallahassee"—as he got ready to make his turn. In any case, the controls jammed and the plane started into a rolling dive.

Because of the situation with the transponder, C.C. wasn't as high as he could have been, but he still had time to switch to an emergency frequency and give a mayday call describing his situation and even making a correction about his location. That particular model of radio also took several seconds to warm up. That is, when you switched from one frequency to another, it might be as much as ten seconds before you were actually able to use it. But C.C. was clearly heard by the tower at Tallahassee. So he had an opportunity *to eject safely.*

But he never made it. The T-38 was in a nearly vertical dive when it impacted in a game preserve near the town of Miccosukee. Analysis showed that it was going about Mach .95 when it hit. C.C. did eject, but too late and too low for his chute to open.

It was probably a typical test pilot response to a crisis: I can get out of this. A lesser pilot would have bailed out sooner.

C. C. Williams had been lunar module pilot on Pete Conrad's crew, so Pete needed a replacement. Pete himself suggested Al Bean, who had been working the Apollo Applications program since the end of Gemini the year before. Pete had been one of Al's instructors at the Navy test pilot school and they got along great. Putting Al on Pete's crew was a logical move.

But that made five funerals in one year.

Joe Engle had a close call later that month. He was coming back to Houston on a Sunday night from a hunting trip in one of our helicopters. I guess he ran out of fuel or something, but he had to try to put the chopper down in a field. He did, but he wracked it up pretty good.

A few weeks later, Major Bob Lawrence, one of the astronauts in the Air Force Manned Orbiting Laboratory program, was killed out at Edwards Air Force Base. He and another pilot, Major Harvey Royer, were doing zoom landings in an F-104, part of the aerospace research pilot's course. Again, something went wrong . . . the 104 hit the runway. Royer ejected and got out, but Lawrence was killed.

The double tragedy was that Lawrence was the first African-American to qualify as an astronaut. When we took some of the MOL pilots into NASA a year or so later, he would certainly have been among them.

Also, on November 15 an X-15 had crashed, killing Mike Adams. Adams had also been in the MOL program, had gotten tired of waiting for flights (which never happened), and transferred to the X-15.

Some of this was just bad luck. We never really found the answer to C. C. Williams's accident, for example, but I always thought it was because of a bum airplane.

Some of it had to do with the kinds of guys we had. They weren't daredevils, but they were confident they could handle anything you threw at them. Most of the time they were right.

While the problems with the CSM and LM were being sorted out, von Braun's team went right ahead with the Saturn V. That big baby was finally ready for launch on November 9, 1967, and a bunch of us flew out to watch.

I'd seen a lot of rocket launches by this time, of course, but nothing was ever as impressive as that first Saturn V. It just rose with naked power, lots of noise and light. A perfect test flight.

That combined with good reports from Downey and Bethpage meant that by Christmas 1967 we could begin to make plans. The first lunar module would fly unmanned in the first couple of months of 1968. There would be two more unmanned Saturn V launches. If all went well, Apollo 7 would follow in late summer, with Apollo 8 and the first manned lunar module in fall and Apollo 9 a couple of months after that.

With that in mind, I announced the crews for 8 (McDivitt prime with Conrad as backup) and 9 (Borman prime with Armstrong as backup).

I warned the crews that they shouldn't get too married to flying a particular type of mission, because I expected changes. And changes I got.

20

THE MOON

We moved into 1968 with some confidence that we were finally back on track. The Block II command module was really shaping up. Everybody was pulling together. Maybe it was easier because we didn't have the Russians breathing down our necks just then. We knew they had to be going through the same thing we were.

Grumman finally had the lunar module ready for flight, which was a hell of an accomplishment. People don't realize it took thousands of people to get one Saturn off a pad in the comparatively benign environment of the Cape. The lunar module had to not only land on the moon, it had to be capable of lifting off with a ground crew of *two*. No pad. No team.

The lunar module consisted of a descent stage and an ascent stage. The descent stage was boxlike and rested on four legs. The ascent stage was the buglike crew cabin. There was a docking port atop the ascent stage and a hatch in the front, through which the crew would exit.

The development was as complicated as anything else in the manned space program. There was constant pressure to make the lunar module lighter while making it more capable. Eventually it wound up being practically a metal balloon with legs. It didn't even have seats inside the cabin. The crew flew it standing up.

On January 22, 1968, LM-1 lifted off aboard a Saturn IB—the same Saturn IB, AS-204, that would have carried Gus's crew. (The mission was officially known as Apollo 5.) LM-1 didn't have four landing legs attached to the descent stage, and its environment control system was incomplete. Other than that, it was the vehicle that would land guys on the moon and get them off.

About four hours into the flight, three hours after the lunar module had separated from the S-IVB upper stage, the TRW-built descent propulsion system (DPS) engine fired. This was the engine that would slow the lunar module down for a lunar landing; unlike other engines to that time, the DPS was throttleable. You could change the amount of thrust. For the LM-1 test the DPS was supposed to start at ten percent of maximum, hold that for twenty-six seconds, then rise quickly to about ninety-three percent in the next twelve seconds. This was the kind of maneuver the lunar module would perform in lunar orbit, to start the descent to the surface of the moon.

The DPS started firing, and shut down after only four seconds. It turned out there was some incorrect programming—a human error—that was caught by the onboard lunar module computer.

We had planned a second burn of twelve minutes, just like a powered descent to the moon, and that had to be scrubbed. Instead the DPS was fired twice, once for thirty-three seconds. Then, after drifting for thirty or so seconds, again for twenty-eight seconds.

At that point, by design, Rocketdyne's ascent propulsion system (APS), which the engine used to get the lunar module cabin and the astronauts off the moon, fired up. It burned for one minute, testing our ability to abort a lunar landing by having the ascent stage separate from the descent stage in flight.

Then we went back to the original program, and made another mistake: the lunar module's onboard computer acted as though the descent stage were still attached! So its steering motors, the reaction control system (RCS), got confused and started firing as if it were steering a much bigger, more massive vehicle. Pretty soon the lunar module was out of RCS fuel.

So when we fired up the ascent stage with the idea of just burning all the fuel in the tanks, the lunar module quickly tumbled.

Nevertheless, this was a good test flight. There were some problems, but no show stoppers. George Mueller canceled the second unmanned lunar module test. We weren't just able to move on to the manned flight, however: a check of LM-2 showed some wiring and corrosion problems. Given the nature of the lunar module, the vehicle practically had to be disassembled in order for it to be fixed. So even though we were skipping LM-2, LM-3 wasn't going to be ready much before the end of 1968 due to these repairs.

The second unmanned test flight of the Saturn V—Apollo 6—was set for April 4, 1968. That happened to be the same day that Martin Luther King was killed in Memphis. (A few days before, LBJ had announced that he wasn't going to run for reelection as president. Which left us wondering who was going to succeed him, and how much support Apollo was going to have.) As far as most of America was concerned, the flight never happened.

Apollo 6 sure didn't go as well as the first one did. While the big S-I first stage was still thrusting, the vehicle began showing some pogo—a strong vibration that nobody had anticipated. (The Titan vehicles that had launched Gemini had bad pogo problems, but they had originally been designed as ICBMs. We didn't think it would be that big a deal with the Saturn, which was always designed as a manned space launcher.) We figured out later that Apollo 4, the first Saturn V, hadn't carried the same mass as Apollo 6 did. In any case, part of the adaptor between the S-IVB third stage and the Apollo CSM came apart because of the vibration.

The first stage shut down as planned and the S-II second stage fired up. Then it really got hairy. Four minutes into a planned six-minute burn, one of the J-2 engines on the stage shut down. A second or so later, so did another one.

If this had been a manned flight, the escape tower on the Apollo would have been commanded to fire, pulling the spacecraft away from the Saturn for a parachute landing in the Atlantic.

But the two J-2s were on opposite sides of the stage; the loss of one sort of balanced the other, and the S-II kept sailing along instead of tumbling out of control and blowing up.

The Saturn's internal guidance system commanded the remaining three engines on the stage to fire longer to make up the loss in thrust, but now the whole vehicle was flying lower and slower than it should have. The S-II burned fifty-nine seconds longer than planned, and still didn't quite get the S-IVB up to the right velocity.

Because of the various problems, however, the S-II had kept itself from pitching over as planned, so its altitude was higher than it was supposed to be. This was even more confusing to the computer guiding the S-IVB, which realized it was higher than it should be . . . and slower. So while it added twenty-nine seconds to the burn, it actually *pointed itself down* toward the center of the earth.

That couldn't go on, of course. Eventually the S-IVB realized it was pointing low, so it pitched back up. Way up. They say it actually went into orbit thrusting *backward*, and wound up in an elliptical orbit rather than the circular one we needed.

We had wanted to restart the S-IVB to simulate injection into a translunar course. When the time came, after two orbits, the S-IVB didn't fire. Instead the CSM was separated and the engine on the service module—the service propulsion system (SPS)—lit up for seven minutes and twenty-two seconds. This burn sent the CSM into a looping elliptical orbit, simulating a high-speed lunar reentry. The command module wound up safely in the Pacific after a pretty harrowing experience.

With a performance like Apollo 6, we all thought we were going to need a third unmanned Saturn V before going on to manned missions. But Mueller, Phillips, and Gilruth figured the pogo problem could be fixed. The S-II engine failure was traced to a wiring problem. And the S-IVB failure was caused by a rupture in a propellant feed line. Those were relatively easy fixes, too.

On the other hand, just processing the Saturn V at the Cape—assembling the vehicle, checking it out, and getting it to the pad—was still taking too long. (We hoped to fly them every two months.) We were confident that 503 could be ready to fly unmanned by late July; a manned mission wasn't possible before November.

But overall there was good news in the Apollo 6 mess: the vehicle had made it to orbit. It hadn't blown up. So they decided to make the third Saturn V, AS-503, a manned flight.

On April 16 the Russians announced that they had rendezvoused and docked two unmanned spacecraft, Kosmos 212 and 213. This was the second time they'd done that, so we assumed they were getting ready to fly manned again.

Later that month, with the Apollo 7 prime and backup crews, I went hunting wild sheep and boar on Catalina Island, just off the coast from Los Angeles, courtesy of Rockwell.

We had different approaches to the practical business of training guys to land the lunar module. Out at NASA, Dryden and Edwards Bell Aircraft had developed a weird-looking machine called the Lunar Landing Research Vehicle (LLRV). It was kind of a flying bedstead, a frame with legs wrapped around a G.E. jet engine that pointed down. Mounted on the frame were half a dozen backup jets and something like sixteen small steering jets. On the front was a seat for a pilot. (In the lunar module, of course, the crew would be standing, but we weren't comfortable with the idea of having pilots fly the LLRV without an ejection seat, so that's what it had.)

The idea was to simulate the last couple of minutes of a lunar landing . . . hovering and coming down softly. NASA test pilot Joe Walker was the first to fly the LLRV, in October 1964, and he found it was easy to learn provided you already knew how to fly helicopters.

So helicopter training was mandatory and we put all our guys through it in 1965 and 1966. We modified the two LLRVs as training vehicles and bought three slightly improved versions called LLTVs. Beginning in late 1967 and early 1968 the guys most likely to command lunar landing missions—Pete Conrad and Neil Armstrong—started flying them.

On Monday, May 6, Neil was making his twenty-first flight in LLRV number one out at Ellington Field. The plan was to climb up to an altitude of five hundred feet, then come down on a target landing site.

As Neil descended to a hundred feet, the LLRV began to pitch forward. Neil tried to stop this with the thrusters, but all that happened was that he started tilting to the right. So he punched out of there. A couple of seconds later the LLRV smacked into the ground.

Neil was fine, but the crash caused us a few headaches. Everybody wondered if we had some serious design flaw in the lunar module. Bob Gilruth put together an investigating board, which included Joe Algranti from aircraft ops, and Bill Anders. It was a few months before they found the cause of the accident, which turned out to be some problem with a helium tank aboard the LLRV that starved the steering jets. But it kind of slowed down our lunar landing training for a while.

In June a crew of Joe Kerwin, Vance Brand, and Joe Engle completed altitude tests with a command module test article. CSM 101, Wally's spacecraft, was ready to be shipped to the Cape. We had eighteen months to fulfill JFK's goal.

I had some attrition in the Astronaut Office that spring and summer. The most public was one of the 1967 group of scientists, Brian O'Leary. He and the other guys in his group had been sent out to various air force bases beginning in March for fifty-two weeks of training that would turn them into jet pilots.

We had already been confronted with some public bitching by the scientist-astronauts, notably Phil Chapman of the new group, and Curt Michel from the 1965 selection. They realized they weren't getting any time to do research and were spending it all traveling around the country learning what NASA did and sitting through lectures on the Apollo spacecraft. We let Michel go ahead and start spending most of his time at Rice University doing research. It was basically his first step out the door.

For the new group, which included Chapman, it sure hadn't been the most exciting few months you'd want to spend; it only happened because we weren't able to ship them off to flight school back in September the way we wanted. The Vietnam War was using up a lot of pilots, and we simply couldn't slot them into Air Force flight schools until early 1968. So these guys had been on a very long orientation course that was probably frustrating as hell.

Anyway, O'Leary was out at Williams Air Force Base in Arizona with Bob Parker and just never seemed to get the hang of things. I guess he soloed, then decided he didn't want to proceed. He called me one morning and told me he wanted out. I had a certain amount of sympathy; he was a young guy who came to NASA straight out of college with a lot of mistaken ideas. He didn't like being ordered around by me or Al Shepard, he didn't like living in Houston, and he didn't like doing grunt work when he wanted to be spending his time on research.

He sure shouldn't be flying a spacecraft if he didn't like flying a Piper Cub. I offered him a chance to move somewhere else in NASA, but he just wanted out.

A couple of years later he published a book called *The Making of an Ex-Astronaut,* which pretty much confirmed my impressions. As far as I was concerned, he never was an astronaut. (Until the scientist-astronauts finished flight school they were only candidates as far as I was concerned.) O'Leary also became a big critic of the shuttle, then a big proponent. Last I heard he was into psychic phenomema.

Another of the 1967 scientists, Tony Llewellyn, dropped out later in the summer, too. He had done somewhat better in flight school, but couldn't hack jets. He didn't want to stick around NASA, either, and became a college professor in Florida.

The nine survivors of the scientist group all became pilots, some of them quite good. Joe Allen was first in his class at Vance Air Force Base; I think Story Musgrave did the same at Reese Air Force Base. (Musgrave went on to log more flying time than just about anyone I know, something like seventeen thousand hours. By comparison, during his career, a guy like John Young or Joe Engle might log eleven thousand.)

I lost one of the pilots selected in 1966 that April, too. A Navy pilot named John Bull had been working with Jim Irwin on lunar module tests. He turned up with a pulmonary disease that made the Navy drop him from flight status, which meant that we couldn't fly him, either. We found him a job at the NASA Ames Research Center in Palo Alto, California, where he did some research flying and went on to get a Ph.D. He was a great guy and would have been one of my early picks for a lunar module pilot. I hated to lose him.

I had to make an actual crew change in early July. Mike Collins was training as the command module pilot for Apollo 9, Frank Borman's crew, scheduled for the E mission. Mike had been having some weird physical symptoms and finally got himself checked. It turned out he had a bone spur between two vertebrae in his neck. He was going to need surgery, and the best prognosis showed that he would be out of action for a couple of months.

The backup crew for Borman-Collins-Anders was Armstrong-Lovell-Aldrin. Frank and Jim had flown together, so moving Jim to the prime crew presented no problems whatever. It left me without an experienced command module pilot on the backup crew, however. So I had to promote Buzz Aldrin from lunar module pilot to command module pilot on Neil's crew. Fred Haise, one of the better guys hired in April 1966, came in as lunar module pilot.

Now the lineup went:

Flt	*Msn*	*Prime Crew*	*Backup Crew*
7	C	Schirra-Eisele-Cunningham	Stafford-Young-Cernan
8	D	McDivitt-Scott-Schweickart	Conrad-Gordon-Bean
9	E	Borman-Lovell-Anders	Armstrong-Aldrin-Haise
10	F	Stafford-Young-Cernan	TBA

Mike Collins sailed through surgery and was back on flying status by early November. Too late for Apollo 8, I thought, though I know Mike disagreed. He plunged right to work in helping get 8 ready, however, and served as a capcom during the mission.

I also regained an astronaut. Al Shepard's Ménière's disease had continued to get worse instead of better. He was still suffering from vertigo and had started discussing possible surgery with doctors from NASA and the Navy. As with any surgery, there was a risk that the problem might be made worse—he could lose his hearing.

But he decided to take a shot. In May 1968 he flew out to Los Angeles and checked into a hospital under the name "Victor Poulis." On the fourteenth he was operated on and a silicone rubber tube was implanted in his left inner ear. From that point on it was simply a matter of observation and recovery.

Even though it had been decided to fly the third Saturn V manned sometime around late November, the third lunar module was simply not likely to be ready by then. Probably not until early 1969.

George Low and Chris Kraft had been thinking about turning the E mission—the high earth orbit lunar module test—into a circumlunar flight since at least April. It would cut out one of the steps and make a lunar landing happen that much quicker. Bob Gilruth got brought into the plot about that time. It was one of those things you talk about, but don't really expect to happen.

During the next few weeks, Low came up with an even wilder possibility: eliminate the E mission altogether and create a so-called C-prime mission.

C-prime would send the Apollo command and service module—no lunar module—on a lunar orbit mission in December.

On Monday, August 5, 1968, Low came back from a vacation and started putting this into high gear. There were a lot of facts to be gathered. Were the navigation and trajectory teams ready for this? Could flight controllers be trained? What about the deep space tracking network?

Low went down to the Cape on that Thursday and heard from Rock Petrone that there was no way LM-3 was going to be ready before early 1969. As for flight crew ops, I thought a crew could be ready in four months, no problem. Hell, this was what we'd been training for.

Friday morning Low got me and Kraft together in Gilruth's office. After hearing us say that we could get our troops ready, Gilruth signed off on the idea right away. Phone calls were made to Sam Phillips and to von Braun, who were willing to go with it. In fact, that afternoon we all flew off to Huntsville and met in von Braun's office.

About the only people who weren't there were Jim Webb and George Mueller. Of course, they were the guys who had to give final approval. And they were both in Vienna at an International Space Conference.

It wasn't until the following Tuesday, August 13, that Phillips got in touch with Mueller. (He had first run the idea past Tom Paine, the acting administrator, at headquarters. Paine had approved it.) Mueller wasn't opposed to the idea, but he didn't jump up and down about it, either. He wasn't clear about the technical gains to be made from sending the CSM alone around the moon. On the other hand, there was a good chance that we might learn something valuable; we would get to fly the Saturn V manned. And he didn't want a big gap in flights between 7 and the lunar module flight.

The big problem was Jim Webb, who just hated the whole idea. He had taken the Apollo fire very personally and could easily picture himself back before some congressional committee testifying about this radical change in plans if something went wrong.

Webb also knew he was coming to the end of his term as NASA administrator. He had been appointed by JFK and continued by LBJ, but knew that either of the two presidential candidates, Nixon and Humphrey, was going to get his own guy. So he was going to be gone in January 1969, anyway.

In Vienna, Mueller went to work on Webb. Paine and Phillips sent him a long cable laying out the technical issues. On the sixteenth Webb finally gave in, telling us to go ahead with the plan. But not to make it public.

I had already charged ahead on my own. On Friday the tenth I called

Jim McDivitt into my office. He was the commander of Apollo 8, and it was time he knew what everybody was planning to do to his mission.

Other Voices

JAMES MCDIVITT

By this time I'd been working on the lunar module flight with Dave and Rusty for a year and a half. I knew something was up when Deke called me in because, for one thing, I knew the lunar module wasn't likely to be ready before January 1969.

Deke explained the situation and said that he wanted me to stick with my original mission—which would now become Apollo 9. But he wasn't going to force me.

It wasn't just a case where, since this C-prime mission wouldn't carry a lunar module, NASA didn't want to throw away our training. Frank Borman and Bill Anders had been training on the lunar module for some time, too. I think it was that Rusty and I knew more about this particular *lunar module than anybody else. So there was a certain logic to keeping us where we were.*

Over the years this story has grown to the point where people think I was offered the flight around the moon but turned it down.

Not quite. I believe that if I'd thrown myself on the floor and begged to fly the C-prime mission, Deke would have let us have it. But it was never really offered.

On Sunday the twelfth I got on the phone right away to Frank Borman, who was out in Downey at Rockwell, and told him to get back to Houston. A few hours later he walked in, and I told him the deal, that Apollo 8 was his if he wanted it. The answer was typical Frank Borman, no hesitation, one word: "Yes."

It was always interesting to me to see the difference between Jim and Frank. You could tell either one of them to take the hill. Jim would look things over and very carefully make a decision. Frank would be out of the foxhole before you got finished talking. But both of them would take the hill.

So I had a new lineup of missions and crews:

Flt	*Msn*	*Prime Crew*	*Backup Crew*
7	C	Schirra-Eisele-Cunningham	Stafford-Young-Cernan

8	C′	Borman-Lovell-Anders	Armstrong-Aldrin-Haise
9	D	McDivitt-Scott-Schweickart	Conrad-Gordon-Bean
10	F	Stafford-Young-Cernan	TBA

Of all the people involved, Dave Scott was the most obviously unhappy about the switch, because he had spent a lot of time getting command module 103 in shape; now he was going to have to switch to 104. (Apollo 7 was flying CSM 101, but 102 had been used for ground tests.)

Bill Anders, who lost his chance to fly a lunar module, was also pretty unhappy. I don't know about the other guys, though I suppose if you looked back, Pete Conrad should have complained. The rotation changed so that he wound up with Apollo 12 instead of Apollo 11.

We announced the crew swap on Monday, August 19, but didn't publicly disclose the new flight plan. We hadn't even flown 7 yet. The most that happened was that Sam Phillips hinted at a press conference a week later that a "new mission" was likely to be inserted on the schedule between 7 and the lunar module test.

Within the month, on September 15, 1968, Webb went to the White House to talk to LBJ. When he came back from the meeting, he was no longer NASA administrator. I never quite got straight on what happened, though the story was that Webb had planned to leave NASA when LBJ left the White House, and that he had reminded LBJ of this at the wrong time. At which point Johnson simply thanked him for all his hard work and took him out to meet reporters to announce his resignation. Tom Paine became the acting administrator.

I do know that on hearing that Webb was gone, Sam Phillips told Rock Petrone, "This makes C-prime possible."

If we needed a little encouragement, the same day Webb was cozied up with LBJ the Russians launched a space probe called Zond 5 toward the moon. Three days later it swung around, then headed back to earth. On the twenty-second it splashed down in the Indian Ocean.

We didn't see pictures of the Zond—which probably wouldn't have done us much good if we had, because we still hadn't seen Soyuz—but a lot of folks thought this was an unmanned test of a manned circumlunar mission. (There were people who didn't believe it, pointing out that Zond 5 had reentered the atmosphere at such a steep angle that the G-forces, as high as sixteen Gs, would have injured or even killed the crew.) It would be twenty years before we got confirmation, but we didn't need it. Most of us thought it brought the Russians that much closer to going around the moon themselves.

That was another thing that Webb knew something about, probably from CIA briefings based on spy satellite photos, though damned little of it was shared with people at my level: he said the Russians were definitely building a big lunar booster. He got so identified with it that it was known for years as "Webb's Giant."

"Webb's Giant" was actually called the N-1, and sure enough, it was about the size and capability of the Saturn V. A mock-up version had been erected on a new launch pad at Baikonur earlier in 1968, which is undoubtedly what the Keyhole spy satellites saw. Korolev and Mishin had been working on this since 1962 or so, and though it was way behind schedule and far from proven, it was real enough. And if we had a big problem, the Russians would be able to make it a real race to the surface of the moon.

They weren't hampered by the N-1 delays in a manned flight around the moon, however. The Soyuz-Zond was launched by a proven Proton booster. And that was ready to go right about when we were.

On October 11, 1968, Apollo 7 was launched with Wally, Donn, and Walt. There was some problem right up to the last minute with wind conditions. If they were too high, there was a chance that an abort could put the crew in a dangerous situation—having the CSM come down on land.

The winds were kind of in and out all morning, but they were clearly over the line. Trouble was, everybody including Wally had a lot of faith in the Saturn IB, and nobody wanted to recycle. So we went ahead with the launch.

The Saturn IB worked great and had the S-IVB/Apollo stack in orbit within eleven minutes. The first job for the crew was maneuvering that stack, since that would be an integral part of any manned lunar mission. Then the plan was to separate from the S-IVB, fly out a few yards, turn around, then go back . . . just the way you would to dock with the lunar module.

One slight problem came up at this point: the four adaptor panels atop the S-IVB were supposed to open wide on hinges. One of them had sort of bounced back. It posed no danger to the crew, but had this flight carried a lunar module, it might have been tough to get it out of there. (On later flights the adaptors were designed to simply fall off.)

The crew spent the rest of the day test-firing the SPS and doing rendezvous maneuvers with the S-IVB.

Toward the end of the first day Wally began to show signs of having a cold. So did Walt. When flight controllers threw some changes in the plan to Wally, he began to show a little annoyance.

One of the big PR things in the mission was a live TV broadcast scheduled for day two, Saturday, October 12. We had carried a small TV on Gordo Cooper's Mercury flight, but this was a bigger deal—a nice test of a system that we hoped would show the first footsteps on the moon in a few months.

But Wally got preoccupied with a second rendezvous planned for later in the day, and he just postponed the broadcast, period. (He had always complained about the scheduling, but hadn't been able to win the battle on the ground. He probably figured there wasn't much we could do to him while he was in orbit, and he was right, but it made my life kind of difficult.) There I was in the MOCR with Chris Kraft and a bunch of people getting pissed off. But there was nothing I could say or do to get Wally to change his mind.

From then on it was kind of a war between him and the flight directors, particularly Glynn Lunney. Wally would be doing something in the spacecraft and start bitching about the "genius" who designed some equipment. He might have been right, but it sure didn't endear him to the guys on the ground to have the astronaut implying they were idiots over the open lines for everyone to hear.

Donn got into the act, too. Walt says he tried to stay out of the way. But all of them were pretty testy up there.

Typically, of course, when they finally turned on the cameras on Sunday, Wally was all charm. "Hello from the Apollo room, high above everything." He even held up a card that said: "Deke Slayton, are you a turtle?" In keeping with the tradition, I turned off my microphone for the reply.

The Wally, Walt, and Donn Show—which they repeated throughout the flight—was so successful that the crew won a special Emmy award.

The problem of colds cropped up for reentry on the eleventh day of the mission. Wally decided that they wouldn't wear helmets with their pressure suits. He was afraid they might rupture their eardrums. We were worried that a pressure leak during reentry could kill the crew; our astronauts had always worn pressure suits during reentry for that reason.

Well, this precipitated another "private" discussion between Wally and me, and I lost again. The best Wally would agree to do was have the crew wear their pressure suits, but with the helmets stowed. Reentry took place as planned, with no pressure leaks and no busted eardrums.

I met Wally on the carrier USS *Essex* and had a few words in private with him. It wasn't so much about his behavior—it was pretty typical Wally—but what he had done to Donn and Walt. I didn't hear this personally, so I can't vouch for it, but Chris Kraft was reported to have announced that nobody on the 7 crew was going to fly again. I made the selections, but I wasn't going to put anybody on a crew that Kraft's people wouldn't work with. Not when I had other guys. And the way things worked out, none of them ever did fly again. (Wally had actually planned and announced his retirement before the flight.)

When it was over, Sam Phillips said Apollo 7 had accomplished "101 percent" of its goals, which was technically true. But it wasn't easy.

The Russians got back in the manned space business a few weeks after Apollo 7, launching Soyuz 3 with a single pilot aboard, Colonel Georgy Beregovoy. He was a forty-seven-year-old test pilot—the first test pilot to fly on a Soviet spacecraft—and his mission was to requalify the Soyuz for manned operations. He was supposed to rendezvous and dock with Soyuz 2, which was launched unmanned, but he couldn't make it work. I think he got within three hundred yards.

This was less of a problem for the Russians than it would have been for us, because they had already demonstrated automatic docking. (That's what they were set up to do, anyway.)

On November 10, Zond 6 was launched toward the moon. It looped around on the thirteenth, then landed back in Russia on the seventeenth. They had tried a more benign reentry that skipped the spacecraft in and out, then back into the atmosphere, which not only allowed them to land the vehicle inside the Soviet Union, it would have subjected a crew to only six Gs, well within tolerances.

As far as we could see, with two unmanned successes and the flight of Soyuz 3, the Russians were ready to go.

After Apollo 7, Tom Stafford and his crew—John Young and Gene Cernan—were free to concentrate full-time on 10, which looked to be the F mission: a full-dress rehearsal of a lunar landing in orbit around the moon. They had been working on this in some fashion since April 1968. I assigned a backup crew of Gordo Cooper as commander, Donn Eisele as command module pilot, and Ed Mitchell as lunar module pilot. Tom's support crew was Charlie Duke, Joe Engle, and Jim Irwin.

The first challenge, of course, was getting *around* the moon. The second one was testing the lunar module. Until those two goals were met, we couldn't predict any mission with real confidence.

Frank Borman, Jim Lovell, and Bill Anders had dug into training for Apollo 8. There were some tricky decisions to be made, such as the usual safety concerns. Apollo carried three fuel cells. If two of them failed, could we still complete the mission? There were two oxygen tanks; how far could we get on one?

Did we just want to have 8 loop around the moon and come back, or fire the SPS and put it in orbit? The first option was the safer one, but the second option was what we would have to do on a landing mission. If we didn't have confidence in the SPS now, when would we get it? So it was decided that 8 would make ten orbits of the moon. And given that everyone from Bob Gilruth to Frank Borman was concerned about the performance of the SPS, we came up with this compromise: on the trip to the moon, the SPS would be fired for a few seconds. If it worked, the mission would proceed as planned. If there were problems, the few-second burn wouldn't keep 8 from simply looping around the moon and heading back to earth.

There were other debates, too. The original plans for the mission called for a splashdown in the Pacific during daylight. But this was based on having 8 make twelve lunar orbits or more. (The launch time was fixed by the position of the earth and moon, and by our need to launch early in the day. The transit time to and from the moon was largely fixed, too. Every other decision flowed from these fixed points.) Frank didn't want to spend any more time in lunar orbit than was absolutely necessary, and pushed for—and got—approval of a splashdown in the early morning before dawn.

Frank wasn't afraid to raise objections with anybody, and I think the mission was better off for that reason.

Following the hardware certification meeting on Thursday, November 7, we had another big get-together on Sunday the tenth, the same day Zond 6 was launched—Mueller, Phillips, Low, Petrone, Kraft, and me, plus representatives of something like sixteen contractors. We asked everybody if they agreed that 8 should go into lunar orbit. Only Walter Burke of McDonnell had any objections.

The recommendation was to go for it. We passed our judgment to Tom Paine, who announced it on November 12.

December 6, the best day for a Soviet lunar launch, came and went without a one. The Soviet tracking fleet, which was always deployed for a manned or manned-related mission, went back to port. The next Russian opportunity wouldn't come until early January. But Apollo 8 was scheduled to lift off on December 21.

It was twenty years before anybody in the West knew why a Soviet launch was passed up on December 6. It turned out that a cosmonaut crew of Alexei Leonov and Oleg Makarov was all trained for a seven-day Soyuz-Zond mission and ready to go. But the Soviet flight operations people had set a requirement of two successful unmanned Zond missions before they would commit a cosmonaut crew. It looked to us as though both Zond 5 and Zond 6 had worked.

But Zond 6 *hadn't worked*, at least, not well enough. The return module had lost pressure during reentry and had landed hard. Either mistake would have killed the crew. Alexei would say years later—long after Apollo-Soyuz—that he and his backup commander, Valery Bykovsky, were willing to take the chance to beat us. But they got overruled.

So without knowing it, we had a good chance to win. But two days later we had another hiccup. Joe Algranti was doing the last checkout in LLTV number one before releasing it for astronaut training.

He was supposed to make a six-minute flight in the vehicle. Four minutes in, at an altitude of two hundred feet, he started having lateral oscillations. Then the LLTV just started to fall out of the sky. Joe ejected safely.

Well, that was another delay, and at the worst possible time. Within a month I was going to have to name the guys who would probably have the first chance to do a lunar landing.

At 2:35 A.M. on Saturday, December 21, I woke up Frank, Jim, and Bill in the crew quarters at the Cape. After a medical check, we sat down to breakfast of steak and eggs. All of us had been through this ritual before, and I'd been through it more than anyone, but I think we all knew this flight was a little different from the ones that had gone before. Even Chris Kraft had said it was the riskiest one we'd ever tried. This hadn't gone unnoticed in the press: there were a lot of people talking about the risk, about the fact that this was only the third flight of the Saturn V. I can't say they were completely wrong.

I took up my seat in the firing room as the crew headed out to the pad. Right on schedule at 7:51 A.M., the Saturn V lit up and began crawling into the air.

Eleven and a half minutes later, 8 was in orbit. For a revolution and a half it stayed in earth orbit while everything was checked out. Then Mike Collins, who was the capcom in MOCR, radioed up to the crew: "Apollo Eight, you are go for TLI." TLI was translunar injection, the firing of the S-IVB. Jim Lovell radioed back a "Roger." They were over the Pacific, west of Hawaii, when the S-IVB lit up. When it shut down five minutes and eighteen seconds later, they had increased their velocity from 17,500 miles an hour to over 23,200 miles an hour. Escape velocity. They were on their way to the moon.

And I was on my way back to Houston.

* * *

Five hours into the flight, 8 was sixty thousand miles away from earth. Since it was, in effect, fighting against earth's gravity all the way, its velocity had dropped to less than six thousand miles an hour.

The test firing of the SPS went off as scheduled, clearing the way for a burn into lunar orbit. The crew then decided to turn in. Frank was supposed to sleep first, then Jim and Bill. But Frank, still wound up, couldn't relax, so he took a sleeping pill. A little while later, he got pretty sick.

It passed quickly and Frank, at first, blamed the sleeping pill. Jim had felt a bit nauseous during the first few minutes of the flight, and Bill felt a little queasy later on. It was probably a touch of what they now called space adaptation syndrome (SAS), which affects about half of the people who go into space. SAS hadn't bothered anybody in Mercury or Gemini—including Frank and Jim—but there hadn't been enough room to move around and get disoriented in those spacecraft.

The rest of the coast to the moon went smoothly. At sixty-six hours into the mission—Tuesday, December 24, Christmas Eve—8 swung around the back side of the moon with a go for LOI—lunar orbit insertion. Then we all sweated the next thirty-three minutes waiting to find out if it had worked. (If the SPS hadn't fired, we'd see 8 ten minutes earlier.)

Twenty-three minutes, twenty-four . . . Jerry Carr, the capcom, was calling, "Apollo Eight, Houston. Apollo Eight, Houston." Right on schedule we heard from Jim Lovell, "Go ahead, Houston." They were in an elliptical orbit, 69.5 by 190 miles.

The crew had a ton of work to do in lunar orbit, beginning with good old naked-eye observations and photography—hampered by the fact that three of the five windows in the command module were fogged over. For one thing, they were surveying potential landing sites. There was navigation work—learning about the way 8's orbit shifted due to unanticipated concentrations of mass, mass cons, under the lunar surface. And just the usual work of keeping the spacecraft operating.

When Frank admitted that they were getting tired, we told them to knock off and get some rest. They perked up for their last orbit around the moon on Christmas Eve, with a live telecast and the three of them reading from the Book of Genesis. You could see the surface of the moon rolling past underneath, with Bill Anders saying, "In the beginning God created the heaven and the earth . . ." It was quite a moment.

Their tenth orbit was the last one. At eighty-eight hours, fifty-one minutes into the mission, early on Christmas Day, 8 swung around the moon for the last time. If the SPS burn went right, they would be back in contact in thirty-eight minutes.

We waited. And waited some more. Then, thirty-eight minutes after loss of signal, we heard Jim Lovell: "Please be informed there is a Santa Claus."

A few hours later, once they were well on their way back to earth, I radioed some greetings of my own. "None of us ever expected a better Christmas than this one."

The guys who were the Apollo program managers—first Joe Shea, then George Low—had always felt that crews should be handpicked for specific missions. I had disagreed with this on the principle that the specifics of the mission kept changing frequently, and so far I had won the argument. My operating rule was to have a pool of guys trained so that anyone could handle anything, and then make selections based on that.

With the success of Apollo 8, it was time to name the Apollo 11 crew. On the planning charts this might very well turn out to be the first manned lunar landing. But no one knew for sure. The lunar module was still a couple of months away from a test flight. There was the Apollo 10 lunar orbit mission, too. Some people were thinking that if 9 went well, we should skip the F mission and have the 10 crew make the landing.

So it wasn't just a cut-and-dried decision as to who should make the first steps on the moon. If I had had to select on that basis, my first choice would have been Gus, which both Chris Kraft and Bob Gilruth seconded. With Gus dead, the most likely candidates were Frank Borman and Jim McDivitt. I had full confidence in Tom Stafford, Neil Armstrong, and Pete Conrad, too. The system had put them in the right place at the right time. Any one of them might very well make the first landing.

Jim McDivitt was still tied up with the lunar module flight, but here I had Frank and his whole crew available, and given that they had already made a lunar orbit flight, of all the astronauts they were clearly in the best position to train up on the landing procedures.

There were two other things to keep in mind. Frank had been away from home pretty steadily for almost two years. He was tired of the grind and not happy about what it had done to his family life. He had already told me he was ready to move on.

Finally, there was no guarantee that 11 would turn out to be the landing. So I figured my best choice was to stick to the rotation and assign Neil Armstrong's crew. I made one change, however: Mike Collins was available and had lost out on the lunar orbit mission. He deserved the first available mission, which happened to be 11.

The crew had Fred Haise, from the 1966 group, assigned as lunar module pilot. Fred was very capable—one of the best people in his group—

but he hadn't been in my plans from the beginning and wasn't part of the eighteen guys I promised the landing to.

Further, Buzz Aldrin had already trained as a lunar module pilot for several months prior to the Apollo 8–Apollo 9 swap. So the crew became Neil Armstrong, commander; Mike Collins, command module pilot; and Buzz Aldrin, lunar module pilot. The backup crew would be Jim Lovell, Bill Anders, and Fred Haise.

Later on, people would talk about this process as if it were some kind of science. Or as if politics had controlled it—the fact that Neil was a civilian. All I can say is that a lot of factors, most of them beyond anybody's control, put these three guys in the right place at the right time. The first person to walk on the moon might just as easily have been Tom Stafford, an Air Force officer, or Pete Conrad from the Navy.

I called Neil, Mike, and Buzz into my office on Monday, January 6 and told them, "You're it."

21

STEPS

Apollo-Saturn 504, the vehicle for Apollo 9, rolled out to the pad on January 3, 1969. This was the first all-up working version of the Apollo spacecraft—command and service modules, operational lunar module, even pressure suits. Of course, 9 wasn't going to the moon. But it was just as tricky.

For the first time astronauts were going to be flying in space in a vehicle that couldn't return them to earth. Not only did you want a very reliable guy flying the lunar module—which we certainly had in Jim McDivitt—but the command module pilot was going to be under a lot of pressure. Dave Scott had really proven himself on Gemini 8, the first manned docking that ended in an abort. He had Jim McDivitt's full confidence, and mine, too.

The lunar module pilot in the crew was Rusty Schweickart, a very sharp guy whom I'd hired from MIT, where he had been working on studies of the upper atmosphere. He had also flown with the Air Force and Air National Guard. In the Astronaut Office he was more of a political liberal than the others—his wife, Clare, was more liberal than just about anybody they knew, and I think her politics caused him a few problems with his colleagues. He was going to copilot the lunar module and test the lunar EVA pressure suit.

These three had been training specifically for the first lunar module mission since November 1966, more than two years. It came to something like seventeen hundred hours of training. They were ready; the only question was the equipment.

Nine would also be the first mission in years in which spacecraft had names instead of numbers. You needed them for radio communications when the command module and lunar module were separated. The command module was going to be *Gumdrop* and the lunar module was going to be *Spider*. NASA public relations people like Julian Scheer hated the names; I guess they weren't dignified enough. But the crew had picked them.

On January 14 the Soviets launched Soyuz 4 with a single pilot, Vladimir Shatalov. The next day Soyuz 5 went up with a crew of three—Boris Volynov, Alexei Yeliseyev, and Yevgeny Khrunov. Four docked with 5 on live Soviet TV. It was a pretty humorous moment: Khrunov piped up "Help, we've been raped!" as 4 slid the docking probe into 5's collar, and that went out over a live feed. (Just once; they cut it from later broadcasts.)

TASS announced that Soyuz 4 and Soyuz 5 made up the world's first "experimental space station," which was sort of a stretch; you couldn't get from one to the other without doing an EVA. It looked to me like they were anticipating the flight of two manned vehicles on 9 in a couple of months and decided to try and steal a little of our thunder.

Khrunov and Yeliseyev made an EVA, moving from 5 to 4. A few hours later the two vehicles undocked, with Volynov coming back to earth on his own while Shatalov returned with the other two.

Even then we realized this was the exact mission the Soviets had wanted to fly in April 1967, testing manned docking and a lunar-style EVA suit. They were at least two years behind their own schedule.

On January 19 the Russians launched another unmanned Soyuz-Zond, aiming for that second successful flight to clear the way for a manned circumlunar mission by Leonov and Makarov. But the Proton didn't make orbit, and the L-2 project basically died at that point. The Russian manned lunar landing project, L-3, was still alive, however. The key test there was the first flight of the giant N-1 rocket.

That was scheduled for February 21. The big thing got off the ground successfully, but a fire broke out between two stages about seventy seconds into the flight, and range safety blew it up.

They weren't ready to catch us yet.

Apollo 9's launch was delayed three days because all three of the crew came down with colds. Nobody wanted to repeat our experience with Wally's guys and their colds, so we waited for them to get well.

Doctors were starting to wonder what was wrong—the 7 crew got sick, all of the guys on 8 were sick to some degree during the flight, and here the 9 crew had colds. We figured it was just that we were working everybody so hard they were susceptible to opportunistic infections. So we tried to start isolating the crew as much as possible in the last few weeks of training. It would continue to be a problem, however.

On Monday, March 3, 1969, the fourth Saturn V took off and gave McDivitt, Scott, and Schweickart a clean ride to orbit. There were some oscillations in the S-II stage, but nothing serious.

Once in orbit, the first job was to check out the docking probe on the nose of the command module. This was a complicated little mechanism that allowed you to latch two spacecraft together, then move from one to the other without doing EVA. The probe and drogue, as we called it, was a tube that extended out from the nose of the command module and poked into a hole in the appropriate hatch on the lunar module. Once the probe was inserted, it would retract, pulling the two vehicles together in the proper alignment. The ring of the lunar module fit snugly inside the ring of the command module, which had a series of latches that clicked shut. Once you were sure there were no pressure leaks, you could disassemble the probe mechanism, store it inside the command module, and swim through to the lunar module.

It was a hellaciously complicated little piece of equipment. If it didn't work the first time, your lunar landing was canceled right then and there. If it didn't work after the landing, you could be dead. The Russians hadn't mastered this, for example, and wouldn't for years.

With Dave Scott at the controls, *Gumdrop* moved away from the LM/S-IVB combination, flipped around, and moved back. The docking went smoothly. Jim and Rusty checked the docking adaptor, taking it apart, then putting it back together. So far, so good. The CSM-LM jettisoned the S-IVB. The last item on the first day's agenda was a firing of the service propulsion system, just to see how this whole 36.5-ton stack would handle it. That went just fine.

Day two was full of SPS burns and maneuvers. Day three was for lunar module checkout.

The first problem that cropped up was with Rusty. When the crew put their pressure suits on, he started feeling pretty sick. In fact, he tossed his cookies right there in the cabin. He felt a little better after that, and he and Jim were able to spend the whole day inside the lunar module.

Day four was supposed to have been Rusty's EVA, going outside on the front porch of the lunar module and then crawling up to the hatch of the command module to retrieve some experimental stuff mounted out there.

Everyone was worried about his condition. Throwing up inside a pressure suit would not only be unpleasant as hell, it might be fatal. We had decided to skip the EVA and just have Jim and Rusty suit up inside the lunar module, then open the hatch. That way there would be some test of the suit in the right conditions.

But on the morning of day four Rusty said he was feeling better, so Jim decided to let him go outside the lunar module onto the front porch. Meanwhile, Dave Scott would do a stand-up EVA in the command module hatch and grab the experiments Rusty would otherwise have retrieved.

What was important about this EVA was that the lunar pressure suit was completely self-contained. All the suits used on the Gemini EVAs had relied on the spacecraft to provide oxygen and communications. The consumables and communication equipment for the lunar suit, which also had to be rugged enough to take the punishment of getting around on the lunar surface, were all in a backpack called the portable life support system (PLSS).

Rusty wound up spending thirty-seven minutes in EVA, not as much as we wanted, but enough to give us some confidence in the suit and the PLSS.

Day five was the big deal.

Jim and Rusty had found it was easy to fall behind the time line—the work schedule—inside the lunar module, so on day five they got in there early, got her powered up and their suits on and the hatch closed. Dave Scott hit the switch that released the probe. It caught for a moment, then let go. A little fly-around so *Gumdrop* and *Spider* could check each other out, then Dave fired his RCS motors and backed away.

Jim fired *Spider*'s descent engine for nineteen seconds to put the lunar module above and below *Gumdrop*. *Gumdrop* was to stay in a circular orbit; the maneuvering was supposed to be done by *Spider*. The next goal was to circularize *Spider*'s orbit fifteen miles above *Gumdrop*. During the first burn, Jim and Rusty had noticed a weird groaning noise when the engine was throttled. It didn't happen the second time, but it was still worrisome.

Spider and *Gumdrop* moved farther and farther apart, to the point where the ascent and descent stages would separate. Boom! Off went the descent stage with pieces of foil flying off it.

Now it was just Jim and Rusty in the ascent stage, dropping farther behind and now below *Gumdrop*. At a distance of about ninety miles, Jim fired *Spider*'s ascent engine for three seconds to circularize his orbit, which was now about ten miles lower than *Gumdrop*'s. Thanks to orbital mechanics, *Spider* then started to gain on *Gumdrop*. In an hour they had closed to within thirty-two miles; there was another burn from the ascent engine, and *Spider* began to rise back toward *Gumdrop*.

Thirty-three minutes after the last burn, *Spider* was sitting next to *Gumdrop*. Dave flew the CSM around the ascent stage, then lined up for Jim to do the docking.

That turned out to be surprisingly difficult. The sun was shining right through the window over Jim's head—the window he was using to look at *Gumdrop*. He got *Spider* close enough, however. Dave punched the button on the docking adaptor, and the two spacecraft were locked up again. "That wasn't a docking," Jim told us. "That was an eye test."

The two spacecraft had operated separately for six hours and twenty minutes.

Jim, Dave, and Rusty came home on day ten, and we were one giant step closer to going to the moon.

Nine had been the D mission. It had worked so well that no one saw the need for the E mission, the high earth orbit test of the lunar module. The next step was the F mission, a lunar orbit test of the lunar module, and that became Apollo Ten. We would keep the CSM in orbit around the moon for thirty-one orbits this time, three times as long as 8. It was going to be a full dress rehearsal of a lunar landing mission without actually going down to the surface.

There was some discussion of having Tom Stafford and Gene Cernan do the landing. I don't think they would have objected, but even though things had gone well, we weren't quite ready to skip the F mission. (Back in 1967, for example, everybody had agreed on half a dozen different docking modes that absolutely had to be demonstrated before a landing could take place. So far we had demonstrated exactly one. There were also tremendous differences in the tracking and telemetry capabilities between earth orbit and lunar orbit.)

There was a practical problem, too: the lunar module for 10, lunar module number four, wasn't capable of landing on the moon and getting off again because it was a few pounds overweight. The effort hadn't been made to get those pounds off the module, because number four had always been aimed for an orbital flight test. The option, then, was to postpone Apollo 10 for a couple of months until lunar module number five was ready. When you added up what we would gain as opposed to what we would lose, the decision was pretty clear, and Sam Phillips was the one who made it: fly 10 as the F mission.

The only other decision to make was what to call the CSM and LM on 10. Tom and his crew came up with *Charlie Brown* and *Snoopy*, from the "Peanuts" comic strip.

Julian Scheer hated those names, too. His people won the battle, but we lost the war. From now on the spacecraft names were going to be dignified, goddamn it.

Tom Paine officially became the new NASA administrator on April 3. He was a Democrat, and Republican Richard Nixon had become president in January. No one knew what that meant. We had never paid a lot of attention to party politics in the space program until then. Nixon seemed supportive at first, that's all I knew. During the spring he set up a task force to do some serious long-range planning for manned spaceflight.

Following 9 it was time to name the crews for 12. The prime crew was an easy choice, right from the rotation: Pete Conrad as commander, Dick Gordon as command module pilot, and Al Bean as lunar module pilot.

The backup crew would be headed toward a downstream mission, Apollo 15. I didn't think Jim McDivitt needed to go through a backup job again, and was thinking of him for Apollo 13 at that point. Besides, I had always thought Dave Scott would be a good crew commander, and after two missions he was ready. He was a logical choice for commander of the 12 backup crew, and so was Al Worden as command module pilot. (Al had worked on the support crew for 9.) Rusty Schweickart would have been a logical lunar module pilot, but that bout of sickness had everybody worried—including Rusty, I think. It didn't seem like a good idea to put him back in as an lunar module pilot at this point. I moved him off to Apollo Applications.

Jim Irwin was the next lunar module pilot in line after Haise and Mitchell and was currently working on the 10 support crew, so I put him with Dave and Al as backup for 12.

P. J. Weitz, Ed Gibson, and Jerry Carr became the support crew. With luck, 12 would be the second manned lunar landing, sometime in October or November 1969.

The crews for 12 were announced on April 10.

Apollo 10 launched on Sunday, May 18, 1969, with a smoother ride than the other Saturn Vs. We had developed a way to deal with the pogo effect in the S-II by shutting down the central engine (of five) for the last minute and a half of the burn, but it didn't seem to help. Tom, John, and Gene had a rough ride, and even rougher one on the S-IVB. John even asked Charlie Duke, the capcom, "Are you sure we didn't lose *Snoopy* on that staging?"

The burn for translunar injection was also rough enough that there was some thought about cutting it short. But the crew rode it out.

One new item on the 10 flight was a color camera, and these guys gave it a workout, showing the transposition and docking between *Charlie Brown* and *Snoopy*, among other things.

The coast toward the moon went fine, and on day four, 10 burned into lunar orbit. One of the crew's jobs was to do an eyeball survey of some of the proposed landing sites for upcoming missions—including the one chosen for 11.

There was some concern that first day about *Snoopy*. The crew hadn't gone aboard until they were in lunar orbit, and as soon as Gene Cernan took apart the probe and drogue, a bunch of Mylar pieces blew into the command module, probably little bits of insulation. It was one of the things that slowed down Gene's checkout a bit, but we decided to go ahead with the separation and rendezvous the next day.

On day four, revolution eleven, as Gene and Tom were doing their final checkout, we discovered a little problem with the way the two vehicles were docked. It developed that *Snoopy* was out of alignment, as if *Charlie Brown* had been rolling slightly at the time of docking. Would we damage the latches by undocking? Nobody knew for sure. Glynn Lunney, the flight director, got advice from George Low, who was in MOCR for the event, and the advice was to go ahead with the undocking.

At that point the two spacecraft swung around the moon, out of communication with us. When they reappeared thirty minutes later, they had separated. There were more systems checks as they passed across the visible face of the moon, then they went out of contact again. When they reappeared the second time, *Snoopy* had already made its first powered descent burn. (Given that our ideal landing site was on the eastern limb of the moon, as seen from the earth, the descent burn had to be made on the far side.)

As John sailed on alone in *Charlie Brown*, Tom and Gene dropped lower and lower, heading toward the surface. "We's down among 'em, Charlie," Gene radioed to Charlie Duke. Tom started seeing mountaintops and even boulders from an altitude that was more suited to flying an airplane than a spacecraft. He described it as looking "very smooth, like wet clay, like a dry riverbed in New Mexico or Arizona."

Their closest approach to the surface was forty-seven thousand feet. In a landing mission, this would be the point at which the DPS on the lunar module would fire again, taking you all the way down to the ground. On 10 Tom Stafford was supposed to fire the DPS, but only to reshape the orbit so he and Gene would have another pass over the site. That burn went fine, though the shock of ignition set off a bunch of warning signals in *Snoopy*'s cockpit.

One thing that got a good wringing out was the communications setup. All during these maneuvers we kept finding we had antennas pointing the wrong way. You could hear Tom and Gene trying to talk to John, John having to ask us what they were saying, that kind of thing. It wasn't just voice communications—though those were important—but telemetry and tracking. We really needed to know where the lunar module was and how it was doing if we were going to put it down on the moon.

At the low point of the second pass Tom and Gene were supposed to drop the descent stage and fire the ascent engine. At separation Tom was supposed to give a little burst on the RCS motors to make sure he got away from the stage.

Right at that point the ascent stage went into a spin and even pitched up and down. "Son of a bitch," Gene said, "something is wrong with the gyro."

It turned out that the crew had left the switch for the abort guidance system (AGS) in "automatic" rather than "altitude hold." As soon as the AGS realized that the descent stage was gone, it automatically began flipping the ascent stage around looking for the command module so it could lock on the rendezvous radar.

A couple of seconds later Tom had it under control. "Something went wild there, but we're all set." It was pretty nerve-wracking. We could hear the tension in Gene's voice as he counted Tom down to the ascent burn.

Since *Snoopy* wouldn't be taking off from the lunar surface, it didn't carry a full load of propellant. (We wanted it to maneuver like a lander when it made rendezvous with the command module again.) The burn was just fifteen seconds, enough to reshape *Snoopy*'s orbit so it would wind up below and behind *Charlie Brown*. A few other tweaks and they were station-keeping.

A good day.

Ten splashed down on Monday the twenty-sixth, having looked pretty smooth. We went through the debrief of the crew, and the operations guys looked over their data. Two weeks later Sam Phillips announced that Apollo 11 would be launched on July 16, and that it would attempt a manned landing on the moon.

22

LANDING

It wasn't until 10 splashed down that everyone's attention turned to 11 as the first possible manned lunar landing, but we had operated with that understanding since early January. Neil and Buzz were spending a lot of time in the lunar module simulator while Mike was concentrating on the command module. With 9 and 10 having priority on those same simulators until mid-May, there was quite a traffic jam. And with all the attention we were starting to get, it wasn't the smoothest training cycle we ever had, either.

For example, up until about November 1968 no one really considered the possibility of having both astronauts leave the lunar module at the same time on that first mission. Sam Phillips, for one, really thought that one guy should get out, grab some rocks, and hustle back inside the lunar module.

But way back in 1965 the scientific community had begun development of an instrument package called ALSEP which was designed to be set up on the surface and left there. Some people were pressing for an ALSEP deploy on the very first landing. In August 1968 Don Lind and Jack Schmitt did some simulated EVA work and discovered it just wasn't possible for one guy to set it up. If you had an ALSEP on 11—and George Low and George Mueller thought we should have it—then two guys needed to do the EVA.

Okay. Then it suddenly became a question of who goes first, the commander or the lunar module pilot? On Gemini missions the EVA was always done by the pilot, not the commander, and some early paperwork showed the same thing for Apollo 11.

But by early 1969, the plans said the commander should exit first, making Neil Armstrong the first one to step on the moon.

Buzz Aldrin's father, Gene, Senior, got into the act somewhere about this point. He was one of those aviation pioneers—he knew Orville Wright and Charles Lindbergh—who just couldn't seem to leave well enough alone. From the moment Buzz joined NASA, his old man was trying to pull strings to get him assigned to a flight . . . then get him promoted to lieutenant colonel. Now he wanted to know why his son was getting the shaft about being the first guy to walk on the moon. I guess he got it in his head that Neil was chosen because he was a civilian.

Buzz raised the question with Neil, who was noncommittal. (Nothing had been set in stone yet.) So Buzz wound up going all the way to George Low, who told him the decision wouldn't have anything to do with who was civilian or military.

It bounced back to me, and I told Buzz I thought it should be Neil on seniority. I felt pretty strongly that the ones who had been with the program the longest deserved first crack at the goodies. Had Gus been alive, as a Mercury astronaut he would have taken the step. Neil had come into the program in 1962, a year ahead of Buzz, so he had first choice.

There was also a technical reason. It turns out that the lunar module was configured differently than the Gemini, in which you had two equally usable hatches. There was just the one hatch on the front of the lunar module, and the way it opened made it easier for the guy on the left, the commander, to get out first. Otherwise you'd have two guys in bulky pressure suits doing some kind of goddamn dance inside the lunar module.

So, as if we didn't have enough to worry about, a few hours got expended on that issue. That was resolved in April, even before 10 flew.

I can't say I saw too many other outbursts between Neil and Buzz. Given the incredible stress they were under, they held up pretty well. In a lot of ways, Mike Collins had it tougher: his training pretty much required him to be alone. If something went wrong, he didn't have anybody else to blame or complain to.

Then there was the "lunar germ" problem. There were some scientists who worried that Neil and Buzz—and Mike—once they were back in the command module, might come home infected with some kind of space virus. Suppose it got loose? The natural response was to come up with some kind of isolation for the crew when they got back, so they could be tested before coming in contact with the world at large.

Of course, there was an immediate problem with some of our operational needs: there was no way I was going to leave the crew inside the Apollo until it got aboard the carrier. An Apollo command module made a great spacecraft, but it was a lousy boat. It usually tipped over on landing (the center of gravity was higher than you'd find in a boat) and had to be righted with some inflatable bags. This was guaranteed to make the crew sick, if they weren't already having problems bobbing around on the ocean after having been in space for a week or more.

So the quarantine was in trouble the moment the rescue guys reached the control module. The solution was to have them douse the hatch area with disinfectant and throw three biological insulation garments—BIGs—in to the crew. Once they had those on, they could open the hatch and get out.

Wouldn't that open hatch allow the lunar germs to escape into the atmosphere? Possibly. The plan was that the atmospheric pressure inside the command module would be lower than outside, so air and bugs would presumably flow *into* the module rather than *out*.

Anyway, once the crew got back to the carrier, they would be sealed up in a kind of trailer called the Mobile Quarantine Facility for the trip back to land. They were supposed to spend three weeks in isolation in the Lunar Receiving Lab, the building where we were set up to deal with the moon rocks. (Which also had to be handled as if they were dangerous.)

Finally, there was the very important question of what to name the 11 command and lunar modules. NASA's chief PR guy, Julian Scheer, had made it clear that he wanted the crew to pick something "dignified." Jim McDivitt and Tom Stafford's crew had gotten a little playful, but Neil, Mike, and Buzz were certainly aware of the historic nature of the 11 mission, and chose accordingly. At Scheer's suggestion, the command module got named *Columbia*, the symbol of the United States and a little bow to Jules Verne, who back in the 1860s wrote about a spacecraft being blasted around the moon from a cannon in Florida called the "Columbiad."

The lunar module was going to be *Eagle*. Jim Lovell had suggested it to the crew, and it was such a good idea it became the figure on their crew patch as well.

By March 1969 Al Shepard had recovered from his surgery and passed the physical; he was requalified for a Class I ticket and eligible for assignment to a flight. Naturally he wanted to fly, and I was ready to put him into the first available job, which at that time happened to be Apollo 13. It didn't quite work out that way.

I had told the eighteen guys assigned to the first three Apollo crews that they would make the lunar landing. I had been ready to break the rotation system, but the way things had worked out, it was still in place through the spring of 1969. I didn't feel any obligation, technical or moral, to keep it in place from that point on. Thirteen looked as though it would be the third manned lunar landing, sometime in the spring of 1970. We had a year to get ready for it, as opposed to six months.

So the first thought Al and I had was to put him on Thirteen as commander with Jim McDivitt as lunar module pilot, and a new guy—probably Stu Roosa—as command module pilot. Stu had been working in the trenches as far back as Gemini and had just spent six months on the support crew of 9. But Jim was weighing a bunch of other offers at the time, including some from the Air Force. (They wanted him to come back and take over the MOL program, for example.)

And, frankly, Jim didn't think Al was ready to fly an Apollo mission especially as commander. He thought he should serve a turn on a backup crew. Further, if Jim was going to fly 13, he wanted his own crew of Dave Scott and Rusty Schweickart. Well, Dave now had his own crew and Rusty was off working on Apollo Applications. So that wasn't going to happen. Then I wanted Jim to take over the Astronaut Office now that Al was back in active training, but he turned that down, too. (Jim was careful and smart, and well thought of by everybody who worked with him. Within a few months George Low brought him in as manager of the Apollo program office at the MSC.)

I had perfect confidence in Al, so I felt safe in growing a whole new crew for Thirteen, with Al as commander and Roosa as command module pilot. Ed Mitchell was the logical choice for lunar module pilot. He was very capable and had done a good job backing up 10.

The two who got sort of left out in the shuffle were Donn Eisele, who I had projected as going to Apollo Applications from the very beginning . . . and Gordo.

The problem was, I hadn't had further plans for Gordo after Gemini 5. It wasn't just that I lost faith in him—I don't think I really did. But he was a hard sell to management because he never really put his heart into training and seemed to spend a lot of his time being more interested in racing cars. He had come in at the last minute to back up Gemini 12 for me, and I'd been able to turn to him to back up Apollo 10. At that point there was an outside chance he could have wound up with 13. But he hadn't done a very good job as backup commander—not good enough to change anyone's mind. So it was time for him to move on.

When Gordo found out that 13 was going to be Al's flight, he really bitched about it. He said he was younger, in better physical shape, had flown more, had been in constant training while Al was in an administrative job. All of that was true, but not much of it was relevant. He never confronted me. It wasn't like we had a fistfight or anything. In fact, I wound up bringing him into my office as one of my assistants for a year, until he retired from the Air Force and left NASA.

I had another unusual thing happen on the 13 crew assignment. For backups I wanted John Young as commander, Jack Swigert as command module pilot, and Gene Cernan as lunar module pilot. (Since Jim McDivitt had turned down the chief astronaut job, Tom Stafford was going to take it.)

When I told Gene about the assignment, he said he didn't want it. Didn't want to fly as lunar module pilot. "Let me get this straight," I told him. "You're turning down a chance to walk on the moon."

"I'd love to walk on the moon," he said. "But I want to do it from the left seat. Nothing against John, but I think I deserve a chance to command my own mission."

All I could tell him was that I would see what I could do . . . but it didn't look good. I assigned Charlie Duke in Gene's place.

I submitted the crews for 13 up the chain to headquarters—Shepard-Roosa-Mitchell backed up by Young-Swigert-Duke. And got turned down.

George Mueller wouldn't approve Al as commander of 13. He didn't think there was time for him to get ready. I think he boxed himself in by saying so publicly, and didn't want to have to defend the assignment. In any case, his word was law. Al would not fly Apollo 13.

I had a possible solution, though. Jim Lovell was backing up Apollo 11, still scheduled for launch, with Bill Anders and Fred Haise. Bill had already accepted a position with the National Space Council, effective in August, so he wasn't going to rotate to another crew. But he had given me notice, so I already had Ken Mattingly working as a parallel backup with him.

Both Lovell and Haise had been through the cycle twice. There was no real reason they couldn't be ready to fly 13—it would be a little tough on them, but not impossible. So I asked Jim if he would like that mission. He was happy to do it.

So I now had crews for 13 that looked like this:

Prime, Lovell-Mattingly-Haise
Backup, Young-Swigert-Duke

Since they would have an additional four months to train, I could put Shepard-Roosa-Mitchell on 14. All I needed was a backup crew, which could then point at Apollo 17.

My first choice for backup commander was Mike Collins. With a little luck, he was going to be command module pilot for the first lunar landing. By backing up 14 he would put himself in line to walk on the moon.

I brought up the idea with him on a T-38 flight to the Cape. He thanked me for the offer, but said that if Apollo 11 went well, he wanted to get out of the astronaut business. He was tired of the grind.

Fair enough. So I made Gene Cernan backup commander of Apollo 14, with Ron Evans as command module pilot and Joe Engle as lunar module pilot. Gene's little gamble had paid off for him. None of these crews was carved in stone, however, and we deliberately held off any public announcement.

On June 12 we were scheduled for a flight readiness review for 11. This was basically a four-way conference call between Houston, Marshall, the Cape, and headquarters to see where we were with all the elements of a mission, from the booster to tracking networks to weather to trajectory to crew training. The week before, we'd had a series of tests with the vehicle at the Cape, including a countdown with the crew. The proposed launch date—again, dictated by our requirements to have the proper lighting at the prime landing site on the Sea of Tranquillity—was the morning of Wednesday, July 16, 1969.

I was at the Cape with Rock Petrone. Sam Phillips was on in Washington. George Low and Gene Kranz (the lead flight director) were with Chuck Berry at MSC. Lee James, the Saturn V manager, was at Marshall. Plus a few dozen staffers listening in everywhere.

Phillips opened by telling us he was ready to push 11 into August, if necessary. The reports showed that the equipment and flight controllers were going to be ready. It was really a question about the crew.

Chuck Berry was concerned about how they were holding up, and I was a little ambivalent: Neil, Mike, and Buzz were really pushing to be ready, and during the past few days I had wondered if they would be. In fact, I had asked them if another month would be better. They said no. They wanted to get on with it, and if Neil, as commander, said they would be ready by mid-July, I was prepared to take his word. So as far as I was concerned, if the equipment was ready, the crew was ready.

That was basically it. We would launch at 9:32 A.M., Eastern Daylight Time, on July 16 using the sixth Saturn V with command and service module 107; *Eagle*, lunar module number five, would land July 20 in the Sea of Tranquillity. Neil and Buzz would walk on the moon a few hours later.

During that last month, as everybody realized we were really going to try this, a bunch of other things came up. For one thing, everybody and his brother wanted to be at the launch. Senators, congressmen, ambassadors.

No matter how we isolated the crew, even they were having to deal with phone calls from long-lost relatives. The press was everywhere. (In this case the preflight isolation, which we'd instituted on 10 because the crews were run-down and catching colds, really helped. The crew had a perfect reason to lay low.) The Reverend Ralph Abernathy was planning to lead a protest at the Cape because money was being spent on spaceflight and not fighting poverty.

President Nixon wanted to have dinner with the crew the night before liftoff. Chuck Berry wondered out loud if that might not break the quarantine. Suppose the crew caught cold and it was the president's fault?

It didn't make a hell of a lot of sense, because while the crew was living in relative isolation, every day they were dealing with a bunch of people who could have been carrying anything. Nobody was walking around with white masks.

But once the dinner was questioned publicly, it was off. Chuck was never particularly popular with some of the guys—he tended to give the idea he was the personal physician to the astronauts, though he was actually the head of the medical office at MSC with a few dozen doctors under him. This did nothing to improve on his reputation.

Tom Paine did come down for dinner the week before the launch. He told Neil, Mike, and Buzz that if they got into trouble, especially during the landing attempt, they should save themselves and come home, they would have another chance on the next mission. He didn't want them taking an unnecessary chance just to make that first landing.

The final medical checks were on Friday, July 11, by the crew's actual physicians, Al Harter, Jack Teegan, and Bill Carpentier. If everything went right, Carpentier would be in postflight quarantine with the crew.

Everybody was fine. Over the weekend the crew just basically relaxed. I flew some acrobatics with Mike Collins in a T-38 out of Patrick Air Force Base to help with inner ear conditioning. There was going to be a last press conference on Monday the fourteenth.

Sunday morning, however, we woke up to the news that the Russians had launched a spacecraft toward the moon.

It was called Luna 15 and it was supposed to go into orbit around the moon on the sixteenth. Meaning it would be in orbit when Apollo 11 arrived. Naturally, we were concerned that it might get in the way—not that it would be bumping into 11, but that there might be radio interference.

I did think it was pretty desperate for the Russians to try something like this in that particular week. I didn't know then just how desperate they were for some kind of space spectacular. The manned lunar orbit program was on hold, and would get canceled before the end of the year.

And on July Fourth, they had tried for the second time to launch their big N-1 booster. This one had failed worse than the first one, blowing up a few feet off the pad and effectively destroying not only itself, but the launch complex. Even before the accident, the Russian manned lunar landing program didn't really have a chance of beating us, unless something bad happened with 11. The "moon race" had been won in December 1968 by Apollo 8, though we didn't know that for several years.

Luna 15 was designed to land on the moon, take some lunar samples, and come back before Apollo 11. It was an attempt to upstage us, pure and simple.

It turned out that Frank Borman had just come back from Russia that weekend. He had made a three-week trip at the invitation of the Soviet Ambassador to the United States, Anatoly Dobrinin. (Since flying 8, Frank had been spending a lot of time in Washington, working with the Nixon White House, and making goodwill tours.)

Chris Kraft got hold of him at home while he was still in Washington on Sunday afternoon, and asked him to look into Luna 15. Frank got on the hotline and placed a call to Mstislav Keldysh, president of the Soviet Academy of Sciences, one of their most visible space people. (A good choice: spacecraft trajectories came from one of the academy's institutes.) It was two in the morning Moscow time, so there was nobody at Keldysh's office to give an answer.

Frank flew back to Houston, then got awakened at six in the morning by a call back from Keldysh's deputy. He didn't have the precise figures, but promised to get them. Sure enough, a few hours later a pair of telegrams were received, one at the White House and one at Frank's house, with all the data. It showed that Luna 15's orbit wouldn't intersect 11's at all.

What was really interesting was that it was one of the fastest, and most significant, exchanges between the two space programs. I think it was one of the first real steps toward Apollo-Soyuz.

Tuesday night, July 15, there was just a quiet dinner in the crew quarters, Neil, Buzz and Mike, with their backups—Jim Lovell, Bill Anders, and Fred Haise—and the support guys, and me. No President Nixon. It was a pretty quiet affair: everybody had to be up early.

Wednesday morning, July 16, 1969, four A.M., I knocked on the bedroom doors of Neil, Mike, and Buzz. They showered and got dressed, then met up with Dee O'Hara for the last medical checks. Breakfast was at five, just the crew, me and Bill Anders. There's always something special about a launch day breakfast. Nobody talks about the work, really, but everybody knows what's coming. Especially this particular day. I think the crew was basically just plain eager to get going.

They got suited up—Neil and Buzz in their fifty-pound lunar EVA suits, Mike in a lighter version, since he didn't need to worry about getting around on the moon. While they were doing that, I went out and told the reporters to get ready. Then I walked the crew out to the van for the ride to the pad. As usual, I dropped off at the firing room, wishing them good luck. Up at the white room they'd be in the hands of Guenter Wendt, the pad crew leader all through Mercury and Gemini. Guenter had been a McDonnell Douglas employee. After the Apollo fire, I had helped to get him transferred to North American Rockwell.

There were something like a quarter of a million people scattered on the beaches, not to mention a few thousand at the press site. The VIPs included LBJ and Jim Webb. Inside the blockhouse was Wernher von Braun with Sam Phillips and George Mueller. (It was Mueller's birthday, in fact.)

I plugged in and sat down, just trying to keep out of the way. More than any other flight before that, things seemed to go smoothly, as if everybody had been pointed at this one moment.

Right on time, at 9:32 in the morning, the big Saturn lit up. Like always, it seemed to take forever to actually get off the ground. But then it just rumbled off into the sky. I think most of us in the blockhouse felt like we were lifting it all by ourselves.

I flew back to Houston to take up my usual position in the MOCR. (I made sure I was there for any critical maneuvers, especially burning into lunar orbit.) It was a pretty smooth flight to the moon—nobody got sick, there were no big technical problems. Seven months earlier we had sweated translunar injection with Borman's crew. Then we worried ourselves silly about flying the lunar module with McDivitt. It had only been two months since we were sweating out Tom Stafford and Gene Cernan in the lunar module around the moon. Because we had done it all once, everybody was focused on the new challenge: going that last fifty thousand feet to the surface, and getting back. Owen Maynard's stepping-stone plan really worked.

Luna 15 provided the only bit of drama. While 11 was still climbing out of earth orbit on Thursday the seventeenth, Luna 15 went into orbit around the moon. We still didn't know what it was supposed to do, but it wasn't going to be in our way.

On Sunday, July 20, 1969, I got over to the Manned Spacecraft Center as early as I could.

Gene Kranz took over as flight director from Glynn Lunney at eight A.M. It seemed like everybody in the world was there, from Tom Paine on down. They had to watch from the viewing room: I was at a console with Mueller, Kraft, Low, and Phillips. We hoped all we would have to do was watch.

Next to me at the console was Charlie Duke, the capcom; he would be the voice link to Neil, Mike, and Buzz. He had worked the lunar module maneuvers on 10, and Neil had personally requested him. He was a sharp young guy, one I had already plugged in to a future crew, but today he was a little nervous, too.

The crew was already up—had been for a couple of hours—and checking out the lunar module. Neil and Buzz got suited up, floated aboard, and buttoned up. They separated, as planned, behind the moon. When they came around, they reported they were undocked.

"How does it look?" Charlie Duke said.

"*Eagle* has wings," Neil answered.

Mike fired thrusters on *Columbia* for eight seconds, to move away. Kranz's team ran through its system of checks for the next forty-odd minutes. Just before *Eagle* and *Columbia*—now separate—went behind the moon again, Charlie Duke was able to tell them they were "Go for DOI," descent orbit insertion. It would take place on the far side.

Again, to this point we were doing things we had proven we could do. Still, that period with no contact was nerve-wracking. Gene Kranz ordered the MOCR doors locked. Everybody who was in was staying in; if you were outside, tough. He delivered a little pep talk to his crew, reminding them it was just like all the simulations. After the landing they would go out and have a beer.

It was two in the afternoon.

If the DOI burn went as planned, *Columbia* would be back in contact first. Right on schedule we got it, Mike Collins telling us, "Everything is going just swimmingly." A minute later we locked on to *Eagle*, but there was a communications glitch. For a few minutes data went back and forth using Mike as the middleman, then things cleared up.

Neil and Buzz were dropping toward the moon, toward a low point of fifty thousand feet, almost ten miles above the surface, which was all the closer Tom Stafford and Gene Cernan had gotten. The next item was PDI—powered descent initiation—the start of a whole new deal.

Neil and Buzz had been flying heads down to the moon. Now *Eagle* rotated so they were on their backs, looking up at the sky. That was so *Eagle*'s radar could aim at the surface. At this point it was lot more important to know how much altitude they had than to see some rocks. In a couple of minutes, of course, they would be vertical.

The "go" for PDI had had to be relayed through Mike. *Eagle's* RCS thrusters fired for a few seconds, to settle the fuel in the tanks. Then the DPS lit up, first at ten percent. (It was so slight at first that the crew didn't realize it was firing.) At twenty-six seconds the DPS throttled up to full power and stayed that way for two minutes of the planned five-minute descent.

Two minutes in, Neil began to yaw *Eagle*—tipping it up so it was eventually facing forward. Lower and lower. There were some data dropouts, not enough to stop us. *Eagle's* landing radar went into operation. First it saw that *Eagle* was a little higher than planned, and fed that to the computer.

Then the computer had a problem. "Twelve-oh-two alarm," Neil radioed.

What this meant was that the computer memory was overloaded. It was not only getting data from the landing radar, it was also processing the rendezvous radar. A "12-02" was the computer's way of telling us it was swamped.

Right at this time, six minutes and twenty-five seconds into the descent burn, the crew was told to throttle back to fifty-five percent thrust. They were supposed to be looking out the window at their landing site . . . but they were thinking about whether their computer was working or not.

This is where all the integrated simulations paid off, because something exactly like this had cropped up then. The rules said you had to abort, but the practice showed you could ride it out—as long as the alarm was intermittent, which meant the computer was, in essence, figuring out how to deal with the data overload. If it was a steady alarm, that was different. Kranz queried his Guidance guys, whose lead on this day happened to be a twenty-six-year-old controller named Steve Bales. Bales said, "We're go on that, Flight."

Neil and Buzz were now just three thousand feet above the Sea of Tranquillity, at what we called "high gate," basically the do-or-die point, where you either better get down safely or punch out of there. This was where Kranz polled his team, and came up with a "go" for landing.

"Understand, go for landing," Neil said. Then there was another damn alarm. "Twelve-oh-one alarm," Buzz reported. Another quick exchange between Kranz and Bales, then Charlie Duke told the crew: "Same type. We're go."

In the *Eagle*, Neil and Buzz were shifting from one computer program to another, anyway. The new one was designed to handle the last few hundred feet of landing on automatic, leaving Neil as the pilot to translate the lunar module right or left.

We were hearing Buzz now, calling out the altitude and the rate of descent. He wasn't talking for our benefit—it was for Neil's. "Lights on, down two-point-five. Forward. Forward. Good. Forty feet, down two-point-five. Picking up some dust." Lights on. Picking up some dust!

This was as bad as being in combat. I did one thing I felt bad about later. Charlie Duke was talking too much when, really, the only thing that mattered now was what Buzz said. So I reached over and smacked him on the arm and said, "Shut up!"

The only peep out of him for the next minute was the warning of "Thirty seconds" to the crew, meaning that's all the fuel they had left.

They were probably less than thirty feet high. Buzz: "Forward . . . drifting right. Contact light." The probes on the feet of the lunar module had hit dirt! "Okay, engine stop. ACA out of detent, modes control both auto. Descent engine command override off, engine arm off. Four thirteen is in." Switches that had to be thrown.

Charlie Duke said, "We copy you down, *Eagle*."

Neil said, "Houston, Tranquillity Base here. The *Eagle* has landed."

Gene Kranz told his troops *Eagle* was okay for "T1," a stay of at least one minute. A good thing, too. The whole MOCR was pandemonium. It took about fifteen seconds to calm down at all. Then it was "T2," another minute.

At the twelve-minute point, unless there was a real emergency, there was no point in having them lift off, because it might take *Columbia* too long to find them. Mike Collins passed by right about this time. He had heard the whole thing. "Good show!"

Neil was saying, "That may have seemed like very long final phase. The AUTO was taking us right into a football-field-sized crater, with a large number of big boulders and rocks for about one or two crater diameters around us." He'd had to steer away from the projected landing site, which is why they'd gotten so close to running out of fuel.

The original time line had called for the crew to rest after landing, making their EVA seven hours after the landing—that is, early on the morning of July 21. I don't think Neil and Buzz ever had any notion of following that part of the plan, and no one blamed them. (It was planned that way because nobody wanted to force the crew into a stressful situation. Getting ready for landing, doing the landing, getting ready for EVA, doing an EVA, all without some kind of break—that was impossible on paper.)

Gene Kranz's White Team had turned over control of the flight to Milt Windler's Black Team at four in the afternoon. When Neil and Buzz asked if they could skip the nap and move up the EVA, Chris Kraft approved, and Cliff Charlesworth's Green Team was brought in early. (They had trained to cover the EVA.)

Six hours after landing, Neil and Buzz were buttoned up in their suits, and the front hatch of the *Eagle* was open. Getting the pressure bled out of the lunar module took longer than planned—a lot of this work took longer than planned, because we'd never had a chance to simulate all of it in the right conditions.

Eventually, though, the TV camera on *Eagle*'s exterior opened up, and there was a fuzzy black and white picture of Neil coming down.

His first stop was in one of the lunar module footpads, checking to see if there were any problems getting back up to the last rung of the ladder. There weren't. "I'm going to step off the LM now."

I watched him move. Then: "That's one small step for a man, one giant leap for mankind." Neil said later he hadn't even decided what he was going to say until just before he went out. That was typical Neil: wait until the last possible moment, then make the right decision.

"The surface is fine and powdery. I can pick it up loosely with my toe." He moved around some more and found it was easier than the simulations.

Then it was Buzz's turn. He came down more quickly than Neil. His first words were, "Magnificent desolation."

Off they went, setting up experiments, collecting rock samples, erecting the flag. They talked with President Nixon and read a plaque mounted on the lunar module descent stage.

A little over two hours later, they were back in *Eagle*.

Mike Collins had the worst possible view of this. He was alone in *Columbia*, circling the moon every seventy minutes. One of his jobs was trying to find Eagle on the surface, and he wasn't having much luck.

I went home to catch some sleep while the crew rested. Everything had gone great to this point, but I always remembered that JFK's goal was to land a man on the moon and *return him safely to earth*.

The guys in MOCR were still sweating a couple of problems that could affect the return. The computer alarms that had cropped up during descent would be even more likely—and more damaging—during ascent. (Tracking *Columbia* for rendezvous required more data than tracking the moon.) A guy at MIT named George Silver, who had worked on the software, got briefed on the problem of the computer going into overflow. He had seen something like this before on 10, where the computer kept recalculating rendezvous angles every few seconds as the angles changed, even though they weren't needed during descent. It turned out the rendezvous radar was still set on "AUTO" instead of "MANUAL." That was sent up to Neil and Buzz about thirty minutes before liftoff.

Also, some of the guys monitoring the *Eagle*'s systems had found something to worry about. Apparently there was some kind of blockage in one of the fuel lines in the descent stage, probably a chunk of frozen helium. The blockage itself wasn't a problem because we weren't going to be firing the descent engine again, but it could cause an explosion all by itself. It turned out that the chunk just melted.

Neil had also managed to clip off a circuit breaker switch inside the *Eagle* cabin while moving around in his suit and PLSS. That particular circuit breaker happened to arm the ascent engine for firing. There were several ways around this, of course, but the best one was what Neil and Buzz came up with: they jammed a felt-tip pen in there.

While Neil and Buzz were trying to get some rest between the EVA and liftoff—and they never got much, really; it was too cold in the lunar module—Luna 15 was commanded to make a de-orbit burn. Instead of landing on the moon, however, it crashed.

At 12:55 P.M. on Monday, July 21, I was in the MOCR again, listening to Buzz do the countdown. "Nine, eight, seven, six, five—abort stage, engine arm ascent, proceed. . . . That was beautiful!" Up they went. "Very smooth, very quiet ride."

"We're going right down U.S. One," Neil said. The ascent burn lasted six minutes, twenty-five seconds.

They still had a rendezvous and docking to make. We had gotten pretty casual about that in the three years since Neil Armstrong and Dave Scott had found themselves tumbling all over the sky.

Once Neil, Mike, and Buzz had burned out of lunar orbit and were on their way home, I got on the line to them. "This is the original capcom. Congratulations on an outstanding job. . . . Don't fraternize with any of those bugs en route, except for the *Hornet*."

Neil said, "Thank you, boss."

Three days later they were in the Pacific.

I don't know if it was, as President Nixon said, "the greatest week in history since the Creation," but it was pretty damned great.

23

CHANGES

Once the splashdown party ended—and I think it went on for two weeks—there was a definite letdown around the MSC. It was especially obvious over the next few months. Not that anybody quit on the job, just that the drive to make the landing was gone. A lot of people started looking around at their personal lives and were pretty appalled. Kids had been growing up without them, their wives had been left on their own. I was one of those who had to reassess.

I was still happy to do my job, especially since it allowed me to fly jets, but I was forty-five years old. The odds of my getting an Apollo flight weren't getting any better.

A bunch of changes started taking place in management beginning with 11. Sam Phillips made plans to return to the Air Force. That was going to leave the Apollo program manager job open at HQ, so Rocco Petrone moved up there from the Cape.

George Mueller was making plans to leave as associate administrator for manned flight. George Low, the Apollo manager in Houston, was going to move to headquarters as an assistant to Tom Paine.

This was a perfect opening for Jim McDivitt, who had a tremendous amount of hands-on experience with the Apollo systems, and who everybody thought was extremely capable. He became Apollo program manager at MSC. It also qualified him for promotion to brigadier general—pretty good for a guy who had thought his military peers were passing him by.

The day-to-day business of being director of flight crew operations went on. On August 2, Ed Gibson, one of the scientist-astronauts selected in 1965, wracked up a helicopter. He was out on the mudflats north of Ellington near La Porte practicing landings when he set down on what he thought was solid dirt. It turned out to be a thin crust of dried dirt over a several feet of mud. The chopper just got wrapped up in that stuff; the rotor clipped off the tail. Ed was lucky to live through it.

But the real problem had just started: he was a hundred yards from dry land in the middle of what was essentially quicksand. He had to slog his way through that stuff. The manager at some oil company site on the edge of this threw down a ladder to him. He made it just in time; Ed had been burning up his adrenaline the whole time and was just about to pass out.

That mishap was a combination of pilot inexperience—Ed had only been through helicopter training at Pensacola a few months earlier and had logged something like eight hours of solo flying time—and managerial screw-up. Again, here was a situation we hadn't anticipated. We figured it was all right to let pilots take these helicopters out for practice, assuming they would use good judgment in where they landed them. From that point on we had stricter operating rules.

The accident didn't affect Ed's astronaut career in any real way: it was pretty clear to me by August 1969 that there weren't going to be many scientist-astronauts landing on the moon, anyway.

It was right around that time that we could make a stab at some interim long-term planning. Interim in the sense that we had a finite number of Saturn Vs, command modules, and lunar modules under construction. Other parts of NASA were doing studies for vehicles that would ideally follow Apollo in the mid-1970s—a twelve-man space station, a scaled-up Gemini transport, even a winged space shuttle. There was also a presidential task force studying this stuff. None of it was funded, however; the only kinds of earth orbital and lunar landing missions we were going to be flying in the next five years were Apollo-type.

The initial follow-on landings were called H missions, which had originally been conceived as backups for a first landing and were designed for short stays on the surface. There were three H missions on the schedule for 1970—13, 14, and 15, to be launched at roughly three month intervals.

We didn't have an I class, because of the inevitable confusion, but there was a J class, which would allow a crew to spend up to three days on the lunar surface and get around with a lunar rover. As of late summer 1969 there were five J-class missions—Apollos 16 through 20—scheduled to fly, two each in 1971 and 1972, with one in 1973.

George Mueller wasn't happy with the schedule because of the big gaps between the later missions. He figured it was difficult to keep ground crews and flight teams sharp. On the other hand, the scientific community needed time between flights to analyze data and make adjustments in the experiment package for subsequent flights. So there was one potential source of conflict.

Apollo Applications, the orbital workshop, was then scheduled to fly in late 1971 or early 1972, between or around the pairs of J missions. Until early 1969 the plan had been to convert an S-IVB upper stage, launched unmanned aboard a Saturn IB, into a habitable vehicle in orbit. This was beginning to look not only difficult, but impossible—at least if you wanted to get any science done during the manned visits.

Wernher von Braun and his people proposed instead that we should modify and fully equip an S-IVB on the ground and launch that instead, so crews would have an honest-to-God space station to work in. This would also be a better stepping-stone to the twelve-man job.

Fine, but in order to launch an S-IVB that way, you had to use a Saturn V—which meant you had to take one from the landing missions. Tom Paine officially approved the idea while 11 was flying, on July 18. It took until the following January for us to announce that Apollo 20 was off the books, but it was actually gone in July. It was just the first of a bunch of cutbacks we had to start dealing with.

Naturally, right about that time I wound up with a whole new group of astronauts.

Twice during the 1960s the Air Force had tried to get its own manned space programs going. First was the X-20 Dyna Soar, which was being talked about even before I left Edwards in 1959. Five Air Force and three NASA guys (including Neil Armstrong, for a while) had been selected for that, all of them test pilots at Edwards. The Air Force guys had been smart enough to keep working on other programs part-time during their years on the X-20. When it was canceled in December 1963, most of them were able to move on to other Air Force jobs.

Beginning in 1965 the Air Force selected three groups of what they called aerospace research pilots for the Manned Orbiting Laboratory program. This was a small space station that was going to be put into polar orbit from Vandenberg Air Force Base. On top of the MOL itself would be a Gemini capsule with a crew of two. Once in orbit, the pilots would open a hatch in the Gemini heat shield, then crawl into the lab for thirty days of experiments.

Actually, it was going to be thirty days of taking pictures and spying. MOL was mostly a manned military reconaissance platform. (That's why it was in polar orbit, where you passed over all of the earth's surface in the space of twenty-four hours.)

Anyway, over three years they got seventeen pilots into MOL, mostly Air Force, but also a couple of Navy guys and a Marine. Flights were originally supposed to start in 1968, then kept getting pushed back because of budget and design problems.

In the spring of 1969 it looked as though all the technical problems had been licked, but flights were still three years off. Suddenly—I think it was in the course of one week in early June—MOL got canceled.

The story was that President Nixon himself had killed it for financial reasons. (MOL had cost over $1 billion by that point.) NASA had the Apollo Applications project going. What did the country need two space stations for?

There were other considerations, however. Unmanned spy satellites had matured so much since 1965 that they had simply overtaken MOL; the intelligence community had unmanned birds ready for launch that could operate for three months at a time and return film of a high quality. The real question was, why did you need people to do this spying?

So the fourteen remaining MOL pilots were suddenly out of a job. (Two guys had resigned by then and another had been killed.) And George Mueller wanted me to hire them.

I told him I didn't need them. Given the flight rates, I didn't need some of the people I already had. Bob Gilruth felt the same way.

"*I* know and *you* know that we don't really need them," Mueller told me. "But we're gonna need the Air Force one of these days, and it won't hurt to make them happy just this once."

I still didn't think it was fair to take fourteen guys like this and hire them under false pretenses. I had work for them, especially in the development of Apollo Applications, but I didn't have *flights* for them.

Mueller said to just make that clear, see if any of them still wanted to come over. I did, and I think thirteen of the fourteen were still interested. That was still too many.

So I looked at their ages and found that six of them were over the age of thirty-six, seven were under. Thirty-six was our upper age limit for pilot astronaut candidates. I said I would take the guys who met the age limit, and on August 9 we announced the "selection" of seven new astronauts:

Major Karol J. "Bo" Bobko, USAF
Lieutenant Commander Robert L. Crippen, USN
Major Charles G. "Gordon" Fullerton, USAF
Major Henry W. Hartsfield, Jr., USAF
Major Robert F. Overmyer, USMC

Major Donald H. Peterson, USAF
Lieutenant Commander Richard H. Truly, USN

All of them went on to make their first flights in the shuttle, some of them several times, but not for twelve years or so. It was a good group.

There was an eighth guy we also took at NASA, Lieutenant Colonel Al Crews. He was already forty years old and was the only MOL pilot to have been involved in the X-20 program, too. After ten years of jobs like that, his career options in the Air Force were pretty limited. He joined aircraft ops with the slim hope that he might eventually qualify for a shuttle crew. That never happened, but I think Al was happy with the move. Last time I looked, he was still flying NASA airplanes at the age of sixty-four.

What's also kind of interesting is a couple of the guys I *didn't* hire. The one who seemed to have no real interest in becoming an astronaut once MOL was gone was Robert Herres. He wound up as a four-star general and vice chairman of the Joint Chiefs of Staff under Colin Powell.

Then there was Jim Abrahamson, who eventually got three stars, headed up the shuttle program in the 1980s, and was the first director of the Strategic Defense Initiative. I ran into him in later years and he told me that, all things considered, he would have traded the stars for a flight in space. I could understand the sentiment, but I'm not sure I believed it.

The crews for 13 and 14 were announced on August 6.

13	(prime)	Lovell-Mattingly-Haise
	(backup)	Young-Swigert-Duke
	(support)	Brand-Lousma-Kerwin
14	(prime)	Shepard-Roosa-Mitchell
	(backup)	Cernan-Evans-Engle
	(support)	Chapman-McCandless-Pogue

The nine "survivors" of the 1967 scientist-astronaut group had by now completed not only flight school, but also several months of survival and parachute training. Of course, I didn't have any more flights for them than I did for the ex-MOL pilots.

I put Phil Chapman to work on the 14 support crew and assigned several others to serve as mission scientists (a support crew position) for the J Apollo flights coming up: Joe Allen, Tony England, and Bob Parker. It was great to have astronauts serving as an interface between the crew and the scientific community. But it also basically took these guys out of the running for the first (and as it turned out, only) Apollo Applications crews.

The other scientists—Holmquest, Lenoir, Musgrave, and Thornton—joined Don Lind in the Apollo Applications branch working with Walt Cunningham and Rusty Schweickart. (Lind had specialized in the lunar module and was pretty disappointed to get moved over to AAP. But with the cancellation of 20, I could see I just wasn't going to have a flight for him.) Three of the original scientist-astronauts, Joe Kerwin, Owen Garriott, and Ed Gibson, were working there, and it's where most of the MOL guys went, too.

One of the first scientist-astronauts, Curtis Michel, decided to say goodbye at this point. He had been part-time for a year, anyway, and couldn't wait around for three more years for a chance to fly.

With the lunar landing accomplished, we had been able to slip the Apollo 12 mission by a month. And make it more complicated.

In 1967 we had landed the unmanned Surveyor 3 in the Ocean of Storms, which happened to be the landing site for 12. The trajectory guys figured that they were capable of putting a lunar module within walking distance of the Surveyor. This was pretty good shooting, considering that we never actually figured out where *Eagle* was on the lunar surface until after the mission.

Not only did this call for some major league number crunching, it also required some pretty good piloting. It was a perfect mission for Pete Conrad and his crew, Dick Gordon and Al Bean. They were all Navy guys, so they came up with all-Navy names for their spacecraft: the command module would be the *Yankee Clipper* and the lunar module would be the *Intrepid*.

The morning of November 12, 1969, there was a thunderstorm at the Cape, not the most promising conditions for a launch. You can launch a Saturn in a storm; a shuttle needs better weather conditions in case it has to return to a landing at the launch site.

My new deputy, Tom Stafford, did the crew wake-up chores. I was part of the crew, including Jim McDivitt and Rocco Petrone, hanging around President Nixon, who was going to watch the launch from the firing room. I can't say Nixon's presence encouraged us to launch in bad weather conditions. The weather was within the mission rules. But it sure made us a little more reluctant to scrub for the day.

Apollo 12 took off right on time. Thirty-six seconds in static washed over communications and telemetry, drowning everything out for twenty-six seconds. The next thing we heard was Pete Conrad telling Jerry Carr, the capcom: "We just lost the platform, gang."

The platform was part of the guidance system in the command module that served as a reference point for everything else. You didn't want to "lose it"—that is, have it fail.

The Saturn was still firing away. Its guidance system was keeping it on track. But something was definitely wrong with the command module.

Then Pete came on the line again telling us that all of his fuel cells were down, that the command module was running on battery power.

The flight director was Jerry Griffin; it was his first time in that job. He polled his troops and got them looking at the problem as the big first stage burned out on schedule. Pete then reported that the fuel cells had come back on, and said, "I'm not sure we didn't get hit by lightning."

That's exactly what had happened. Lightning had struck the vehicle twice, once at six thousand feet, then at fourteen thousand. The discharges didn't do any particular physical damage, but they did momentarily overload all the electrical circuits in the vehicle. The interruption caused the platform to go down.

Once 12 was safely in orbit, there were some hairy decisions to be made. They couldn't go anywhere without the platform, which had to be realigned by star sightings. That was a time-consuming process, however, which Dick Gordon handled with seconds to spare.

Also, we wanted to be damned confident that the spacecraft was safe. We didn't know at the time how much damage this theoretical lightning strike had caused. For example, what if it had damaged the recovery system parachutes? (We did talk about that, and realized that if that had happened, the crew was dead, anyway. So why not let them land on the moon?)

We burned up the phone lines between Houston and the Cape for the next hour, but the decision was Jerry Griffin's. He had the command module systems guys going through their checklists, and except for the wild card of this bolt of electricity having run through there, everything looked fine. Griffin told Chris Kraft, his boss, that he thought 12 should go for translunar injection, and that's what they did.

Four days later it was Pete Conrad's turn to crawl down the ladder of a lunar module and step onto the surface of the moon. "Boy," he said, "that may have been a small one for Neil, but it was a long one for me." Well, Pete was a few inches shorter than Neil, and *Intrepid*'s legs hadn't compressed quite as much as *Eagle*'s had.

Looking backward from the lunar module, he said, "Guess what I see sitting on the side of the crater? The old Surveyor." It was only a couple of hundred yards away.

Pretty good shooting.

* * *

After Pete, Dick, and Al returned, I assigned their backup crew of Dave Scott, Al Worden, and Jim Irwin as the prime crew of Apollo 15. Pete Conrad and Al Bean wanted to move over to Apollo Applications—which was about to be officially named Skylab—but I still had Dick Gordon available to recycle. I made him commander of the 15 backup crew, with Vance Brand as his command module pilot.

If everything went right, this crew would fly Apollo 18, the eighth lunar landing mission. Figuring we would have a pretty good handle on the business by then, I decided to give Jack Schmitt a shot at lunar module pilot.

The support crew would be three scientist-astronauts, Joe Allen, Karl Henize, and Bob Parker.

Other Voices

THOMAS STAFFORD

From the time I went to work for Deke as chief astronaut in the summer of 1969 through early 1971, when Al Shepard came back on the job, I worked pretty directly with him. In fact, we were probably in contact three or four times a day.

Even though the landing had been accomplished, we had a hell of a lot of work to do. The lunar landing missions were getting more and more complex, and they were scheduled to take place every three or four months for the next few years. We couldn't afford to let down.

Deke didn't. We worked long days during the week, and weekends, too, all through this time, because that's what it took. We all paid a price, but we never really stopped to ask ourselves if it was worth it. We just figured it was.

Not everybody was as willing to do the job as we were. At the end of 1969 we had to fire an astronaut, Donn Eisele. He had been a pretty good worker up to Apollo 7, but then his personal life changed—he got divorced and remarried—and his professional skills just went to hell. He just lost interest in the job, and Deke and I sat him down at the end of the year and told him it was time for him to move on. And to be gone within sixty days.

Sam Phillips, who was in the process of going back to the Air Force, took pity on Donn and arranged to get him transferred to NASA Langley in some capacity while still staying on active duty. (At the time Donn still had a couple of years to go before retirement.)

It wasn't for personal reasons; John Young was in the process of getting divorced and remarried right about that time. It was just that Donn quit doing the job.

In the fall of 1969 the presidential task force headed by Vice President Spiro Agnew had published its recommendations for the future of American manned spaceflight. It was pretty ambitious stuff, calling for the development of a reusable space shuttle and the construction of a permanently manned orbiting space station. A space tug was also part of the program—this was basically a vehicle designed to operate only in space, from the station up to geosynchronous orbit, or possibly to the surface of the moon.

One of the options even called for a manned mission to Mars in the mid-1980s. I didn't believe this was particularly likely, not on the schedule the task group laid out, anyway. There was a hell of a lot we didn't know.

There was a bonus to all this long-term planning, however. Somebody realized that the closest thing on earth to a long-term space station or lunar base was a scientific installation in Antarctica. So a few people from different NASA centers got signed up for a trip down there, and I was one of the reps from the MSC, along with Bob Thompson (the Skylab program manager) and Dave Scott.

At that time—January 1970—you didn't have a lot of casual visitors to the South Pole because it was just damned difficult to get a plane in there. And it was pretty clear to me that regular visits were a necessity; you could have the best-motivated people in the world, and they'd still have problems if they were all cooped up in orbit or on the moon for months at a time. It made me realize we didn't know nearly enough about human factors to be going to Mars.

Other Voices

ROBERT THOMPSON

I had known Deke from the first days of Langley, where my office was right next to one of the Mercury simulators. I remember getting a little annoyed with all the racket—the simulator banged every time it moved—and especially with one guy in particular who was always there. It turned out to be Deke.

In the early 1960s I worked on recovery systems, so our paths didn't cross. And when I started working on Apollo Applications, my primary contact in flight crew operations was Deke's deputy, Tom McElmurry.

But I got to know Deke better during the South Pole trip, and it was on the way back, when we stopped on the South Island of New Zealand, that Deke said, "How about stopping off on the way back and go hunting?" I wasn't an avid hunter, but I wasn't about to turn down the chance to have a little fun. So he set it up.

So there we were in the Alps of the South Island, pretty rugged mountains. It turned out we were going to be hunting together in a little two-man Bell helicopter that had the doors off. The pilot sat right in the middle of the helicopter and Deke and I were strapped into the corner of seats on either side with our feet out on the skids. There we were, flying around through the mountains this way looking for animals.

I can still see Deke dangling out of this helicopter on a rope, trying to retrieve a mountain goat out of a canyon. I never told anyone back at MSC that Deke wasn't quite as religious about applying his flight safety rules in his hunting as he was in our configuration and control boards.

Apollo 13 was scheduled for launch on April 11, 1970. If things went right, Jim Lovell and Fred Haise would land near the crater Fra Mauro on April 15 and spend two days there, with Ken Mattingly overhead in the command module.

We had managed to overcome the early problems we'd had in Apollo with crews catching colds and getting run-down in the last few weeks before launch. Partly it was because we had become aware of the problem and made them slack off, and partly because the missions were stretching out to one every four or five months.

On Sunday, April 5, I was at the Cape when I got a call from Chuck Berry. He told me one of Charlie Duke's kids had come down with the measles. I didn't think this was particularly worrisome: Charlie Duke was lunar module pilot on the backup crew. He wasn't going to be flying 13.

Then Berry pointed out that rubella—measles—was pretty damned contagious and that it had a two-week incubation period. During that time Charlie could have exposed the prime crew.

Well, we all started chasing around. It turned out that Jim Lovell and Fred Haise had already had measles. They were immune.

But their command module pilot, Ken Mattingly, had not. The next day Berry said Ken should be grounded for ten days, the incubation period for measles. Since the launch window for the Fra Mauro site was about five hours long, that effectively meant he would not be eligible for launch.

I had two options. The first was to swap the prime crew and backup crew, which was pretty much out of the question. The Russians had always said they trained two pilots or crews for each mission and that neither one knew until the last day which would fly. Bob Gilruth had wanted to do the same thing with us back in Mercury, until he got talked out of it. And the Russians hadn't done that in practice, either.

The prime crew was just better trained, period. They had first demand on simulators, on everything. The backup crew was basically there as insurance in case of a disaster, somebody getting killed in a plane flight—which had happened. You didn't want to have to postpone a mission for several months.

The second option was to swap the prime and backup command module pilots only—Swigert for Mattingly. But that one required some thought, too. Jim and Fred and Ken had developed their own language and their own way of working together. So had John Young and Charlie Duke and Jack Swigert. There had been some cross-training, and certainly some of these guys knew each other fairly well, but it was not the same thing.

The one factor in favor of a switch, if you had to make one, was that Ken was the command module pilot. He would be spending a good chunk of the mission operating alone.

I had to call Ken in and tell him he was probably going to be off the flight. It wasn't pleasant, especially for me, because I could sympathize. I told him I would rotate him onto the next available crew, Apollo 16, but he and I knew that might not mean much: one more budget cut and there might not be an Apollo 16. He thought I had ruined his life.

The guy I had to convince was Jim Lovell, who wasn't at all happy about losing a member of his crew a few days before launch. (We could have just as easily postponed the mission by a month, but I guess we were scared off by the expense. Looking back, it was a dumb decision.) He liked Jack Swigert fine, of course, but wanted to test him.

So we put Jack in the simulator with Jim and Fred for a twelve-hour test on Thursday, April 9. My branch chief Riley McCafferty ran Jack ragged.

Jack had been on two support crews, 7 and 10, and by April 1970 had been training as a backup for the better part of a year. It was obvious to all of us that he knew his stuff, and when the test was over, Jim signed off on the switch.

The 13 launch was at thirteen minutes after one Central time, which is what MOCR used, on Saturday afternoon, April 11—1313 in military time. (Some people who were superstitous made a big deal out of the number of thirteens relating to this mission.) Ken Mattingly watched from MOCR, where he did some capcom work.

Just so we didn't get any ideas that a Saturn V launch would ever be routine, the middle J-2 engine aboard the S-IIC second stage cut out early. The remaining four had to fire longer, and so did the single J-2 on the S-IVB third stage.

Nevertheless, 13 made it to orbit in fine shape, then went for translunar injection. The command module, which was named *Odyssey*, pulled away from the stack, came back and docked with the lunar module, which was *Aquarius*. Then Jim, Jack, and Fred settled down for the climb up to the moon.

Fifty-five hours into the mission, just after nine in the evening of Monday, April 13, the crew had just finished up a TV guided tour of *Aquarius* when something happened. The flight controllers had asked Jack Swigert to do a routine "cryo stir"—that is, to flip a switch on his console that would turn on a fan inside one of the oxygen tanks in the service module. The cryo stir made it possible to get an accurate measure of how much oxygen was left in the tank.

We later figured there were bare wires inside the tank—its insulation had been cooked off by mistake during some ground tests weeks before launch. The cryo stir caused a spark that started a fire inside the tank. When the pressure was too much, it just blew . . . taking the side of the service module with it.

The first anyone on the ground knew was that Jim Lovell suddenly came on the radio saying, "Houston, we've had a problem," just as some red warning lights appeared on the EECOM (electrical, environmental, and communications) console.

Pretty soon the fuel cells powering *Odyssey* began to fail. Oxygen pressure began to fall. No matter what Gene Kranz, the flight director, and his troops tried to do, nothing seemed to help. *Odyssey* was dying. It took about an hour before everyone realized that there wasn't going to be a lunar landing, that the goal was now to get the crew back alive. Gene ordered the crew to get the lunar module powered up, something that usually wasn't done until after reaching lunar orbit.

I hustled over to MOCR as soon as I heard. Over the next few hours, I was joined by Gilruth, Kraft, and McDivitt in addition to three astronauts: Walt Cunningham, Dave Scott, and Rusty Schweickart. Charlie Duke—Mr. Measles—set up shop in the lunar module simulator, where he could try out procedures before they were sent up to the crew.

Kranz went ahead and handed control over to the next team a little more than an hour after the accident, figuring he was going to need his troops alert and rested over the next few days. Glynn Lunney's team took over and handled the lunar module power-up, which was a pretty complicated procedure. One of the things that had to go right was that the lunar module's inertial guidance unit had to be configured properly—or the crew would have to do what Dick Gordon had done on 12, take star sightings so the vehicle would be able to maneuver.

The first pass at the lunar module lifeboat scenario wasn't too promising: it would take more hours to loop 13 around the moon and back than the lunar module would last. We talked about having the crew fire *Aquarius*' descent propulsion system and send it directly back to earth.

The problem was that 13 was still really hauling ass and the DPS might not be able to accomplish a direct return. It might slow it down just enough to make sure it didn't loop back toward earth, or worse yet, might hit the moon.

Prior to the explosion, 13 had fired its service propulsion system to reshape its trajectory so it wasn't technically a free return. Getting back on that free return became the priority. In order to do that properly, of course, *Aquarius* had to be powered up, the guidance platform aligned, etc.

It wasn't until almost six hours after the accident that Jim and Fred were able to do the thirty-second-long burn with the DPS which meant that, one way or another, they would be headed back toward earth.

We didn't have much choice except to let the crew go around the moon and head back. The problem would be keeping them alive while they did.

One of the tricky things about getting the crew through that first burn, and anything else, was that they were getting tired. They had been ready to start a sleep period when the accident occurred. With the shortage of power, the command module in particular got cold, which also made it hard to sleep. I'm not sure any of the crew got any real rest during the remainder of the mission.

(I didn't get much, either. I had been battling a cold, so I wasn't feeling particularly good. Someone took a picture of me dozing at the console at the one-hundred-hour point in the mission. My own copy has an inscription from Jim, Jack, and Fred "thanking" me for my attention. . . .)

It was one mini-crisis after another. No sooner did we think we had a problem on stretching the consumables—air and water and power—over the four days it would take to get the crew back than we realized we had a potentially lethal problem with the buildup of carbon dioxide in the spacecraft. We were using the lunar module's system for "scrubbing" the air, but it only had two lithium canisters—enough for a lunar landing mission for two guys, not enough for three guys for four days.

While the crew systems people were worrying about that, there was more debate on firing the DPS again and somehow getting the crew home before Friday. It still looked as though a big burn like that would take all of the available fuel and not leave you enough for adjusting the orbit when they got close.

(We actually didn't want the crew back too soon, since we had to have recovery ships in position. That would also take a couple of days.)

What we decided to do was find a way to get them back a few hours early. This was going to be a four-minute, twenty-three-second burn at pericynthion, closest approach to the moon, plus two hours: P C + 2.

Seventy-seven hours into the mission, less than a day after the explosion, 13 looped around the back side of the moon. At least they were heading home now.

Jim and Fred tried to make sure they were oriented. It was hard because the service module was still venting—the spacecraft was traveling in a sea of little particles that looked just like stars through an eyepiece. And the crew was really getting tired. Chuck Berry and I, who rarely agreed on anything operational, were both worried about them.

The P C + 2 burn went so well, the trajectory guys figured all 13 would need was a final tweak the last few hours before reentry.

There was a dust-up in the MOCR afterward, however. Max Faget was worried about the external heating on 13, afraid that parts of it were being left in the shade long enough to freeze. Gene Kranz agreed with him. They wanted the crew to put the spacecraft into a passive thermal control (PTC) mode—to get it rotating slowly so it would be heated evenly.

The problem was, going to PTC would take the crew two hours minimum, and they were so tired right now I wasn't sure they could do it. Chris Kraft was more worried about conserving power than going to PTC.

It was a real battle for a while there. But Gene Kranz was the flight director, and he was pretty adamant that if some pipes on *Aquarius* froze and broke, it wasn't going to matter if the crew was rested. He won, and Jim and Fred—following instructions from Kranz's guys—got into PTC quicker than anyone thought they would.

The next day, Wednesday, as they headed back toward the earth we read them a procedure for adapting lithium canisters in the command module for use in scrubbing the air.

That gave us time to start worrying about other things. Had the command module heat shield been damaged by the explosion? Would *Odyssey* have any life at all once *Aquarius* was jettisoned? For that matter, just what were the steps for jettisoning a lunar module in a situation like this?

John Young and Ken Mattingly were in the simulator helping to create a checklist for that. Ken came to MOCR to read it up to the crew on Thursday.

With communications dropouts and a tired, dehydrated crew, it was a tedious business. Later that night, when we should have been done with it, somebody insisted we wake up Jack and give him more updates. This woke up Jim and Fred, who weren't sleeping much, anyway. (The temperature inside the spacecraft was close to freezing.) Eventually all this finally got to Jim. "Unless the changes are really essential, don't bother sending them up."

Time for the original capcom to step in. "I know that none of you are sleeping worth a damn because it's so cold," I told Jim, "and you might want to dig out the medical kit there and pull out a couple of Dexedrine tablets apiece." The medics had been suggesting that for a while, but I was reluctant, because when the Dexedrine wore off, the guys were likely to be even more tired and depressed. But we were getting close to reentry now.

Jim said he might try it, but didn't until a few hours later, when it didn't make any difference.

Since we were close, however, the controllers figured we might as well use some of the power to warm up the spacecraft.

At 138 hours into the flight, on Friday morning, with the command and lunar modules still attached, the crew separated the service module. Jim Lovell had the best view of it as it drifted away. "There's one whole side of that spacecraft missing!"

Three hours later, *Odyssey* was all powered up, running on batteries, with the crew inside. Then *Aquarius* was separated. "Farewell, *Aquarius*, and we thank you," Jim Lovell said.

Jack Swigert radioed down a thank-you—very classy, under the circumstances. Then *Odyssey* went into the communication blackout as it entered the atmosphere.

Normally the blackout would last three minutes. We passed that, and then some. Another long minute. Then the tracking people reported a beacon, meaning the command module had survived. No word on crew yet. . . .

Joe Kerwin was the capcom. "*Odyssey*, Houston, standing by." A couple more seconds passed. Then: "Okay, Joe," said Jack Swigert.

Another three minutes. Then we had a picture from the recovery team—we could see the three chutes.

They'd made it.

* * *

From the moment 13 splashed down, all future Apollo missions were on hold, pending an investigation. Tom Paine appointed Edgar Cortwright, director of NASA Langley, as chair. Neil Armstrong represented the Astronaut Office.

The board reached its conclusions fairly quickly about a combination of human error and design flaw. Bad as the situation had been, it could have been worse. If the tank had exploded after the landing, the crew would have never made it back.

After 13, I was able to make good on my promise to Ken Mattingly (who never had come down with the measles) to put him in the next available crew. I assigned him as command module pilot for 16, with John Young as commander and Charlie Duke as lunar module pilot.

The backup crew, pointed at 19, was going to be Fred Haise as commander, with Bill Pogue as command module pilot and Jerry Carr as lunar module pilot. (Jim Lovell was retiring and I didn't want to send poor Jack Swigert around the moon again.) The support guys were Tony England, one of the scientist-astronauts, and two of the MOL astronauts, Hank Hartsfield and Don Peterson.

Those assignments never got announced. As the summer wore on, it was clear that there wasn't going to be enough money for the last two landings, 18 and 19. They were finally canceled in September. Fourteen would be the last of the H missions, and it could not be launched before early 1971.

Fifteen would follow later in 1971; it would now be a J mission and would have Dave Scott, Al Worden, and Jim Irwin on the crew. Sixteen and 17, the last two J missions, would take place in 1972. Then we would move on to Skylab.

Eighteen and 19 had already been in jeopardy because of the fiscal year 1972 budget problems, but the close call with 13 frankly scared Bob Gilruth. I don't think he tried to get 18 and 19 canceled, but he didn't fight it. He was right, in a way. The system we had was great for making the first landing, but when you were talking about longer stays on the lunar surface . . . shit, you were far away from home and help. He believed we should be on the moon, but he also believed we should have a safer way of operating there. Apollo wasn't it.

I really thought it was unfair to ask somebody to do a backup job that was a dead end unless they were coming off a similar mission. Dick Gordon, Vance Brand, and Jack Schmitt were sort of stuck dead-ended on 15. However, the crew for 17 hadn't been assigned and wouldn't be for some months, even though Gene Cernan's crew would have the edge going in. Dick's team hung in there in the hope that I might break the rotation.

I was in a position to do something about 16, however. Good old Fred Haise was willing to stay as backup commander, but I moved the two guys on his crew who hadn't flown—Jerry Carr and Bill Pogue—over to Skylab, which was still almost three years away from a launch. They'd have plenty of time to get in line there. I would replace them with Stu Roosa and Ed Mitchell from the 14 crew after their flight.

24

RETURN TO FLIGHT

Other Voices

JOSEPH ALLEN

I was one of the scientists who learned to fly after becoming an astronaut in 1967. Now, the Air Force has a very practical and effective way of training pilots—the cookbook method. It runs you through a basic set of steps, each one building on the next, and is so efficient that it can take just about anybody, from a jock with an aeronautical engineering degree to an English major from Yale, and turn him or her into a pilot. As long as you follow the cookbook.

This method drills you in following certain operational rules as if they were the word of God. In the T-38 jets, which were what I flew at NASA, there was a strict airspeed rule. Don't let the plane go slower than 270 miles an hour, or it'll fall out of the sky.

This rule is relatively easy to follow most of the time you're flying: a jet like the T-38 just sort of overpowers nature. Hit the gas and you go. Wind and rain aren't going to slow you down.

The trick is that in landing a T-38 you eventually have to fly a lot slower than 270 miles an hour. Setting up to approach the runway, you have to put yourself into a careful turn that bleeds off airspeed while still maintaining lift. We were taught that the only warning we were going to get that the T-38 was about to stall—to fall out of the sky—would be this little burble or flutter. If you got that, you were in big trouble, especially since in approaching the runway you were fairly close to the ground.

I don't think I ever flew an approach in a T-38 without having my eyes glued to the airspeed indicator, making sure I wasn't going too slow. I barely even looked out the canopy, because I didn't want to get that burble. It must have worked, because I logged several thousands of hours in the T-38 without killing myself.

I still remember the first time I landed a T-38 with Deke, however. He was flying and I was in the backseat—his "second" pilot. First of all, we came zooming in across the runway, just to sort of say hello to the tower, I guess, then out and into the turn to set up for landing.

We start into the turn, and right away the plane starts to burble. *It's wobbling all over the place and I'm reaching for the handle on the ejection seat thinking I'm dead.*

Then, just like that, we leveled out . . . the burble went away . . . and we made a smooth landing.

I was still waiting for the color to return to my features when I asked Deke, as casually as I could, because he hadn't seemed bothered at all, "Say, how fast were we going during that turn?"

"Damned if I know," he said.

That's when I realized how different it was to have learned to fly when Deke did—in prop planes that are much more subject to weather and wind. Deke had learned how to fly by touch, not by a cookbook, and that's how he did things.

Being restricted to flying with another pilot had one bonus—I flew with everybody in the Astronaut Office and in aircraft ops, so I had a pretty good idea how guys handled themselves. That was the only good thing; otherwise the restrictions were a total pain in the ass. I felt like a second-class citizen, and it was during 1970 that it finally started to bother me.

There were probably other reasons. My mom was sick that year, and finally died. My dad wasn't in the greatest shape, either. I was dealing with a teenager at home, and Marge was wondering how long I was going to keep doing what I was doing, given that the clock was ticking and the flights were running out. NASA was going through cutbacks, too.

I was on an antelope hunt out in Wyoming in July 1970 when I had a fibrillation. There was nothing particularly unusual about it, except that I realized it was the first one I'd had in months. I'd never gone that long without the condition—not without some kind of medication—so I tried to figure out what was different.

The fibrillations had stopped back in early April, just before the 13 launch. I had been trying to fight off a cold down at the Cape and had gone to Chuck Berry, who had loaded me up with vitamins. And I had stayed on them. The hunting fibrillation came after a day when I was awake late, then up early, a little more stressful than most.

So, just as little experiment, I started taking the vitamins again, and keeping detailed notes on my intake and my reactions. All through the preparations for Apollo 14 and after, I never had a recurrence.

It wasn't the kind of thing you could use as hard medical evidence: I couldn't run to Chuck Berry and say, hey, look, I'm taking vitamins now—give me a Class II ticket. But it got me thinking that I might still have a more realistic chance of getting one. If only I could find a doctor who would go to bat for me.

The fixes had been made to the Apollo command module to make sure an accident like 13 wouldn't happen again. Fourteen was set for launch in late January 1971 on a mission to visit Fra Mauro, the original target for 13.

Al Shepard had been real smart in his use of the extra time for training. He and his crew—Stu Roosa and Ed Mitchell—were really ready to go. Al had even whipped himself into such terrific physical condition that nobody was worried about the fact that he was forty-seven years old. (We'd come pretty far from the days when NASA thought an astronaut would be washed up at forty.)

The only question came from William Loeb, publisher of the *Manchester Union-Leader* up in New Hampshire. On the pretext that he was "concerned" about Al Shepard, a native of New Hampshire, Loeb wrote to the White House to raise some questions about Al's physical condition. Hadn't he been grounded for all those years? What had been corrected? Wasn't he too old to be risked on a mission like this?

Poor Chuck Berry had the White House climbing all over him looking for a way to answer Loeb. I think Chuck had to write about a ten-page letter justifying Al's assignment on medical grounds alone. It was a real pain in the ass, and totally unjustified.

We were down at the Cape the last couple of weeks before launch, the prime crew and backups—Gene Cernan, Ron Evans, and Joe Engle. All the crews were spending time in the simulators. Al and Gene would occasionally do some helicopter flying to tune up for a lunar landing.

On January 23, four days before 14 was scheduled to take off, Gene Cernan was flying a helicopter a little too low over the Indian River, which ran between the Cape and the mainland. He smacked into the water and got knocked out. When he came to, the chopper was sinking and the water was on fire.

He kicked his way out of the cockpit, dived under the flames, and got himself to shore. He was pretty damned lucky: I think the worst injury he got was a bump on his head and some singed eyebrows.

Other Voices

GENE CERNAN

I was getting cleaned up in the crew quarters when Deke came in. After making sure I was okay, he said he was going to have to talk to the press in a few moments. "So," he said, "exactly how high were you when the engine quit?"

"Deke," I said, "the engine didn't quit. I just flew the son of a bitch into the water."

His expression didn't change. "Maybe you didn't hear me. When exactly did that engine quit?"

"Like I said, Deke. It didn't quit. I just screwed up."

Deke just smiled, shook his head, and shrugged. "Well, if that's the way you want it."

That was typical Deke. He knew and I knew I had screwed up, but he was damned if he was going to blame me publicly.

Fourteen lifted off on the morning of January 31, after a forty-minute delay for weather. (We weren't going to take any more chances with lightning.) The ride up was as uneventful as these things get, and in a few hours 14 was headed for the moon.

Then there was a hiccup. The 14 command module, *Kitty Hawk*, separated from the S-IVB/LM stack in order to move out, transpose, and come back to dock with the lunar module *Antares*.

Stu Roosa, flying *Kitty Hawk* from the left seat, nosed the docking probe into the cone atop *Antares* and hit the switch to fire the latches. But nothing happened. Stu tried again, this time burping the RCS thrusters to give a little shove. Still no joy.

Obviously, if they couldn't dock with *Antares*, the crew wasn't going to be landing on the moon. They tried to dock twice more, with no luck, and backed off while the flight control guys looked things over. There were no obvious problems and we were wondering if there wasn't some loose nut or chunk of ice jammed into the probe somewhere.

The plan they came up with was this: for Stu to drive the *Kitty Hawk* probe into the *Antares* cone and hit the switch for a hard dock, bypassing the normal step of a soft dock. That worked, and they were finally in business.

Three days later the *Antares* set down on the moon near the crater Fra Mauro. "We're on the surface," Ed Mitchell reported. "That was a beautiful one." We had learned enough about guiding the lunar module, and Al was confident enough, that he deliberately landed a hundred yards short of the actual aiming point. He and Ed were supposed to do an EVA in that direction and he thought it would save them some walking.

They did two EVAs over the next day, hauling a wheeled trolley called the modularized equipment transporter (MET) and setting up the usual gaggle of scientific instruments.

The real highlight of the EVAs as far as I was concerned came at the end of the second one. In *Life* magazine a dozen years back, Al had said that one of his ambitions was to one day be able to golf on the moon. Well, there he was, just before climbing back into *Antares*: "What I have in my hand is the handle for the contingency sample return, and just so happens I have a genuine six-iron on the bottom of it. In my left hand I have a little white pellet that's familiar to millions of Americans. I drop it down. Unfortunately the suit is so stiff I can't do this with two hands . . . but I'm going to try a little trap shot here."

He took one swing, got more dirt than ball, then tried it again. Fred Haise, the capcom watching on TV, said, "Looked like a slice to me."

Not to Al. "Straight as a die," he said. "Miles and miles. . . ."

One thing about 14 that got more publicity than anything else was Ed Mitchell's "experiment" in ESP. He took along a pack of something called Zenner cards that had certain symbols and shapes on them, and by prearrangement with some parapsychologists on the ground, tried to transmit these symbols to them half a dozen times during the flight to and from the moon (but not on the lunar surface). Ed said later it was statistically significant; other people disagreed.

I thought it was worth a look. Hell, NASA doesn't know everything. And in spite of the fact that we'd had a little argument a few months before launch—Ed didn't want to stick around and do the backup job on Apollo 16, so I told him in that case he didn't need to fly 14, either—Mitchell was one of the smarter guys we had in the Astronaut Office.

Tom Paine left NASA in March 1971. Tom had really started to get frustrated by the Nixon administration, which kept saying it was behind the space program, but really didn't put any money or time into it. Tom really wanted the administration to commit to the development of a big space shuttle with a manned orbiter and a manned reusable booster, but even in NASA he didn't have everyone behind him. Forget the Nixon administration, which took one look at the project costs and said no way. Worst of all, I suppose, Paine was a Democrat.

George Low became the acting administrator, and he put a lot of his energy into finding some compromise that would keep the United States in the manned space business. One of them was a cheaper space shuttle that would have a manned orbiter launched by reusable solid rocket motors.

Eventually a good Republican was found to become administrator—James Fletcher, a physicist who had been president of the University of Utah. He really had his work cut out for him, however. The Nixon administration didn't have the money for any of those ambitious programs Agnew's task force had laid out to follow Apollo. But they still weren't ready to get completely out of the manned space business.

It would have helped a lot if the Russians had kept up with their moon program. It was still alive, though barely. After the big blowup at Baikonur in July 1969, their N-1 rocket was back on the pad for its third test in June 1971.

They had had a few other manned flights since then. Three Soyuz spacecraft were in orbit at the same time in October 1969, for no particular reason that I could see. Two of them—Soyuz 7 and Soyuz 8—were supposed to rendezvous and dock while the third, Soyuz 6, was supposed to rendezvous with the other two and film the docking.

I guess there was some technical value in trying to control three manned spacecraft at the same time. God knows we had enough problems with a command module and a lunar module. But it still looked like a public relations stunt, some kind of bogus "space first." Soyuz 7 and 8 didn't manage to dock, anyway.

In June 1970, right when Neil Armstrong was making a visit to the cosmonaut training center, the Russians launched Soyuz 9 with Andrian Nikolayev and Vitaly Sevastyanov. They stayed up for eighteen days, a space record at the time. That kind of mission had some use, especially if you were going to build space stations—which the Russians were saying they were going to do.

Nikolayev and Sevastyanov came back in pretty bad shape, however. They had to be carried from the spacecraft. In a few days, though, they were up and walking around. But it was pretty clear everybody had to learn a lot about keeping people healthy on these long missions.

The Russians finally did launch their space station, called Salyut or Salute, in April 1971. Following a few days of checkout on Salyut, Soyuz 10 went into orbit with a crew of Vladmir Shatalov, Alexei Yeliseyev, and Nikolai Rukavishnikov.

They docked with Salyut and immediately ran into problems. They couldn't get the docking hatch to open. I guess there were some problems equalizing pressure between the Salyut and the Soyuz. Docking systems gave the Russians fits: they hadn't been able to develop one at all for their lunar program. Now these guys had to give up and come home. (That version of Soyuz had such limited consumables that it could only operate on its own for forty-eight hours.)

Six weeks later they were ready to try again. Soyuz 11 lifted off with a crew of Georgy Dobrovolsky, Vladislav Volkov, and Viktor Patsayev. They docked and got the hatch open, and moved aboard Salyut.

The plan was for them to spend thirty days up there, and they did pretty well with it. There were some technical problems, including a small fire aboard the station. (Russian spacecraft used a mixed gas atmosphere under normal pressure, so a fire wasn't automatically a catastrophe.) There were also some personality clashes between Dobrovolsky and Volkov. (Volkov had flown before, while Dobrovolsky, the commander, had not, and Volkov figured he knew more about what was going on. This was because of that goofy Russian system in which the commander was from the air force cosmonaut team while the flight engineer was from another one.)

The cosmonauts also had a ringside seat at the third N-1 launch on June 27. They were observing from orbit as it went up for about a hundred seconds, then failed and had to be destroyed.

Three days later Dobrovolsky, Volkov, and Patsayev packed up to come home. Just prior to reentry, after the Soyuz retros had fired, as the orbital module separated, a valve opened by mistake in the descent module. (It was supposed to open just before landing). All the air started whistling out.

They tried to close it—even tried blocking it—but couldn't. The landing was automatic, and the rescue teams knew something was up because there had been no contact with the crew. But it must have been a hell of a shock for them to open the hatch and find the crew dead.

We all got a little worried. Given that Nikolayev and Sevastyanov had been pretty weak . . . here you had this crew *dead*. Was there something about being weightless that long that could kill you?

The Russians were starting to open up a little. They announced that the crew had died because of a technical failure, not a physiological problem.

When we got word, as we had with Komarov in 1967, we offered to send a representative to the state funeral. This time the Russians accepted, and Tom Stafford got on a plane.

With 14 behind us, I had to assign a crew for 17. I wasn't hurting for candidates. There were the backups for 14—Gene Cernan, Ron Evans, and Joe Engle. And the backups for 15—Dick Gordon, Vance Brand, and Jack Schmitt.

The scientific community was convinced that Jack Schmitt had to fly. After all, he was a geologist, wasn't he? I even had Dave Scott telling me

I should use Dick's crew, which included Jack, just because he thought they were better than the other guys.

I didn't have anything against Jack Schmitt. He had done a hell of a lot of good work as a geologist getting crews ready for lunar missions. He had dug in there and gone through the training, becoming a jet pilot and all of that. And I had ranked him right where I thought he should be—behind Joe Engle.

Joe had some flaws as an astronaut—for one, he seemed pretty uninterested in rendezvous—but he was a terrific stick and rudder guy. He'd flown the X-15 and earned Air Force astronaut wings before NASA ever got hold of him. I thought it would be nice to send a geologist to the moon, but it wasn't my business to be nice. I had at least one guy I thought was more qualified.

So the choice was obvious to me. I submitted Cernan-Evans-Engle as the crew to NASA HQ. And had it rejected. It was a little like 13 all over again.

So I bit the bullet and dropped Joe Engle off Gene's crew and replaced him with Schmitt. I hated having to explain that to Joe, but he sort of realized it was out of my hands. He took it better than I would have.

The 17 crew was approved with Gene as commander, Ron Evans as command module pilot, and Jack Schmitt as lunar module pilot. I planned to recycle the 15 crew—Dave Scott, Al Worden, and Jim Irwin—as their backups. I gave Gene Gordon Fullerton, Bob Parker, and Bob Overmyer as support guys.

Al Shepard went back to being chief of the Astronaut Office. Tom Stafford moved up to become my deputy in flight crew operations.

My last big foray into crew assignments came right about this time, for Skylab. We were expecting to fly at least three missions, and hoping for even a fourth visit and a theoretical fifth beyond that. So I needed at least that many crews; none of them would be automatic dead ends. Fifteen guys, nine of whom could realistically expect to fly.

The problem was, there were about twenty working on the program. Pete Conrad, Al Bean, Walt Cunningham, and Rusty Schweickart had already flown missions. Pilot candidates were P. J. Weitz, Jack Lousma, Bill Pogue, Jerry Carr, Bruce McCandless, and some of the ex-MOL guys. And there were a whole gaggle of scientist-astronauts—Ed Gibson, Owen Garriott, Joe Kerwin, Story Musgrave, Bill Lenoir, Bill Thornton, Don Holmquest.

There had been some talk of having each crew consist of a pilot-commander and two scientists, but that wasn't going to fly: Skylab was a whole new deal for us. I figured the first crews, at least, should have at least two people who had some experience in troubleshooting.

I especially wanted an experienced guy commanding the first mission, so Pete Conrad was an easy choice. (If it hadn't been Pete, it might have been Tom Stafford, who was still eligible for flight, but Tom really wasn't enthusiastic about going around and around the world for a month.) P. J. Weitz had been working AAP for a while and had done a good job, so he got pilot on the first crew. And everybody wanted the first science pilot to be a physician, since the mission would be lasting twenty-eight days. Joe Kerwin was the best candidate.

Here's what I laid out early in 1971:

Skylab 1	(prime)	Conrad-Kerwin-Weitz
	(backup)	Cunningham-Musgrave-McCandless
Skylab 2	(prime)	Bean-Garriott-Lousma
	(backup)	Schweickart-Lenoir-Lind
Skylab 3	(prime)	Carr-Gibson-Pogue
	(backup)	same as Skylab 2

No sooner were these announced internally than Walt Cunningham dropped out. He had been working AAP/Skylab since the end of 1968 and had wanted to fly again. In spite of the flight operations opinion that he shouldn't, I wasn't going to rule him out. But it was a numbers game.

He accepted that, but didn't think he wanted to stick around for another two years just to serve as a backup, so he resigned and went into business.

I moved Rusty Schweickart to backup for the first Skylab, then started looking for another backup commander. The best candidates were Dick Gordon and Vance Brand, since they would be coming off 15 in a few months. Dick had let me know that he was retiring after 15, however, so I turned to Vance.

Except for a theoretical joint flight with the Russians, this was the end of the road as far as crew assignments went. And I told the guys in the Astronaut Office so. Eventually it got into the press that I had said most of them "might as well pack their bags," because our manpower exceeded our needs by a factor of three.

My statement didn't exactly lead to a mass exodus, though two scientist-astronauts, Phil Chapman and Tony England, did resign. (England came back to NASA as an astronaut in 1979 and wound up flying on the shuttle in 1985.) Other guys took different jobs in government or NASA—Joe Allen went to NASA HQ.

Some of the others resigned because they were ready to make changes, whether it was the end of twenty years in the military or just personal. (Ed Mitchell, for example, had gotten divorced and wanted to start over.)

* * *

Apollo 15 was the first of the J lunar missions, capable of twice as much time on the lunar surface. The crew was Dave Scott, Al Worden, and Jim Irwin, who had been training for it for eighteen months. (There had been some discussion of replacing Jim Irwin with Jack Schmitt, but since he was now assigned to 17, Jim's shot at a lunar landing was safe.)

Maybe they'd been training too long. At one of the last press conferences before the launch, Jim Irwin complained about some technical problems with his lunar module, which was going to be named *Falcon*. (The crew had also picked *Endeavour* as the name for the command module.) There were always some last-minute glitches, but we usually didn't raise them with a bunch of reporters.

Fifteen was launched on Monday, July 26, 1971—it was Jim Fletcher's first manned flight as NASA administrator, and he was there to watch. The liftoff and eventual translunar injection went well. A couple of problems cropped up on the way out, nothing that couldn't be worked around, and three days after launch they burned into lunar orbit.

The landing site for 15 was in the Hadley Apenine Mountains, chosen especially because it promised to give clues to the possible origin of the moon. Since it was in the mountains, however, Dave and Jim really had a challenging landing. They had to burn almost straight down at the end there. The descent engine kicked up a hell of a lot of dust, so much that they couldn't see the actual touchdown. They hit pretty hard and even slid a few feet before settling down. "The *Falcon* is on the plain at Hadley," Dave radioed at touchdown.

They had three days on the surface, three big EVAs including use of the lunar rover. Better yet, the TV mounted on the rover was remote-controlled from earth. Dave and Jim could be off setting up experiments and taking samples and the operator, Ed Fendell, aka Captain Video, could follow them or take in some other sights. Joe Allen was the mission scientist and capcom for the EVAs, and I know having the TV used that way was a big help for him.

It also got us my favorite astronaut joke. Dave Scott was trying to jam a core sample into the ground and not having much luck. He really reared back to ram it into the moon, which got Joe Allen, watching on TV, to say, "Careful you don't puncture it."

At the end of the third EVA the crew did some fun and games. Dave Scott dropped a hammer and a falcon feather at the same time, "proving" Galileo's theorem about objects falling at the same velocity no matter what mass they were. There was a little stamp cancellation ceremony.

The crew also deposited a small figurine called the "Fallen Astronaut," and a plaque, in memory of the Americans and Russians who had died in spaceflights.

The rest of the flight was pretty straightforward. Al Worden did an EVA on the trip from the moon to the earth, retrieving some experiment packages from the service module. (Another option with the J mission).

The EVAs were a hell of a physical drain on the crew. We wound up depleting their potassium levels, something we corrected for 16 and 17. Jim Irwin wound up having a history of heart problems, and he tended to blame some of it on the stress of those EVAs.

There was on one potentially huge problem: during deployment of the parachutes that were supposed to slow *Endeavour* for splashdown, only two deployed fully. The third one came out, then collapsed. We guessed later that RCS fuel, venting prior to splashdown, burned through some of the lines. It meant a slightly harder landing.

The *real* hard landing for 15 wouldn't come for another year.

Chuck Berry knew about my vitamin therapy and apparent lack of heart fibrillation. He also knew I was looking for a doctor would would sign off on my suitability for flight.

He attended a medical conference in Turkey, of all places, and ran into a specialist from the Mayo Clinic named Hal Mankin. Mankin listened to Berry's description of my situation—that every doctor said I was fit for flight, but no doctor would certify me for it—and agreed to test me himself.

The catch was, this was final: if I flunked, it was the end of the screen. They were going to ground my ass permanently. If I passed, however, Mankin said he would sign the certification.

So in December 1971 I flew up to Rochester, Minnesota, and checked into the Mayo Clinic under the name "Dick K. King." For a week they ran me through the usual tests—angiograms, whatever. Not only was there no sign of any coronary disease, which had been one of the worries back in 1962, there was no recurrence of the fibrillation.

Other Voices

CHARLES BERRY

As soon as I got the verdict from Dr. Mankin, I called my friend, Pete Segal, who was Federal Air Surgeon—the chief medical officer at the FAA—and got Deke qualified for Class I.

I was concerned about last-minute objections. The seven doctors who, at one time or another, had given Deke a thumbs-down still needed to be consulted, however informally. I recall that six of them were perfectly willing to approve his new status, and the seventh, though he grumbled, said he wouldn't stand in Deke's way.

I also got approvals from Kraft, Gilruth, and Low, in time to release the news of Deke's return to flight status at the same press conference where the Skylab crews were announced. We emphasized that Deke was requalified and eligible for any flight, including the joint flight with the Russians, which was then under discussion.

By an odd coincidence, the announcement was made on March 13, 1972, ten years to the week from when I had been pulled out of my Delta 7 simulator and ordered up to Washington.

I didn't make a big deal out of it, but NASA alerted somebody at the Houston *Post* that I was back on full flight status, and there was a photographer out at Ellington Air Force Base the day I took a T-38 out for my first solo spin since 1962.

The same week I was allowed back in the cockpit alone, NASA announced its choice for the upcoming space shuttle design. (James Fletcher had had to go directly to Nixon in January to get it sold.) It was a model using a winged one-hundred-ton reusable orbiter launched atop a stack of a throwaway external fuel tank and two reusable solid rocket motors. According to the studies NASA commissioned from a company called Mathematica, Inc., each orbiter would fly a hundred missions over its lifetime.

Those missions were to carry all U.S. unmanned satellites into orbit, thus eliminating the need for expendable launch vehicles like the Titan, Delta, or Atlas. (A good quarter of these missions were likely to be military, and so the orbiter had been designed according to Department of Defense specs.) The study also projected as many as *fifty* shuttle flights a year.

In the original 1969 Space Task Group plans, the shuttle had been part of a three-way system that included a space station and a space tug. The tug was gone and, worse yet, so was the space station. What we were left with was the shuttle, a truck that had really no place to go.

But it was the only truck we were likely to have.

Apollo 16 was aimed at the area around the crater Descartes, a J mission using the rover. It had a crew of John Young, Ken Mattingly, and Charlie Duke. John was one of the unsung heroes of the Astronaut Office, a real hardworking guy who did whatever you asked him to, no problems. He had been on the first Gemini, had commanded his own Gemini, then had gone to work for Tom Stafford on Apollo 10. The only thing that held him back was that he was not comfortable with public speaking; he tended to freeze up and give one-word answers. I knew what that was like, because that had been my problem for years. I don't know that he ever got over it . . . but then again, I don't know that it made much difference. He's still flying twenty years later.

Ken Mattingly, of course, was the guy I'd had to yank off the 13 crew. He was finally getting his chance. And good old Charlie Duke, who had been through the first lunar flights of the lunar module on 10 and 11 as capcom.

The launch got postponed a month, from March to April, because Charlie Duke came down with pneumonia. There wasn't even any discussion of replacing him; that was one lesson we'd learned on 13.

Sixteen lifted off on Sunday, April 16, and ran into a bunch of little problems, any one of which had the potential to screw up the mission. First there were some leaks in the S-IVB stage even before translunar injection; those had to be worked around.

After transposition and docking Charlie Duke got sent into the lunar module—*Orion*—earlier than planned because the crew had seen some particles flaking off the exterior. (They turned out to be flakes of white paint that had gotten too cold and brittle.)

Once they were in lunar orbit three days later, getting ready to separate *Orion* from the command module, *Casper*, there were other glitches: an antenna aboard *Orion* wouldn't work right in its automatic mode, so the crew had to be ready to steer it manually, adding to their load. There was a regulator problem in the *Orion*'s RCS system.

Finally they got squared away and separated, and the biggest problem of all arose. Ken Mattingly, flying alone in *Casper*, had had to postpone firing the SPS engine because of some unplanned oscillations. Since that engine had to be fully operational for us to proceed with a lunar landing, everything ground to a halt.

John and Charlie couldn't just sit there in orbit indefinitely: after five revolutions their orbit would have migrated too far for them to reach the landing site.

It took two whole revolutions for all of us in MOCR to be sure the SPS oscillations didn't mean there was something seriously wrong with that engine. We gave John and Charlie a go-ahead for powered descent initiation, and read up all the new data. John put *Orion* down on the moon within a few yards of the target point.

Since he and Charlie were having an exceptionally long day, they went to sleep before starting their EVAs. Neil and Buzz had been too excited, but we were getting a little blasé about being on the moon by now.

On the morning of Friday the twenty-first John climbed down *Orion*'s ladder, followed by Charlie, for the first of three EVAs. They really did a good job, visiting distant craters and giant boulders and setting up the experiment packages, though John accidentally broke one when he tripped over a data cable. He was pretty upset about it.

One of the challenges was just dealing with the mechanics of working on the moon: there was dust inside *Orion*'s cabin and spilled orange juice inside a pressure suit.

John, Ken, and Charlie splashed down on the twenty-seventh. And there was just one more landing to go.

Returning to flight status in the spring of 1972 didn't leave me a hell of a lot of options as far as getting into space. The last Apollo lunar landing crews—16 and 17—had already been in training for months. The Skylab crews had just been publicly announced on January 18, 1972.

The space shuttle was just getting congressional budget approval. We already had a small group of astronauts helping with development—Hank Hartsfield, Bo Bobko, and Don Peterson working under Jack Swigert. But the shuttle wasn't targeted for manned flights until the end of 1977 at the earliest.

Fortunately, the Cold War had thawed enough to allow the United States and the Soviet Union to consider a joint manned space mission. Discussions had been taking place in a serious way since the end of 1969, though not with much seriousness on either side until late 1970.

Oddly enough, one of the things that moved the joint flight closer to reality was a fictional movie called *Marooned*, based on a novel by Martin Caidin. In the original version, published in 1964, a Mercury-Atlas 10 astronaut is rescued by a Soviet cosmonaut. The movie had been updated (and Caidin wrote a new novel version as well) showing how a Skylab crew might be saved by a Soviet Soyuz pilot.

The movie never made much money in the United States, but it apparently impressed the Soviets that Americans were ready to consider international flights—especially to demonstrate the concept of space rescue.

George Low visited Moscow in October 1970 and in 1971, and several of the Soviets' people visited the United States. That's when serious talking started. The formal agreement was signed by President Nixon during his visit to Moscow in May 1972.

A couple of weeks later we had our first big briefing for this project at the Manned Spacecraft Center with Glynn Lunney, who had been a flight director during Apollo and who was going to be the program manager. There had been talk of an Apollo-Salyut docking, but with the deaths of the three cosmonauts the summer before, the profile had been changed to a docking in orbit between an Apollo and a Soyuz—the Apollo-Soyuz Test Project (ASTP).

I figured I had a little jump on a possible crew assignment: I had already started taking Russian lessons. With me were several other guys, all of them thinking the same thing, Tom Stafford, Bo Bobko, Bob Crippen, Jack Swigert, and Vance Brand among them.

I was still flight crew operations director, of course, but I asked Chris Kraft, who succeeded Bob Gilruth as director of the Manned Spacecraft Center in early 1972, to handle the crew selection for ASTP, since I now considered myself a candidate.

Sometime after the official agreement, in May 1972, Kraft asked me for my recommended assignments. With the understanding that it was up to him, I made them:

ASTP	(prime crew)	Slayton-Swigert-Brand
	(backups)	Bean-Evans-Lousma

I had fifteen guys to choose from, but thought Jack Swigert deserved another chance to fly, since he had been thrown into 13 at the last minute, with all that happened. Vance Brand was already assigned as backup commander for the last two Skylab missions and had done a good job backing up 15. He was extremely well trained and deserved a chance, too.

The backup crew came out of the rotation: I picked guys who were still in Skylab or Apollo training at the time, but who would be free early enough to provide support. Ron Evans was assigned as command module pilot for 17, the last Apollo, which would fly in December 1972. Al Bean and Jack Lousma were going to be free following the second Skylab, no later than September 1973.

Unfortunately, this crew didn't turn out much like my recommendations. The first problem was Jack Swigert.

Right about this time, the summer of 1972, we were dealing with a regular goddamn scandal. It turned out the the crew of 15, Dave Scott, Al Worden, and Jim Irwin, had carried something like 440 unauthorized postal covers—specially-stamped envelopes—along on their flight. After the landing, they let a hundred of these things be sold to dealers with the idea that the money would go to a trust fund for their kids. That was completely against NASA rules. There were two hundred *authorized* postal covers aboard—Dave and Jim had even stamped a couple of them right there on the moon. But they had put these others in their personal kits without getting authorization, and were making money on the deal.

Worst of all, the sculptor who had made the Fallen Astronaut figurine had sold some copies of that, and there was more money going to the crew.

I had gotten wind of it just prior to the launch of 16. In fact, I confronted Dave and Al about it the day after Jim Irwin and I flew to the Cape prior to the launch. (Jim was off fishing and couldn't be reached.) They told me what the deal was, and I got pretty goddamn angry.

So I was through with Scott, Worden, and Irwin. After 16 splashed down, I kicked them off the backup crew for 17, replacing them with John Young, Stu Roosa, and Charlie Duke. Since all three of the offenders were active duty Air Force officers, the Air Force punished them, too, with letters of reprimand. Jim Irwin had been planning to retire following 17; I told him it was fine with me if he left right now. Since he had been spending a lot of his time since 15 on religious activities, he became a full-time preacher.

I told the Air Force that they could have Al and Dave back, if they wanted them. Al wasn't interested in putting on the blue suit again, so he made some calls to NASA people other than me and wound up with a job at the Ames Research Center. Dave got placed at the Manned Spacecraft Center as a civilian.

It was particularly tough on Dave Scott, since up to the stamp thing he had been a pretty good bet to become a general. I know he wanted to go on and fly the shuttle, too. But he was the commander of the crew, and if anyone was to blame, he was the guy.

In August the three of them had to testify before the Senate. They came clean and took their lumps, but I was still pretty pissed off about it. We had spent years trying to operate above board, within guidelines. We even had to review everybody else's dealings.

It turned out that the same dealer in Germany who'd sold the 15 stamps had an old project with most of the Apollo crews, getting them to sign blocks of stamps commemorating their flights. The money was something like $2,500 each.

Not all of the guys on the crews did, but some went for it. One of those who did was Jack Swigert. But when we first asked around, he denied that he'd signed any envelopes. A few weeks later he changed his mind and admitted that he had, though. That was enough for George Low, who was having to deal with this. Jack Swigert wasn't going to be on ASTP.

In the last half of 1972 it seemed there was a constant stream of key people leaving NASA. Jim McDivitt resigned as head of the Apollo program to go into business. Wernher von Braun, who had been at HQ for a couple of years, did the same thing. Bob Gilruth had left the MSC director's job earlier. It wasn't just the big guys, either. Employment at the Cape went from twenty-six thousand down to thirteen thousand around that time, too.

At the same time we were looking at *four* manned flights in the next year or so, including the longest ever attempted.

The first one to get ready was the last Apollo J mission, 17. This was Gene Cernan's crew, with Ron Evans and Jack Schmitt. The landing site was in the Taurus Mountains near the crater Littrow—Taurus-Littrow—and, since everybody figured this was the last chance at a manned lunar landing for the foreseeable future, the flight was packed full of scientific experiments. It would be the last manned Saturn V—number 512—and the last lunar module, number 12. The crew chose *America* as their name for the command module and *Challenger* for the lunar module.

Orbital mechanics dictated that we launch 17 at night. All through the evening of Wednesday, December 6, 1972, we struggled through hours of glitches and holds—all the way into December 7, Cape time—when 17 finally lit up.

It was literally like a second sunrise. I hear it lit up the whole East Coast of the United States as far away as North Carolina.

The flight was smooth from launch all the way into lunar orbit. A good thing: in order to reach Taurus-Littrow, 17 had to be put on a trajectory that really didn't allow for a free return in case of trouble. We were all banking on everything working right.

On the eleventh, day four of the mission, Gene and Jack climbed into *Challenger*, buttoned up, and separated. "Okay, Houston, the *Challenger* has landed," Gene said. They had had a good jolt when they set down, though. "Boy," we heard him tell Jack Schmitt, "when you said 'Shut down' and I shut down, we dropped, didn't we?"

They went on to do the longest and most complex EVAs yet, over twenty-two hours' worth. It was great to have professional geologist Jack Schmitt there when the crew found orange soil on the moon, too.

They finished up with a nice ceremony about the end of the "first phase" of lunar exploration, and three days later, on the nineteenth, splashed down in the Pacific.

Apollo was over.

Right after 17, Chris Kraft called me into his office. There I found him with Tom Stafford and Vance Brand. "You guys are the crew for Apollo-Soyuz," he said. It was good news and bad news, because Chris went on to explain that Tom was being assigned as commander of the mission: I was the docking module pilot.

We were also assigned a four-man support crew, one more than normal because we would need to have a representative in Moscow during the flight: Bobko, Crippen, Dick Truly, and Bob Overmyer.

It was a little deflating—hell, here I was, the senior American astronaut—but I understood the reasoning. If you're going to fly a high-profile international docking mission, you want an experienced pilot commanding the thing. And Tom was one of the best.

Not only that, but he had been sent to the USSR in June 1971 for the funerals of Dobrovolski, Volkov, and Patsayev. That particular assignment wasn't anything he lobbied for; he just happened to be available. Nevertheless, he had already met several cosmonauts and Soviet space officials. I accepted the decision. What the hell . . . flying is flying. And I was finally getting a chance.

25

APOLLO-SOYUZ

The American crew for the Apollo-Soyuz Test Project was announced on January 30, 1973. It turned out to be the last crew to be assigned for an American spaceflight for the next five years.

At the press conference a couple of days later I had a chance to publicly thank Bill Douglas and Bob Gilruth for finding a way to let me keep flying planes following my grounding in 1962. (Bill Douglas paid a price, I think. He complained so much about the bureaucratic bungling that cost me Delta 7 that he really hurt his Air Force career.) I also gave a nod to Chuck Berry and Hal Mankin of Mayo Clinic for getting me requalified. Then I thanked Chris Kraft for making the decision to put me on the crew.

I made plans to give up the director of flight crew operations job as soon as necessary. It wasn't necessary right then in January 1973, however. The ASTP launch wasn't going to take place until July 1975—two and a half years down the line. We didn't even have a Russian crew named yet, though there were rumors it was going to be Shatalov and Yeliseyev. I could keep up with my Russian language lessons and still do my job before any real intensive training began.

There were going to be at least three manned American missions during 1973. After years of postponements and being treated like a stepchild to Apollo, the Skylab program was finally ready for launch. The plan was to orbit the five-story-long workshop itself on a Saturn V number 513 on May 14, 1973, followed a day later by the first crew—Pete Conrad, Joe Kerwin, and Paul Weitz. They would activate the station and spend twenty-eight days working aboard it.

The next crew, Al Bean, Owen Garriott, and Jack Lousma, would shoot for fifty-six days aboard Skylab beginning in July. A third crew of Jerry Carr,

Ed Gibson, and Bill Pogue would do fifty-six days and maybe a little more, depending on results from the first two missions, beginning in November.

The three manned missions were to use command modules 116 through 119 and Saturn IBs 206 through 209. (There were four CSM/Saturn IB combinations available for Skylab—three for the visits themselves, one set for a possible rescue mission. It was the first and only time we had a realistic chance of doing something like that.)

There were four main objectives to Skylab. One was just to prove out the basic engineering of surviving in space for a month or two months. Nobody had managed to do that yet. Then there were a bunch of experiments in solar astronomy (using the Apollo Telescope Mount on the front of Skylab), earth resources, and medical.

Some of these medical experiments, as they developed over the years, had been pretty wild, if not downright humorous. For example, at one point the doctors thought it would be great if they could have a biopsy from a Skylab crew member during the mission. Well, I don't know about you, but a making a biopsy is a bit trickier than giving yourself an injection. (We were even training some guys to do emergency dental extractions.)

Somebody came up with a tool that would be sort of a do-it-yourself biopsy kit. Slam it right down on your thigh, press a button, nothing to it.

Well, I insisted that this kit be demonstrated—with one of its designers as the subject. Sure enough, the guy just yowled with pain as this thing tried to rip out a hunk of his thigh. Back to the drawing board.

They also wanted to run a heart catheter into one of the crew members, to monitor heart functions during liftoff, zero-G, and landing. Fine in theory, but not so fine in practice. Maybe it was my own history with cardiologists—I hadn't been impressed by too many of them over the years—but I was worried about the effect on the crew member. So we arranged another demonstration, and damned if the subject didn't put himself into cardiac arrest. Fortunately, it wasn't fatal.

They finally did try the heart catheter, but not until a Spacelab shuttle flight in 1991.

LBJ died of a heart attack in January 1973 and on February 17 the Manned Spacecraft Center was renamed the Lyndon B. Johnson Space Center.

The Saturn V carrying Skylab was scheduled for liftoff on May 14, 1993. It looked kind of funny sitting out there on the pad without an Apollo CSM on top of it. Kind of sad, too, since it turned out to be the last Saturn V ever launched. The first crew was supposed to be launched aboard a Saturn IB the next day.

Skylab took off right on time at one-thirty in the afternoon. We were all watching and listening from the press stand as things went fine through the first couple of minutes.

Right as the stack moved through Max Q, a little over a minute in, telemetry showed something funny: the meteoroid shield had supposedly deployed. This was a shell around the skin of the workshop that was supposed to be opened slightly, giving the vehicle a jacket of sorts. But only after the whole shebang was in orbit.

Then there was a report that one of the two big solar panels had deployed, too. Since everything else seemed to be going fine, no one worried about it too much right then. We all thought it was a bad signal. The S-IC burned out and the S-II lit up. Ten minutes after liftoff, Skylab was in orbit.

As Skylab made its first pass around the world, all its automatic systems went on line, powering up and deploying—except for one of the big solar panel booms. And the meteoroid shield.

Skylab needed about eight kilowatts of power to operate. There was a set of solar power panels on the ATM, and they had deployed. But they were only capable of providing about forty percent of what the workshop needed. One panel seemed to be gone, and the other was just putting out a trickle of power.

And then the temperature started to rise inside the workshop. (The meteoroid shield not only protected from collisions with small space debris, it served as a shade for the skin of the workshop.)

That's when we realized we had a sick bird. Bill Schneider, the Skylab program manager, announced that the launch of Conrad's crew had been postponed to May 20 at the earliest.

One team began trying to figure out what kind of missions we could fly with the station crippled the way it was. Another bunch began to figure out what could be done to fix it.

Within a few days everyone concluded that the crew needed to erect a new shield around the workshop itself. There were three candidates, but the winning solution was a huge parasol. The crew could basically push it out the side of the workshop through a small air lock designed for scientific work, unfurl it, then pull it back up snug against the exterior. The tent pole could be taken apart piece by piece as the whole thing was pulled back.

This parasol—the sun shade—had to be designed, built, and tested, and the crew trained, all in the space of a few days. (We couldn't just let Skylab operate with its current internal temperature, 140 degrees, because eventually all the food and film aboard would get cooked.)

Rusty Schweickart and Story Musgrave spent a lot of time in the neutral buoyancy simulator—the big pool where astronauts practiced EVA in weighted pressure suits—testing the deployment, and found that it would work.

We also figured that some debris was keeping the surviving power boom from deploying fully. It was showing some power generation, not much, so we knew it was still there. It was possible the crew might do an EVA and cut away the debris. That called for some special tools.

Three days before launch, then, the crew tried out the various techniques, dividing up the work. Pete Conrad and Paul Weitz would be ready to deploy the parasol; Joe Kerwin would be trained for a backup method (a flat shield that would be deployed from the ATM on the nose of the workshop); Weitz would also be ready for a stand-up EVA to try to free the stuck power boom.

The challenge was also to get the equipment built and configured so it would fit inside the command module—and be delivered in time. The two shades weren't installed in the command module until two A.M. Friday the twenty-fifth. The crew was awakened less than a couple of hours later.

The damn thing is, it all worked. Following launch, the crew flew up close to Skylab. P.J. leaned out of the command module with Joe Kerwin holding on to his legs. He could see that there were pieces of metal bent around the surviving boom, but he couldn't tug it free. They buttoned up and tried to dock; that didn't work right for a couple of hours, but eventually they got aboard.

Within two days the crew had erected the parasol and temperatures had begun to drop back to what we wanted. That allowed the Skylab mission to go ahead, even though it was with reduced power.

On the fourteenth day of the mission, June 7, Pete and Joe did an EVA to free the solar power wing. It pretty much put the workshop back to full capacity and made it possible for us to go ahead with planning for the next two visits.

It was a pretty wild month.

The Russians announced their ASTP crew on May 24, 1973, the first day of the Paris Air Show. We had expected a crew of veterans, with the smart money on Vladimir Shatalov and Alexei Yeliseyev, and with Andrian Nikolayev and Vitali Sevastyanov as another possibility.

But the crews were something different. Prime crew—first crew, in Soviet terms—was going to be Alexei Leonov, the first man to walk in space, with flight engineer Valery Kubasov, who had flown on Soyuz 6 in 1969.

Second crew consisted of Anatoly Filipchenko, another veteran (Soyuz 7), and civilian engineer Nikolai Rukavishnikov (Soyuz 10). This wasn't a backup crew per se, but a true alternate crew. The Soviets planned to fly a fully manned dress rehearsal mission, and Filipchenko-Rukavishnikov were going to fly that sometime in 1974.

The third and fourth crews—which were a combination of backup and support crews—consisted of four rookies, the first time the Soviets had ever identified unflown cosmonauts before a mission. The third crew was commanded by Vladimir Dzhanibekov, with civilian engineer Boris Andreyev. The fourth crew was commander Yuri Romanenko with engineer Alexandr Ivanchenkov.

Tom Stafford flew to Paris and made a joint appearance with Leonov. Vance and I were tied up with Skylab.

We didn't know it at the time, but the reason Leonov and Kubasov were assigned is that their planned sixty-day Salyut mission had just been scrubbed. On May 10, 1973, the USSR launched a massive satellite eventually known as Kosmos 557. It was actually a civilian Salyut space station intended for manned occupation. Leonov and Kubasov had been scheduled to fly to it.

But Kosmos-Salyut developed severe problems during its first orbit, losing propellant and never really coming under control. The Soyuz launch was postponed for a year, and as I understand it, Shatalov, the cosmonaut training chief, called Leonov into his office and said, "I want you to fly the Soyuz-Apollo mission."

"But I don't speak English," Leonov said.

"You have two years and two months in which to learn," Shatalov told him . . . which pretty much dates the conversation to May 1973.

Two months later, in July, the eight cosmonauts in the crews, plus Shatalov and Yeliseyev (who were indeed going to fly with us, until the Salyut failure), came to the United States. We met them in Long Beach, California, where we were working at Rockwell.

I'd seen a Russian cosmonaut before—I even met Leonov, in fact, in Athens in 1965. Beregovoy and Feoktistov had come to the United States in late 1969, and so had Nikolayev and Sevastyanov back in 1970. Those were social visits, but they confirmed my gut feeling, which was that the cosmonauts were basically a lot like us—pilots and engineers—and their pilots were actually pretty polished. Leonov was jolly and friendly, always joking, a lot like Wally Schirra. I never got much of a read on Filipchenko, but Dzhanibekov and Romanenko were pretty well-rounded guys, very at ease with public speaking and all the other political crap we had to go through.

The engineers were actually harder to get to know. Valeri was a nice guy, but very quiet. Rukavishnikov had a pretty dry sense of humor, but still deferred to Alexei and Anatoli. The backup and support guys spent more time with Andreyev and Ivanchenkov.

There weren't any unusual preparations on our side, no security briefings. Everyone assumed that we would be discreet . . . especially since NASA was a civilian agency, we actually didn't have much to hide from the Soviets. It certainly wasn't like we were spying on them, though obviously we were interested in what we learned.

At that first meeting none of us was very conversant in the other's language . . . and since our plan was that they would speak English and we would speak Russian, it made communication difficult. They had translators with them—KGB, we assumed—and we were never alone with any of them. Dzhanibekov and Romanenko were the best at English at that stage: maybe that was why they were assigned to the mission.

The Russians went home on July 21.

The first Skylab mission ended on June 22 with a splashdown in the Pacific. (On June 18 the crew had broken the Russian space endurance record; Shatalov had sent us a telegram of congratulations.) The crew wasn't in terrific shape—Joe Kerwin was really miserable for the first few hours back—but it was nothing they didn't get over in a couple of days.

The crew had proved that we could do effective work for a month. The next crew, Al Bean, Owen Garriott, and Jack Lousma, was going to fly twice as long. (They were also going to deploy a new sun shade and replace some failed equipment on the workshop).

They were launched on July 28. Their only immediate problem was a leak from one of the RCS thrusters aboard their service module. It didn't prevent a docking, but it made us start thinking about the effects of two months in space on that equipment.

So we hauled out some of our plans about possibly flying a five-man CSM to pick up the crew, if necessary. (We could fit two more seats inside the command module and launch it with a crew of two, Vance Brand and Don Lind.) It was just a contingency plan. With Skylab we had the luxury of taking a wait-and-see attitude.

In late September the Russians resumed manned flights. Soyuz 12 was launched with a crew of Vasily Lazarev and Oleg Makarov. They were supposedly testing a Soyuz that was now modified to carry two guys in pressure suits. It only lasted two days and really didn't seem to prove anything. I guess it was the vehicle Leonov and Kubasov should have flown to their Salyut; they needed to have the manned test flight before they could go on to other missions.

The first big American ASTP party—fifty people—arrived in Moscow on October 1, 1973. Our first destination was the cosmonaut training center ("TsPK" was its Russian abbreviation) named for Yuri Gagarin. The Gagarin Center was an installation of the Soviet air force complete with is own commandant, a Major General Nikolai Kuznetsov. It was located about thirty miles northeast of Moscow at a village called Zvezdny Gorodok, "Star City" or "Star Town." Star Town and the Gagarin Center had been built beginning in the summer of 1960; by the time we visited, it had three thousand permanent residents. Not only the air force cosmonaut detachment, but their families, various training officials, instructors, and support people.

(There was a whole separate civilian cosmonaut team based at the Korolev Bureau in Kaliningrad, which was another Moscow suburb north of the city. The Korolev Bureau designed and built the Vostok and Soyuz spacecraft, and also ran the flight control center nearby. Kubasov, Rukavishnikov, Andreyev, and Ivanchenkov were from the Korolev Bureau; they literally worked for another organization and did not live at Star Town. They only commuted in for training sessions.)

We were never alone at the TsPK. For one thing, we learned pretty quickly that there were quite a few air force cosmonauts, about forty-five of them at the time. On our first day at the center we were shown the gym—there was always a gaggle of guys there.

I have to admit that I was mildly surprised at Russia during that first visit. Things were better than I had thought they would be. Partly this was because Bob Overmyer, who had already made a trip to the USSR, brought a metal suitcase with him full of snacks—nuts, peanut butter, what have you. I said, "Come on, Bob, don't they have food in Russia?"

"Yeah, but not food like we eat. They had caviar for breakfast. . . ."

Well, caviar for breakfast didn't bother me, let me tell you. I also liked their soups and breads. . . . I got along fine.

We lived at the Hotel Intourist downtown. In the basement they had a big bar which was just overrun with Finnish and Swedish businessmen. We spent a little time drinking and partying down there with them.

I remember that on each floor the hotel had what they call a "*klyuch* lady," a key lady, whose job it was to let you into your room with this big goddamn key. Didn't matter what time of day or night it was, one of them was on the job. We tried to sneak around her, but the rooms were bugged, so it didn't make much difference: they knew when you got in and when you went out.

We assumed we were under surveillance all the time. It never bothered us; it was just a fact of life. For example, in addition to having some drinks at the Intourist, we went over to the Marine bar at the U.S. Embassy a couple of times. One of our party decided to leave and walk back to the Intourist. It was late at night, and he'd had a couple of drinks, so naturally he got lost. Well, once it was obvious that he was lost, a car just showed up by magic, and the driver offered him a ride back to his hotel. Which he took.

Vance and I found some time away from the people who were tailing us, but it happened by accident. Independently we each decided to go shopping in GUM, the big department store in downtown Moscow, and we ran into each other. By watching out for each other, we managed to lose our tails in the crowd and sneak a little free time . . . not that we did anything with it. It was like a big game.

Because I'd been off in Moscow for ASTP, Al Shepard had taken over my day-to-day work in flight crew operations. The Skylab 4 crew—Jerry Carr, Ed Gibson, and Bill Pogue—had been launched to the workshop on November 8, 1973. And within a couple of days there was a good old crisis.

It turned out that the whole crew was pretty sick during the first couple of days—with Bill Pogue being the sickest. (That was the last thing you'd expect: he had flown with the Thunderbirds.) The fact that they were sick wasn't the problem . . . the problem was they didn't tell anybody. And, in fact, they tried to make sure nobody found out about it.

They forgot, of course, that Skylab was bugged. All the Channel B tapes—the recordings of the extraneous conversations inside the station—were automatically stored and fed down to Houston. There they were in the goddamn transcripts the very next morning talking about tossing the airsick bags overboard and not saying a word, how everybody in NASA would do the same thing.

Well, the shit really hit the fan. Here was NASA's very own Watergate. Of course, what do you do to discipline a crew in flight? Wally Schirra had made his own decisions on Apollo 7, whether anyone liked them or not. It wasn't as though we could send a cop up there.

So Al had to deliver a chewing-out to Jerry Carr in particular—he was the commander. The guys were contrite, and everybody moved on.

I think the incident colored their dealings with mission control, however. The guys on the ground tended to overload them with work, and the crew complained about it for the next eighty days.

While the crew for 4 was in orbit, the Russians launched Soyuz 13 with a crew of Pyotr Klimuk and Valentin Lebedev. They were probably the youngest space crew ever flown—both of them just thirty-one years old. (I found out later they had originally been assigned as backups to two more senior guys, but had gotten swapped about ten days before launch.)

Soyuz 13 carried an astronomical sensor called Orion and the crew aimed it at Comet Kahoutek, which was on its closest approach to earth. There was some talk of hooking them up by radio with Skylab, but we weren't set up for that yet. Klimuk and Lebedev came home after eight days in space.

Carr, Gibson, and Pogue splashed down on February 8, 1974. After eighty-four days in space they were in better shape physically than either of the other two crews, meaning we'd obviously gotten a handle on the exercise and health maintenance routines. A couple of weeks later we ran them through debriefing, with all the ASTP crew members attending.

There was some talk about possibly sending a crew to make a twenty-day visit to Skylab—we had the standby CSM and Saturn IB. But there wasn't any money for it. We figured the next visits to the workshop would be by shuttle crews.

With the end of Skylab 4, the only future American manned spaceflight was ASTP. We were next on the runway.

I gave up the title of director of flight crew operations as ASTP took up more of my time. Chris Kraft suggested that I might want to move on to the shuttle program and manage the approach and landing tests (ALT), which was an idea I liked. But with ASTP training I had very little time for that: during late 1974 and early 1975 my former deputy in the flight crew directorate for Skylab, Tom McElmurry, handled ALT for me.

Al Shepard, who had returned to the Astronaut Office after Apollo 14, in March 1971, moved up to become deputy director, replacing McElmurry. Kenny Kleinknecht succeeded me as director, but only briefly. He got crosswise with NASA management because he took some hunting trip that was paid for by a contractor, and by the end of 1974 he had been replaced by George Abbey.

Kraft was the guy making these decisions: neither Kleinknecht nor Abbey were astronauts. Kraft's idea had always been that since flight directors were in charge of missions, flight *crew* operations—that is, the pilots—should be under him, too. And now they were.

Other Voices

KENT SLAYTON

The time when Dad was training for Apollo-Soyuz was probably one of our happiest. He had never sat around complaining openly about not being able to fly—the only thing he ever said to me about it was that it was just one of those bad breaks. You have to learn to deal with them.

One time we were working out in the yard, and Mom got worried that neither of us was talking. She encouraged Dad to be more open, to give me some fatherly advice. Dad thought it over for a few seconds, then said, "Son, always take a nap and use the head whenever you get a chance." Then we both started laughing. I would have to admit that Dad wasn't much of a philosopher. But he was clearly a happier guy when he got requalified.

We had some of the astronauts and cosmonauts, and even some Russian space people, out to the house a couple of times. We also had more time for some family trips; we went on a safari in Africa with Jim Lovell and his family.

The only bad thing I remember is that Grandpa Slayton died the same year Dad started flying again. It was right before a hunting trip that Dad and Uncle Howard were supposed to take. So they had the funeral and, according to his wishes, Grandpa was cremated and his remains scattered on the farm.

Then Dad and Uncle Howard went off to Wyoming. Some families might have thought that strange, but not ours. Grandpa would have wanted it that way.

In terms of getting ready for ASTP, the only new mission-critical item was the docking module, which was my responsibility. Not only did the docking module make it possible for Soyuz and Apollo to link up, it also served as a sort of regulator for the different environmental systems: Apollo was a 5 pounds per square inch pure oxygen system, while Soyuz was at 14.7 pounds per square inch with mixed gas. Inside the docking module we had 10 psi with a mixed nitrogen/oxygen atmosphere.

We also planned a series of joint scientific experiments, and some free-flying between the two spacecraft. I spent a lot of time in the simulator training to fly the command module. I'd be damned if I was going to wait all these years to get into space and then not have a chance to actually fly the spacecraft.

That part of the training went smoothly. The language stuff was a real bear, however. None of us had done any further Russian language study since our visit to Russia in October; seeing that the Russians were getting much better in English made us realize we'd better get serious, and right about this time we got several additional instructors.

I never really got comfortable with the Russian language to the extent that I could have conversations. Partly this was because we concentrated on technical exchanges—knowing the words for docking and pressure, things like that. I could barely get along in a social setting.

When I added up the hours later, I found I spent more time studying Russian than doing any other kind of training for Apollo-Soyuz.

As I moved out of the flight crew operations job and concentrated on training for the mission, I found myself with a whole new kind of lifestyle. I had always done a lot of traveling, so that didn't change—most of it had been Houston to the Cape and back. And, with the restrictions, always with a second pilot.

So I really enjoyed being able to go out and grab a T-38 by myself. During the training for ASTP, however, I came close to wiping out. I somehow managed to put a T-38 into a flat spin over the Gulf, something that was supposed to be impossible. Impossible or not, there I was, whipping around and around.

I got her straightened out in plenty of time, but it was a close call, the closest I'd had in years.

On our next visit to Russia, in July 1974, without any advance warning we were taken directly from the airport to Star Town . . . driven right through the gates in our KGB staff cars, to a brand-new hotel that had been built on the site. The Hotel Cosmonaut, it was called, and it had been built just for us—American astronauts and training officials. (They used it later for other "foreign" visitors like the Interkosmos pilots who flew Soyuz missions between 1977 and 1981. And for the French and Indian pilots, too, at least on their initial visits.)

The rooms were pretty plush by Russian standards, with all kinds of furniture—too much, in fact—but also typically Russian: there weren't curtains on the showers, for example. Doorknobs would be missing.

They built us our own restaurant and our own bar right in the hotel, to make things easier for everyone—it certainly kept us from associating with the residents of Star Town.

It got so that whenever we wanted something, all we had to do was speak . . . the walls had ears. One day we decided to test it, and complained loudly that we didn't have anything to do. "Too bad we don't have a pool table."

The next day, by God, there was a pool table in our bar downstairs. It was probably the only pool table in Russia: it had square corners and balls that were too big for the pockets. You could play one game all night.

Then there was the time Bob Overmyer was at a meeting at the flight control center over in Kaliningrad. He decided to move his chair and tried to do so: the thing wouldn't move, so he yanked on it . . . and some wires came out of the base. Everybody just sort of looked the other way.

We called this the speak-into-the-lamp mode. We actually used it to get rid of a couple of technical people on the Russian side who weren't particularly productive: just complained about them among ourselves, when we knew someone would be listening. And we never had to deal with them again.

Don't get the impression we were locked up, however. We could go out anytime we wanted: there was a staff car available to drive us wherever we wanted to go. (It even trailed Vance and me when we went running out on the highway.)

Nevertheless, we didn't have a lot of dealings with Star Town. It was actually a military facility belonging to the Soviet air force. The two officials we dealt with were former cosmonauts, Vladimir Shatalov and Georgy Beregovoy. Shatalov was the director of cosmonaut training, a lot like the wing commander at an air force base. Beregovoy was the director of the Gagarin Center itself, having taken over from Kuznetsov: he was like a base commandant.

During that July 1974 visit they had a flight going on, Soyuz 14, which had docked with their Salyut 3 space station. The crew was two military guys, Pavel Popovich (from Vostok 4 back in 1962) and Yuri Artyukhin. No one said much about what was going on up there, which made sense later when we figured out it was a military flight, basically a Russian version of the Manned Orbiting Lab.

It was during this trip that we ran through simulations of different flight events with the American crew speaking Russian. We also spent time in the Soyuz simulator at Star Town.

The first time I ever really got some private time with a cosmonaut was during the Russians' trip to Houston right after this, in September 1974. We were out at the Johnson Space Center one day, and when we took a break I put Leonov, Kubasov, and Stafford in my Camaro and said I was taking them to a sporting goods store downtown. I added that I didn't have room for anyone else, and off we went.

Shortly after that, thanks to Tom Stafford and a pal of his at Learjet, six of us were invited for a antelope hunt out in Wyoming. A Learjet doesn't hold any uninvited guests . . . so it was Tom, Vance, and me together with Alexei, Valery, and Vladimir Shatalov out on a ranch.

Vance, who was our best Russian speaker, found himself baby-sitting Shatalov, while the rest of us were off hunting and drinking vodka. We had a great time . . . although I can't say that Alexei and Valery acted any differently without their "minders."

During the Russians' visit a new manned flight was launched, Soyuz 15, on August 26. The crew of Gennady Sarafanov and Lev Demin rendezvoused with Salyut 3 on the twenty-seventh (Russian rendezvous techniques required them to spend twenty-four hours playing catch-up with the target), but couldn't dock. They were back on earth the next day, and the Russian press talked about the mission as if it had been a complete success—a "test" of automatic docking systems. Well, this was complete horseshit and we knew it: something had gone wrong with the docking.

It wasn't of immediate concern to ASTP—Apollo would be the active vehicle with Soyuz as the target, and we were confident in our system. But it raised some questions in the public's mind about Russian reliability.

Tom and Shatalov had a discussion, which wound up with the two of them holding a press conference. Shatalov admitted that something had gone wrong with the automatic docking system, but that was as far as he went. (Sarafanov and Demin were to have spent two or three weeks aboard Salyut 3.)

The real ASTP dress rehearsal was supposed to be a manned flight by Filipchenko and Rukavishnikov aboard a model of the Soyuz that could operate autonomously for eight days, and which would also carry the androgynous docking adaptor.

The Russians kept mum about the actual launch date—their "tradition"—and so we didn't know exactly when it would take place. (NASA's attitude was if you tell us, we'll tell anyone who asks. We couldn't be in a position to keep Russian secrets from anyone.) On the other hand, we planned to track the new Soyuz to test our ability to do that.

What happened was that Vladimir Timchenko, one of the Russian flight controllers, called Glynn Lunney's office shortly after the launch, 6:35 A.M. Houston time. Lunney wasn't in, but a guard took the message. A couple of hours later he and Timchenko talked; we got the orbital parameters and began tracking the flight.

Filipchenko and Rukavishnikov thumped down in Kazakhstan right on schedule on December 8. A good flight with no problems.

Right on the heels of this, on December 26, the Russians put a new Salyut station in orbit—Salyut 4. Two weeks later a crew was sent up to it,

Soyuz 17, with Alexei Gubarev and Georgy Grechko. We wondered about the Russians' ability to control two missions simultaneously, but were given to believe that operations aboard Salyut 4 would be over by July.

Leonov and Kubasov and company came back to the United States on February 7, landing in Washington to show the flag, then down to the Kennedy Space Center. We spent a day taking them through procedures down there, including a visit to CSM number 111, which was still in the Vehicle Assembly Building. Then it was back to Houston, with a detour to Disney World.

We spent the next two weeks in flat-out training—mission simulations with the prime and backup crews—and figured we were getting pretty comfortable with the language and procedures.

Our last Russian trip before launch was scheduled for April 14–30, 1975. Before that could happen, another political and technical problem cropped up.

Cosmonauts Gubarev and Grechko had come back to earth in mid-February after a thirty-day mission aboard Salyut 4. The next flight, scheduled to last sixty days, was supposed to begin on April 5. Once again, we didn't have any advance warning about this—especially since it wasn't directly related to ASTP the way Soyuz 16 was.

Well, they had a big problem. The Soyuz took off all right, with cosmonauts Vasily Lazarev and Oleg Makarov, but when it came time for the big core booster to shut down and separate, all it did was shut down. There was the Soyuz and its upper stage, still sitting on top of the booster, only the booster is spent. It's nothing but dead weight.

The Soyuz had a launch escape tower, same as we had in Mercury and Apollo, but that dropped off earlier in the launch phase. All they could do was separate the Soyuz from its upper stage and go through reentry. In essence, they did a suborbital mission not much different from Al Shepard and Gus Grissom.

The crew wound up pulling about eighteen Gs, however. I guess it was kind of tough on them. And they came down in the damn mountains near the Chinese border, miles away from any rescue teams, on the side of a mountain. They started sliding toward a cliff and were saved from going over only because the parachute snagged in some trees.

Anyway, the Russians knew the crew was safe, but it still took a couple of days to get them out of there. Nobody really knew what the deal was until the seventh, and now some politicians were really yelling about Russia's safety record.

Konstantin Bushuyev, the Russian program manager, talked to Glynn Lunney on the eighth, assuring him that the booster used on "the April fifth anomaly" was an older model, and that the problem couldn't recur on ASTP. We were willing to bet that it wouldn't, but not because we believed that excuse.

We arrived in Russia on our third and last visit on April 14. During that first week we got our first visit to Kaliningrad, to the Russian flight control center—TsUP. It had just gone into operation that summer; all the early Russian manned flights had been controlled out of Yevpatoriya down in the Crimea. (They had to be down there at the start because Moscow was too far north to be in direct contact with an orbiting spacecraft. Once they got a network of tracking sites built, it didn't matter where they put TsUP, so they made it handy.)

It was an impressive place, especially since it was new. There was a lot of construction going on in Kaliningrad and at Star Town. Even though we were getting out of the manned space business, the Russians weren't.

The real big event was that we got an unplanned and unprecedented trip to the Russian launch center at Baikonur, out in Kazakhstan. It started with a request from Glynn Lunney, the head of ASTP on our side, to Viktor Legostayev, his Russian counterpart, though it came from the crew. (Tom Stafford really pushed it.) We wouldn't fly in space in any vehicle we hadn't been in on the ground. It was that simple.

So one day they came to us and said, tomorrow we're going to Baikonur. The morning of the twenty-eighth it was out to the airport and into a plane. I don't think it was any accident that it was night when we got to Baikonur, which was about as far from Moscow as Houston is from Los Angeles. The Russians made no bones about the fact that this was a military facility, and they had no intentions of opening it up: we were there to see hardware relating to their "international" programs. National programs were off limits.

We spent the night in downtown Leninsk at the same hotel Russian cosmonauts lived in before their flights. The next morning they loaded us onto a bus for the drive north, across the Syr Darya River, to Baikonur. It was a huge facility spread over miles of steppe, with no real launch complexes visible from the highway, though you would see tank farms and support equipment. Everybody was military. There were some flowers blooming, but you could tell it was closer to desert than anything else, more like Edwards than the Cape.

One of our first stops was the MIK, the spacecraft assembly building. Unlike us—we stacked our vehicles vertically—the Russians assembled their spacecraft horizontally, then drove them out to the pad on a train, elevating them into place. This gave them the ability to launch a lot of vehicles quickly, since the bird didn't go to the pad until two days before liftoff. With the Atlas and Titan we had the pads tied up for weeks. Even the Saturn V, which got trucked out to the pad from the vehicle assembly building, still sat there for weeks at a time.

It was a very efficient system. The whole thing was a little cruder than ours—with much less emphasis on keeping stuff "clean." (The MIK was pretty much a huge barn. I got the impression that sometimes they worked on vehicles with the doors open.) But it worked. They had a bunch of Soyuz vehicles in there, too, in various stages of assembly. And here we were lucky to have a single goddamn Apollo to fly.

We got to crawl into our Soyuz and made ourselves somewhat at home. On the way back to Leninsk we stopped and planted trees, a traditional activity of each cosmonaut before his flight. A few days later we were back in Houston getting ready to move down to the Cape.

26

ORBIT

Tom and Vance and I went into medical isolation on June 24. From this point on, we were allowed to deal only with a restricted number of people, to keep us from catching some last-minute cold. I needed all the spare time I could get to keep working on Russian, among other things.

We spent the last week of June and the first week of July in Houston going through simulations in addition to a countdown demonstration test at the Cape. We moved down there on the thirteenth for the launch.

There was a new controversy about ASTP out in the real world, most of it coming from Senator William Proxmire. He had raised concerns about the safety of the Soviet spacecraft before this—pretty late in the game for a guy who had done everything he could to keep the American manned program from getting any money over the years.

But there was a legitimate concern about tracking facilities, because following the April 5 aborted Soyuz flight, the Russians had gone ahead and launched a new manned mission on May 24. This one carried Pyotr Klimuk and Vitaly Sevastyanov on a sixty-day visit to Salyut 4. Fine, but sixty days from May 24 carried them right into late July—and overlapped ASTP.

It was a questionable decision; I guess the Russians were forced into it by projections of Salyut's operational lifetime. Besides, they had had a hell of a time with their space station programs: the first Salyut got visited by one crew that died; the next three Salyuts either failed during launch or after reaching orbit, and never got visited at all; only one of the two scheduled crews had gotten aboard Salyut 3; same thing to this point with Salyut 4.

The only factor which took a little of the curse off the decision was that they still had two working flight control centers—the new one in Kaliningrad and the old one in Yevpatoriya. So theoretically they could control two flights simultaneously.

There was also another potential problem—but I didn't know about it until long after.

Other Voices

CHARLES BERRY

I had always worried that somebody would find some reason to reverse the decision to requalify Deke for a flight and assign him to a crew. By the time of the ASTP launch, however, I wasn't even working for NASA anymore—I had left after the end of the Skylab program.

Some of the people in the medical branch had insisted on a flight rule for the countdown—if Deke fibrillated on the pad, they were going to go to a hold at four minutes and take him out of the spacecraft. I guess they had inherited some concerns about his suitability for flight, even though by that time a number of physicians had signed off on him.

Well, Chris Kraft was just livid at this idea. He called me and asked me what this was all about—he thought Deke was fully qualified to fly. So I had to go over to the JSC and meet with the doctors, and when I left, there was complete agreement that this rule wasn't necessary . . . that Deke wasn't going to fibrillate on the pad . . . that even if he did, he was a slow fibrillator and would not be adversely affected.

Chris did take the precaution of asking me to go to the Cape for the ASTP launch, even though I was no longer a NASA flight surgeon. "I only want to talk to you," he said.

Sure enough, two days before the launch there was another big fight, because this rule had crept back into the planning. Chris shipped the two flight surgeons—Hawkins and Winter—home. Literally got them away from the Cape.

Tom Stafford knew about the argument, but we kept it from Deke until after the mission.

Deke later gave me a gift, which I still have in my office. It's the cardiac monitor from his ASTP medical harness, the one tracking his heartbeat during the mission, and it's mounted on a piece of tracing paper showing his heartbeat . . . steady, no fibrillations.

* * *

One of the benefits of ASTP was that it got the Russians to open up a little. Aside from letting us visit most of their facilities, they agreed to broadcast the Soyuz launch live.

On July 11 they rolled the prime Soyuz—which we now figured would be Soyuz 19 unless the Russians had some other mission up their sleeve—out to the same pad that Gagarin was launched from. Two days later they moved the backup Soyuz to the backup pad. (They really believed in redundancy.)

Because of the unique orbital dynamics, both launches were scheduled to take place in midafternoon local time on Tuesday, July 15. Midafternoon at Baikonur was the middle of the night in Florida, so we actually slept through the Soyuz 19 launch.

Since our wake-up time was ten in the morning, I was already awake when John Young knocked on the door. It was unusual to be on the other end of that little bit of business—the first time since 1962 for me.

Tom and Vance and I had a gentleman's lunch of steak and eggs with John, Ron Evans, and Jack Lousma. We heard then that Soyuz 19 was safely in orbit and that everything was looking good for us.

We got suited up, got in the truck, and drove out to the pad. I have to admit I felt pretty good walking across the swing arm to the spacecraft . . . what the hell, it was only thirteen years overdue. I never planned on being the world's oldest rookie astronaut, but I wasn't going to complain.

We strapped in, Tom in the left couch, Vance in the middle, me on the right. The only hiccup in the countdown was an umbilical that got hung up. We didn't want to have to stand down and recycle for a launch a couple of days later, because Soyuz 19 would have to come home. (That's what that backup Soyuz was doing on the pad.)

The umbilical problem cleared, and right on time we lit up and took off.

I'd debriefed every Gemini and Apollo crew, so I wasn't surprised by much that happened. The noise at liftoff was greater than I imagined it . . . we had eight engines running back there, and they got even louder as we moved through Max Q, then things began to smooth out.

Whoppo! Shutdown was pretty abrupt. You went from being pushed back in your couch to hanging in your straps. We were in zero-G.

None of us got sick during the mission, unlike about half of the later shuttle astronauts. In fact, there was very little space adaptation syndrome during Gemini, Apollo, and Skylab. My theory was—and is—that you can train yourself out of SAS. That was one of the reasons we had the fleet of T-38 jets, so pilots could take them up and do spins and other maneuvers, to get themselves used to being in zero-G. (Once I even had to publish a memo to that effect, because the T-38s were basically being used as airliners, not training vehicles.)

The first order of business was to move the CSM away from the S-IVB stack, turn around, and come back for the docking module. Tom did the flying and did it perfectly. Then we got out of our suits and got down to work.

One of the interesting things was that in order to be in the same orbit as Soyuz we had headed northeast from the Cape, and we would be flying over territory no American astronaut had ever seen—including chunks of the Soviet Union. I kept trying to sneak peeks out the windows and wished there was a bubble turret on the spacecraft, so I could just sit up there and watch the world go by. But there wasn't a hell of a lot of time for sightseeing that first day. We had two SPS burns to make, first to circularize our orbit, then to put us into an elliptical one which would allow us to gradually close with Soyuz over the next two days.

There was some little problem with the hatch leading into the docking module—we literally couldn't get it open. Since it was the end of a long day, Tom told mission control that we would deal with it first thing in the morning, which we wound up doing.

We folded up Vance's couch to give ourselves more room and eventually went to sleep, two guys in couches, one underneath, which didn't work out so well. From that point on I slept in the docking module itself, except for one night in a couch.

Wednesday was taken up with getting the docking module ready. There were experiments. And I had to get ready to fly the Apollo around Soyuz: I'd spent hours in the simulator training to do orbital mechanics by ear.

On the morning of the seventeenth we began to close in. There was another SPS burn to drop our apogee, which allowed us to speed up as we moved toward Soyuz. We were able to see Soyuz through the sextant, so I got on the VHF radio and talked directly to the other crew—Valery Kubasov—for the first time.

Tom started the terminal phase of the docking, the final approach, as both vehicles passed over the Pacific. The three of us were buttoned up in the command module, in case of a collision. Leonov and Kubasov were actually in their pressure suits. We were sending TV pictures down through a communications satellite, so everyone in the world could watch.

A couple of minutes before 11 A.M., Houston time, Thursday July 17, 1975, there was a Russian Soyuz sitting outside the window a couple of hundred yards away. Dick Truly was the capcom, and told us, "I've got two messages for you: Moscow is 'go' for docking, Houston is 'go' for docking.

It's up to you guys. Have fun." We were just coming up on the coast of Portugal.

Alexei had to roll the Soyuz sixty degrees to line up for docking. Then Tom edged forward. At 11:09 Tom said, "Contact." We heard Alexei a moment later: "Capture!" Then he said, "Apollo and Soyuz are shaking hands now."

All the various crew exchanges had been scripted in advance—these two Americans go to Soyuz while this Russian comes to Apollo, that sort of thing. Given the different pressures in the two spacecraft, you couldn't just open the hatches on both ends of the docking module and go through.

Tom and I were scheduled for the first visit, so I unbuttoned the docking module while the other guys were getting last-minute updates on who was going to be talking to which president when. There was one last-minute problem—I opened the docking module and got a whiff of what smelled like burned glue. A few minutes later it was gone; it had probably come from one of the experiments onboard.

Alexei and Valery opened the Soyuz hatch that led into the docking module at 2:10 P.M. Seven minutes later Tom opened the docking module hatch, and there we were, looking at Alexei and Valery floating in a bunch of umbilicals. "Come in here!" Tom told them. Then Tom and Alexei shook hands and tried to hug.

We got a message from Soviet President Brezhnev, then a ton of honest-to-God questions from President Ford. It was kind of tricky, since we kept having to hand headsets from Tom to Alexei to Valery to me to hear the questions and give the answers. Then we exchanged flags and signed some ceremonial documents.

Other Voices

ALEXEI LEONOV

In 1965 we had a first meeting—Charles Conrad, Gordon Cooper, Deke Slayton, me, and Pavel Belyayev. During a party, Deke Slayton said it was necessary for our two countries to cooperate in space. During this party I made a speech, and I proposed that we should have a drink sometime in the future—aboard our joint Soviet-American spacecraft.

While Valery Kubasov and I were getting ready for our flight, I prepared several special tubes of food—only these said "Russian Vodka," "Stolichnaya Vodka," "Old Vodka." During the flight, I invited our guests—Tom Stafford and Deke Slayton—to drink before eating. And when Deke said, "Alexei, ten years ago you promised me a drink of vodka in our international spacecraft," I took out these tubes.

Tom said it was not possible; there were so many people watching on TV, including the president. I said it was a Russian tradition. Tom and Deke drank . . . but inside the tubes was soup. Deke complained, but I told him it was the thought that counts!

After Apollo-Soyuz, for many years, I was in charge of training cosmonaut crews for their missions, much like Deke Slayton. Before every flight I had to make a very long report—a long speech—to the State Commission about their readiness. But I remember that Deke told me how he reported on American crews and their readiness: "They're ready." Two words.

So one time, instead of my long speech, I simply told the State Commission: "They're ready."

Academician Semonov said to me, "You sound like Deke Slayton."

Tom and I moved back into Apollo around 6:00. The next morning it was Vance's turn to go over to Soyuz. Then Tom brought Alexei back to Apollo, so we spent most of the morning that way: Tom, Alexei, and me in the Apollo, Vance and Valery in Soyuz. There was a TV tour of the Soviet Union, as seen from orbit, narrated by Alexei and Valery.

After lunch Tom and Alexei swapped with Vance and Valery. We got through a televised press conference. Alexei got to show off some of his sketches. Tom did an impromptu speech—in Russian—sort of an address to the Russian people. "Let the things that went on yesterday in our flight, and today, be a good thing for both our peoples." We hoped that Apollo-Soyuz would lead to more cooperation between the United States and the USSR, and it certainly helped—but it took fifteen years.

On Saturday the twenty-first, after two days of joint operations, we undocked from Soyuz and backed away. I was at the controls in the left seat, Vance was spotting in the right seat, and Tom was handling the computer down in the equipment bay. The plan was for me to create an eclipse, as seen from Soyuz, so it could be photographed for a solar experiment. Then I did a manual redock . . . which went pretty well, though I was getting blinded by the sun for the last few feet. I also tweaked the hand controller the wrong way, once we had captured Soyuz again, causing the two spacecraft to shake a little.

Not only was this a test of flying ability, it was also an exercise in fuel management. I had 225 to 250 pounds of fuel alloted to me . . . not much. But I think the exercise taught us something about possible future space rescues.

The Soyuz with Alexei and Valeri moved off at that point. They had two more days in space, ending with a good thump-down on the twenty-third.

We had a ton of scientific experiments and earth observations to keep us busy for the rest of our flight. Everybody had made up a wish list, since this was the last time Americans were going to be in space for quite a few years, until the space shuttle started flying.

One of our experiments was related to the shuttle. The three of us had to make some precise measurements of our height after five days in orbit. People tended to get taller as their spinal column stretched out. This wasn't a problem for Apollo crews, but shuttle pilots needed to be able to use their feet to control the vehicle during landings. The worry was they might not fit in their original couches.

It turned out to be a nonproblem. I got some questions during the flight about how I was holding up, and answered that I really wasn't doing anything my ninety-one-year-old Aunt Sadie couldn't have done.

For reentry Vance and Tom swapped seats, with Tom in the middle following the checklist. My job from the right seat was to get as many pictures as possible and keep track of the altitude, calling it out as we went.

One of the important items on that checklist was to close down the RCS system once we hit forty thousand feet. The reaction control motors were the tiny rockets spaced around the command module that let you orient it: they were fueled with toxic chemicals.

Anyway, either Tom didn't call for RCS close, or he did and Vance just didn't hear it. The small drag chute deployed with a big *whap*, and suddenly we had a cockpit full of yellow gas. We knew pretty quickly what the problem was, because the RCS was still firing.

It got shut down as we rode down on the drogue. That was a rough ride; I hated it. Then the three main chutes came out and we slowed down. We were all hacking and coughing by the time we hit the water.

Then it was my turn to screw up. We were in the water a few minutes, still hacking, when they dropped the frogmen. One of them appeared in the window, and like a dumb shit I gave him the thumbs-up sign. Everything's okay. Well, of course it wasn't. But everybody outside thought it was, so there was no special effort to get us out of the command module. To make things even worse, as usually happened with the command module, it tipped over. We were hanging upside down for a few minutes while the flotation devices inflated.

That was bad. We were all three hacking, with no fresh air. I think Vance might have passed out. Tom was scrambling around down in the bay getting gas masks.

Eventually we were upright with the hatch open, fresh sea air pouring in. We climbed out and got lifted up to the choppers and off to the carrier. I didn't think a whole lot about what we'd just gone through.

It wasn't until the press conference on deck, when we were talking to President Ford, that we even bothered to mention the problem. It came up when we summed up the flight as very smooth, a piece of cake "except for the last four minutes." I guess no one knew how much trouble we'd been in.

They stopped the conference and hauled us downstairs. It was the first indication anybody had had that we might have inhaled that crap.

The doctors started pumping cortisone into us. A good thing, too. We hadn't felt too bad once we got out of the command module and onto the ship . . . but about three-quarters of an hour later, suddenly we all felt like we had pneumonia. A lethal dose of that gas was four hundred parts per million. They estimated we had inhaled it at three hundred parts per million. Pretty close.

For the next ten days we steamed toward Honolulu, with the doctors giving us chest X-rays every few hours. I wasn't feeling too hot there for a while.

One interesting thing happened, though: the X-rays turned up a spot on my lung. So the first thing I did when I got back to Houston was check into the hospital for lung surgery. Fortunately, the lump turned out to be benign.

But if they'd found it before ASTP, I never would have been allowed to fly.

27

SHUTTLE

After getting back on my feet after the surgery, I took Marge and Kent with me, and Tom and Vance and their families, on the traditional postflight public relations trail. (The single bonus about being ineligible for flight all those years meant that I missed all of that crap.) We went back to Russia, then home to the United States.

Marge enjoyed the trip, I know. She seemed real happy that I finally got a chance to fly, and had even taken a shot at old Larry Lamb, the first doctor to screw me up, in the papers on the day of the ASTP launch. Once we were back home, it was back to work.

Chris Kraft had hired me to run the shuttle Approach and Landing Test (ALT) program even before the ASTP flight. I had no interest in going back to flight crew operations, since there weren't going to be any flights in the immediate future. Tom McElmurry, who had been my deputy in flight crew ops for Skylab, moved over to the ALT program before I did and was basically up and running things by the time I came aboard in November 1975.

The goal of the program was simple: to prove that unpowered landings of the shuttle orbiter were possible while also testing the basic aerodynamics of the vehicle itself. The shuttle design had been "frozen" in 1972—after about forty-nine other designs had been considered. All this was based on work with some smaller rocket-powered vehicles called the M2-F2 and the HL-10, which NASA had flown in the 1960s.

Max Faget had wanted the orbiter to have a pair of small, short, straight wings, which were really all it needed to go from reentry to landing.

But the Air Force insisted that the orbiter had to be capable of a fifteen-hundred-mile cross range. That is, from the point it reentered the atmosphere, it had to be able to reach a landing site fifteen hundred miles away in each direction.

The reason for this was that the Air Force required manned launches out of Vandenberg Air Force Base in California for polar orbiting missions. A once-around abort profile without that cross range meant there was no place to land, other than in the Pacific Ocean. (Hell, the downrange abort landing site turned out to be Easter Island.)

The Air Force also had some single-orbit polar reconnaissance or quick satellite deployment missions in its plans. The idea here was that you could launch from Vandenberg and, somewhere over the South Pole or Indian Ocean, deploy a satellite in an orbit where it would be damned hard to track . . . and reenter ninety minutes later back at Vandenberg.

Forget that the Air Force never flew anything like this unmanned. Forget that the Air Force was literally forced to come up with a wish list that included stuff like this because NASA needed its support. (The Air Force got dragged into the shuttle program by the heels. It never would have touched it otherwise.)

The important thing to remember is that in order to have that fifteen-hundred-mile cross range you have to have a big, heavy delta wing on the orbiter. (The delta wing gave better lift than Max's straight wings.)

This impacted the shuttle program in a big way, because the weight of the delta wing meant there was less weight available for other systems—jet engines, for example. And it raised the amount of heat the orbiter would have to survive, which forced us to develop a whole new shielding system.

Although it was the solid rocket boosters that later became everyone's worry, the early concern around NASA was with the idea of an unpowered landing. There were a lot of nonflying types who wanted to stick jet engines on the orbiter for that purpose.

I was one of those who didn't think it was necessary. I knew that a dead-stick, unpowered landing was workable—I'd done a few of them myself. And adding the engines to the orbiter cost you a tremendous amount of weight and drag while getting you very little in return. That is, if you came out of reentry in bad enough shape that you couldn't reach an airfield, you probably had bigger problems: you couldn't carry enough fuel or have enough engine power to make much difference. All you needed to do, really, was get to where you could circle a runaway. Once you reached that point, getting down—with or without power—was not a problem.

Anyway, we formed a task force within the agency to demonstrate this concept. The original plan, baseline 1975, was to fly about a dozen unpowered landings with the first orbiter, which at that time was supposed to be named the *Constitution*, at the NASA Dryden Flight Research Center, located at Edwards Air Force Base, beginning in early 1977. As usually happened, we had planned about twice as many flights as I thought we would need.

There was a lot of simple ground-crunching to be done. We not only had to demonstrate a landing, we had to set up a whole system for ferrying the orbiter—a hundred-ton glider the size of a DC-9—from one place to another. Modifying a Boeing 747 to carry this thing was a bigger challenge than most people realize. I know of one Air Force pilot who saw the shuttle carrier aircraft (SCA) arrive at the Cape one day in 1979 hauling the orbiter *Columbia*, and being more impressed by the carrier than by the shuttle itself. We also had to design and build a structure to literally lift the orbiter into the air so it could be mounted on the SCA, and things like that.

The task force consisted of about six hundred people, many of them contractor personnel from Rockwell, since they were building the orbiters in Palmdale, just down the road from Edwards. There was a smaller number of people from Boeing, who were dealing with the SCA. The rest were NASA people from JSC and KSC, engineers and future flight operations guys.

We also had a pair of astronaut teams, one consisting of Fred Haise and Gordon Fullerton, the other of Joe Engle and Dick Truly. Haise had been a NASA test pilot before becoming an astronaut: he was familiar with the demands of a program like ALT. Joe Engle had flown the X-15 at Edwards during the mid-1960s. Fullerton and Truly were from the MOL group.

On the surface it looked like a handpicked team, and maybe it was: these were the first crews—even though they weren't actually assigned to a flight—selected by George Abbey and not by me since 1962. (I had been sounded out about any objections I might have had, and I had none. I knew these guys very well, and between 1975 and 1977 got to know them even better, since we spent a lot of time flying back and forth together.) We also picked up Bob Overmyer, who had worked on ASTP, as the support crewman. Bo Bobko, another guy from ASTP, did some support and capcom work on ALT, too.

The only bad thing about the ALT period was that my personal life started getting kind of rocky. I think Marge wanted me to move on to something else—she thought that after I got the flight I would find something to do that would keep me at home more. Instead, I wound up doing the ALT program and basically spent the next two years living in a motel in Lancaster, California, out in the desert. Kent was off at school now, too, and that was tough on her.

As far as getting the job done, being out in the desert, communicating with Houston by telecon, was great; there was nobody jumping down our ass. And I've always liked the desert. There were a lot of old friends around from my time at Edwards. It was a great bunch of people and a great time: we had a reachable goal and a deadline, two things that you have to have.

The only real problem was between our task force and the director of NASA Dryden, a former astronaut named Dave Scott.

Dave, of course, was a very capable Air Force test pilot who came to NASA in 1963 with the third group of astronauts. In fact, he was the first of that group to go into space, aboard GT-8 in 1966, and had a good career: command module pilot on Apollo 9 in 1969, then commander of the fourth lunar landing, Apollo 15, in 1971. In some ways, though, Dave was more openly political than most guys, a real Boy Scout, quite intolerant of what he saw as failings in other people. That's why it was a real surprise when he got involved in that stamp business. Chris Kraft and I had to stomp on him pretty hard. After that Dave got involved with ASTP, doing some technical liaison with the Russians. He was very capable, and when ASTP ended he was named director at NASA Dryden.

That's where we started butting heads. He felt that as director up there he should have a greater say in the program. He also complained that we were using his people and not telling him, all that. I guess he complained directly to John Yardley, who was the associate administrator at headquarters, the head of manned spaceflight.

There was one big meeting, and it was clear pretty quickly that Dave just didn't know what was going on. The Dryden people who worked with us had signed off on everything they were supposed to. From that point on, Dave was on the downhill side with NASA. He wound up leaving in 1977, and getting involved with one questionable business deal after another. (He seemed to have a weakness for anyone who would throw green at him.)

Other than that we had no problems: Tom Stafford, who had returned to active duty as a major general following ASTP, was commander of Edwards for most of the time I was there. He was succeeded by Phil Conley, who was an old Air Force pal of mine.

(One interesting note: during the ASTP training, when the first shuttle simulator was being put together at JSC, we invited Alexei Leonov to have a look at the thing. What was odd was that he had no interest in it whatsoever. We couldn't figure out why, unless he was under some sort of guidance to stay away from U.S. shuttle stuff so he wouldn't be put in the position of commenting on Russian shuttle programs. They had a limited spaceplane program going at the time; the Buran shuttle program was probably just getting started in 1974.)

The ALT program began to come together in September 1976. On the seventeenth we rolled out the first shuttle orbiter at the Rockwell plant in

Palmdale. The original name had gotten changed, thanks to a write-in campaign by a bunch of *Star Trek* fans: the first orbiter was now called the *Enterprise*. On January 31, 1977, we trucked it thirty-six miles up the highway to Edwards-Dryden, and we started to really get serious.

The astronauts trained for the flights on modified F-104s—if you shut down the engine, an F-104 would land just like a shuttle orbiter. We also began to modify a couple of T-38s (opening the speed brakes) for powered simulations of unpowered shuttle landings.

We did several captive-inactive flights in February and March. That is, we simply mounted the *Enterprise* (with a tail cone, for aerodynamic stability) on the SCA and flew it around to test the combination. In June we did three captive-active flights, with the crew aboard. None of these gave us any reason to change anything.

Then, between August 12 and October 26, we flew five free flights, three with a special tail cone on the orbiter to help it fly a little better. They went fine, well enough to give everybody a warm feeling.

The only problem was on the fifth, the one flown with the tail cone off and targeted for a runway landing. Fred Haise and Gordon Fullerton were in the cockpit. As the *Enterprise* came out of its preflare, raising the nose for landing, the crew realized it was going way too fast, over three hundred miles an hour. Fred popped open the speed brakes, which actually increased the speed. So he dropped the landing gear and pitched down to get to the runway. The wings wobbled a bit . . . the wheels hit . . . *bam*! The damn thing bounced back into the air. It took another couple of seconds for it to straighten out and settle down.

We found out we had a little lateral instability in the orbiter. I also think Fred was worried about getting to the runway, got uptight, and overcontrolled.

Then, while the *Enterprise* was still out at Dryden, there was a leak inside the vehicle from one of the auxiliary power units. We spent a couple of months literally just cleaning that up and learning that that was a problem you really had to avoid.

But we verified the orbiter aerodynamics to a high degree of fidelity. We tested some of the thermal protection system (there were patches of different coatings on the *Enterprise*). When the ALT program ended, we were officially two years away from a first manned orbital flight test. It stayed that way for another couple of years.

The ALT task force was broken up in the fall of 1977; the JSC and KSC people went back to their centers, to make the transition to the orbital flight test program. A lot of the contractor people left, too, though there was still a permanent crew at Dryden who would prepare to deal with the orbiter on landing, getting it ready for return to the Cape.

Bob Thompson was the shuttle program manager throughout the ALT program. As we wound down, he asked me to stay with the shuttle program and press on to the orbital flight tests.

Other Voices

ROBERT THOMPSON

I had become shuttle program manager in Houston since 1970 and had grabbed Deke for the ALT program as soon as he was available. In typical fashion, Deke went off, spent about one day looking things over, and came back and said, "I'll take it."

"What do you need?"

"I need an office, I need a secretary, and I'll need my old buddy Tom McElmurry to coordinate between myself and your office, and I'll need access to the T-38s." I don't know that the job actually required the program manager to fly the T-38s, but I was getting off easy. I'd thought I was going to have to give him a lot more in the way of resources.

Moving Deke down to the Cape to be manager for orbital flight test was a natural move. That was the next phase of the program, and he was a valuable deputy.

The way the shuttle program operated during its development phase was like this: We were moving ahead building the orbiter, external tank, and solid rocket motors approved in 1972. Every phase of development within those parameters—every change, every question—had to be brought to my attention for the weekly change board meeting, which I held each Friday. (One of Deke's jobs was to keep the change board up to date.) Here is where we fought the battles between various points of view, made technical and financial decisions. What was effective about the system was that it had continuity: I had been running the program since 1970, and if some new person tried to reinvent the wheel, I could tell him why and how things had evolved to where they were.

It was quite a challenge. The money kept getting cut, the schedules kept changing, and we were trying to do a bunch of things that no one had ever done before.

Deke did a good job. That one argument with Dave Scott, which I refereed, was basically a case where Dave was new to his job as director at Dryden, and not too sure what all these people from other centers were doing all over the place. Deke was also new to his job; it took someone higher up to smooth things out. There was nothing personal in it.

No sooner was I back in Houston than Marge and I finally decided to split up. Kent was coming home from school for his twenty-first birthday—April 1978—and bringing a girlfriend. I guess Marge didn't want the girl to get the wrong impression, that we were together.

I bought a condo over near the center and moved there. I still saw Marge from time to time. She was diagnosed with cancer not too long after this, and it was really tough on her.

One thing that was bad was that I didn't see much of Kent during those years. He was busy with school and his life, and helping his mom, and I was off at the Cape. It wasn't a great time.

The last group of new astronauts came to NASA in August 1969; those were the Manned Orbiting Laboratory pilots we were essentially forced to take. The last selection had been two years prior to that, for the second group of scientist-astronauts. None of these guys had flown in space, and as the ALT program turned into the OFT program, it didn't look as though any of them would soon.

There were still about twenty-seven people in the Astronaut Office in 1977, most of them assigned to various shuttle development duties. Somebody at NASA figured it was time to start thinking about recruiting the next generation, even though Chris Kraft was saying that the current bunch was more than adequate to support ALT, OFT, and the first few operational shuttle flights. That is, we didn't anticipate the need for new astronauts until 1982.

It took about two years to get somebody through the initial phase of training and a technical or support assignment. Back that up from 1982 and you'd have 1980 as the earliest date you needed new people reporting. Given the nature of the shuttle, the new people you would need would have to be test pilots with experience in high-performance aircraft.

Then there was the issue of nonpilot astronauts. The shuttle was designed to carry crews of more than two, and it was quite true that these other people didn't need to be pilots at all, though familiarity with aircraft operations would make their transition easier. Here's where NASA really went overboard: they started a massive recruitment campaign designed to bring women and minorities into the program.

That was fine—I had nothing against women and minority candidates becoming astronauts. But it meant having to deal with over a thousand applications, and how do you do that fairly? They asked me how I would go about it, and I said I could think of quite a few places where qualified people could be found: the military test pilot schools (which were now accepting flight engineers) and flight test centers, NASA centers themselves, certain contractors. My approach would have been to hire from those places, since you'd have a smaller, preselected pool.

That was about the last I had to do with the process. It went ahead without me, and a group of thirty-five pilots and mission specialists was announced in January 1978. There was still some last-minute political bullshit, because the original selection had twenty pilots and fifteen mission specialists. Oops—only one woman made the cut, so five pilots were dropped from the list (they got selected a couple of years later) and replaced with five women mission specialists.

They were a good bunch and several of them did support work on the OFT program—Loren Shriver, Dick Scobee, Hoot Gibson. But as it was, none of the folks from this group got into space until 1983.

The orbital flight test program was originally going to be a series of six two-man flights, with the first taking place in the fourth quarter of 1979. They were to lead to an increasingly ambitious schedule of "operational" shuttle launches that would kick during 1980–81.

In March 1978 Abbey assigned four two-man crews to the OFT program. The first one consisted of John Young and Bob Crippen. The second was the ALT crew of Joe Engle and Dick Truly. The third was Fred Haise and Jack Lousma. The fourth was Vance Brand and Gordon Fullerton. They were called the A, B, C, and D crews. The plan was they would fly OFT 1 through 4 in that order, and then the A and B crews would do OFT 5 and 6, if necessary.

At one point I wound up in some discussions about crew scheduling for the operational missions, especially since the manifest called for as many as fifty-two flights in one year: one per week. The feeling around JSC and headquarters was that a lot of astronauts would be needed in addition to the thirty-five new guys (which is what the 1978 group named themselves).

My proposal, which was strictly back of the envelope, called for six two-man crews, no more, who would fly six or seven missions a year, one every couple of months, as a team. A crew might fly a satellite deployment mission one time and a Spacelab mission the next, the same way an airline crew would fly Atlanta to Newark, then switch to Newark to Los Angeles. If we were truly going to have an operational system—a cross between an airline and a military transport squadron—then we had to treat it like that.

Obviously, you'd have a couple of extra guys—backups, if you will—and eventually you'd replace people. But you didn't need a hundred astronauts: for my plan, for fifty flights a year, you needed less than twenty.

There was the matter of mission specialists and payload specialists. Mission specialists were intended to be career astronauts who performed different functions on orbit, supervising scientific experiments, performing EVAs, things like that. Like the scientist-astronaut group, as originally envisioned, they did not have to be pilots.

Payload specialists were simply scientists or engineers who had a real need to be included in a single crew or mission, something like a Spacelab mission which might require them to train for five years.

I thought the mission specialists—one per crew—could fly a minimum of once per year, maybe twice. Maybe more: if you had three satellites that were to be deployed with the inertial upper stage, for example, why not fly the same mission specialist three times? So you might have another twenty or so mission specialists in addition to the pilots. Payload specialists would come and go as needed.

My proposed shuttle Astronaut Office would have had more like fifty people—each one flying at least once a year, some as many as six or seven times. The current office has about a hundred, flying a maximum of eight missions a year. And it's rare that anyone flies once per year.

What happened was the astronaut office loaded up on people in 1978–80, while the flight rate kept getting cut and flights were delayed. Once crews of more than two started flying in 1982, somebody suggested or realized that having an extra pair of hands on the flight deck was a great idea. Another rationale was that half of the shuttle astronauts were sick during their first couple of days in orbit, so the more people you had in the vehicle, the more likely you were to have a couple of people who could function. (Of course, you could wind up with five, six, or seven people who were *all* sick, too.)

Now it's just as likely that a shuttle crew will have six or even seven astronauts. This makes sense for a Spacelab mission—if you forget that two of the seats were supposed to go to Spacelab payload specialists—but I don't know what all these people are doing in a five-day satellite deployment. I think it's a case of adapting the crew size to the size of the Astronaut Office, rather than having the number of astronauts the missions require.

The large crews totally overwhelmed the hygienic facilities aboard the shuttle. As recently as January 1993 they were still testing a new $23 million toilet.

* * *

The moment I became manager of the OFT program, though, I knew there was trouble. The damned thermal protection system just wasn't ready to go. Partly this was due to the continued stretch-outs of the program for budget reasons. There was no formal goal, which we'd had in Apollo. We didn't have to accomplish a manned orbital shuttle flight before the end of 1979, so whenever Congress would trim money from the NASA budget, everything would get pushed into the future. Where it wound up costing more than the amount that got saved.

The other problem was just technical: the orbiter had been designed (with its big delta wing) with the assumption that a lightweight, survivable thermal protection system would be developed. It was . . . but two years later than it was needed. We had tested some of these materials on the ALT flights and found that you could make all kinds of stuff that would keep the orbiter from burning up . . . but the material itself would crumble if you looked at it wrong. Or would simply fall off the vehicle.

When *Columbia*, the first orbiter scheduled for an actual launch, was rolled out in March 1979, it was shipped to the Cape with its tile system unfinished. The very act of flying *Columbia* to the Cape aboard the SCA broke hundreds of tiles. I knew because when it flew over Houston I took a NASA photographer up in a T-38 and saw the damage with my own eyes.

So there we were in the spring of 1979, trying to figure out how to improve this system so we could fly. I figured it was a two-year delay, and that's what it turned out to be.

There were also severe problems with the space shuttle main engines. These babies were not only supposed to be some of the most powerful engines ever built, they had to be throttleable *and* reusable. This was one area where the lack of money really hurt—the testing program was just starved; they couldn't build the number of test articles they needed, so when they had a problem, it really set them back.

The third area of concern, after the tiles and the engines, was the flight control software. The shuttle had a fly-by-wire system—when the pilot pushed on the stick, it wasn't physically connected to steering motors or control surfaces; it was sending a command to a computer that managed everything. Where the Apollo computers were the best we could get for mid-1960s technology, the shuttle computers were circa 1970 . . . the Stone Age when you look at them now, but getting all the codes and commands written and tested and debugged was another major job, and that also caused delays.

As manager of the OFT program I was back in the NASA mainstream, because the shuttle program was most of NASA at the time. This was a whole different ball game from ALT. Instead of six people, I was in a program involving thousands. I spent most of my time at the Cape running the change board, keeping track of all the engineering details. Which meant every working day at noon I was in charge of a telecon involving fifteen different locations.

Here's an example of the kind of thing that would come up: It was in the planning for operational flights, the matter of ejection seats. We had two for the OFT flights: the commander and pilot wore high-altitude pressure suits similar to those worn by SR-71 spy plane pilots. They had the ability to get out of the spacecraft during a launch emergency.

Someone—I don't remember who—proposed that we keep the ejection seats for commander and pilot when we moved on to operational missions and crews of three, four, or more astronauts. Vance Brand wound up commanding STS-5, the first operational shuttle launch, with the first crew of four. *Columbia* carried its ejection seats on that flight, but they had been disabled. Vance joked that he was afraid mission specialists Bill Lenoir and Joe Allen (Allen wasn't even on the flight deck during launch, but had a seat below in the mid-deck) would find the lack of an escape mechanism "demoralizing."

I found the whole idea flat-out unacceptable. Everybody's going to have an ejection seat or nobody's going to have one. And it wasn't technically possible to put more than two, maybe three in the orbiter. See what I said earlier about crew size.

Then there was a lot of time spent on a possible Skylab revisit. The workshop had been going around up there in orbit since early 1974 with no visitors. NASA had always assumed the shuttle would be flying long before Skylab's orbit deteriorated to the point where it would reenter.

There were two mistaken assumptions in that: first, the shuttle got delayed. Second, unusual solar activity affected the upper atmosphere, increasing drag on Skylab as it went around the world sixteen times a day. It became pretty clear that the workshop was going to reenter sometime in 1979. And, of course, it didn't have engines of its own to boost the orbit.

The Skylab visit was programmed for the third OFT mission with Fred Haise and Jack Lousma, who were supposed to rendezvous their orbiter with Skylab and dock a small engine to it. They would back off, and the engine would fire to boost the workshop into a safer orbit.

We finally bit the bullet on that one in early 1979; it just wasn't going to happen. Fred Haise left the astronaut team not long after—I think he just got tired of waiting for the shuttle to fly. Jack Lousma moved up to commander of the C crew with Gordon Fullerton moving up from D pilot to C pilot. Bob Overmyer became the pilot on the D crew.

In the summer of 1979 a book called *The Right Stuff* was published. The author was Tom Wolfe and he told his version of Chuck Yeager's career and the early days of Mercury. It turned out to be a big seller, and suddenly I was getting asked to autograph copies of it, along with all the other photos and stuff I was also having to sign.

I gave it a quick read and thought it was actually pretty good—he certainly captured the spirit of the times, though there were some inaccuracies. One big problem was his portrayal of Gus Grissom as some kind of screwup, which was never the case.

I'd been interviewed probably a few thousand times by that point in my career, so I can't say that I never met Tom Wolfe. But I never talked to him for *The Right Stuff*, though I would have been happy to correct his impressions of Gus. I figure Wolfe must have talked to John Glenn and Scott Carpenter, and probably Wally Schirra, though it's clear his main sources were Yeager himself and Pete Conrad.

The book became the one thing everybody wanted to talk about; most people in the program didn't like it as much as I did. It did bring the surviving Mercury guys back together, in a way. Scott and Gordo Cooper had both left NASA unhappily—Scott in 1967, Gordo in 1971—and probably a little resentful of the way things had turned out. I hadn't seen much of them over the years. Gus was gone, of course, and Betty had wound up suing North American over his death, which had alienated her from some of the guys. Wally and Al were both in business, and I was closer to them. John, of course, was a U.S. senator.

Thanks to the book and the later movie—which was as bad as the book was good, just a joke—we were all lumped back together in people's minds again, whether we wanted it or not. I can't say it bothered me; it was kind of like old Army buddies. Old disagreements didn't seem so important anymore, and I think we all got on good terms.

Over the next few years we even got into some charity work, setting up the Mercury Seven Foundation, things like that.

There was a bad flood in the Houston area at the end of June 1979, and the Friendswood house wound up with several feet of water in it. We got Marge moved out of there while things were cleaned up, but we lost a lot of personal stuff, including mementos from the Mercury years, tons of family photographs just ruined. It was sad, but we didn't dwell on it. Clean up and move on.

Other Voices

BOBBIE OSBORN SLAYTON

I had been working at NASA since 1970—technically for a company called LTV, then for Rockwell. My job was helping to run three conference rooms in Building One.

I had been married and divorced and was supporting my two kids, Jim and Stacy. One day I saw Deke coming out of a meeting. I knew who he was by name; I'd grown up in the area and had always been warned to stay away from astronauts . . . that they were a wild bunch. As he walked past my desk and out the door, I commented to one of the contractors that Deke was absolutely the best-looking man I had ever seen. (He reminded me of Clint Eastwood.) Fifteen minutes later Deke was back in my office asking me to go out for a beer. I told him no; he asked me several other times, and I still had the same answer. Finally he asked what the problem was . . . and I said I didn't like beer. We both broke up with laughter and I decided he wasn't much of a threat to my job or reputation.

Our relationship wasn't too serious for a couple of years. We were both dating other people. In 1982 Deke moved in with me, and in October of 1983 we were married. We had ten tremendous years together. He was a wonderful, loving husband and a kind and generous father to my children.

We were finally ready for the first OFT launch in April 1981. It had been almost six years since ASTP. In that time the Russians had gone charging right ahead. They were operating Salyut 6 by now and had managed to keep crews working aboard it for up to six months at a time. They were still using a version of the Soyuz, but they made the system work. (They were even able to resupply Salyut with unmanned versions of Soyuz called Progress. And Salyut had its own engine system, so it could keep itself in orbit.)

The crew for the first OFT flight, now called STS-1 (Space Transportation System), was John Young and Bob Crippen. John had been around NASA for nineteen years at that point and had already flown four times. He was the kind of guy you could count on to stick around and work hard as long as there was flying to do.

We had had a couple of arguments during the OFT development. John was famous for writing memos—Young Grams, I called them—that would raise all kinds of alarms about various issues, some of them big, some of them small. I was all for hearing his side of a problem, and so was Bob Thompson, but John didn't want to go through the change board process.

He preferred to send around a memo, I guess to get it on the record. One of the big questions was about shuttle landings back at the Cape. John didn't think we were ready to do that. Well, he should bring it to the change board, I said. I didn't think there was any point in saying the shuttle was an "operational" system if you had to land the thing out at Edwards, then spend a couple of million dollars flying it back to the Cape on top of the carrier aircraft. It was the kind of thing that should have been thrashed out. And I think memos like this hurt John's career after *Challenger*.

John's pilot on the crew, Bob Crippen, had come over from the Manned Orbiting Laboratory program in 1969 and had worked on Skylab and ASTP before going into the shuttle. He had done a lot of work on the shuttle's computer system. I think he knew more about it than anybody else in the Astronaut Office.

For these flights my station was the launch control center. Walt Kapryan was the launch director—something he'd been doing since Apollo—but I had the final "go"–"no go" decision. For example, when the issue was icing on the external tank, I had to go out to the pad and look at it with my own eyes. Four thousand people can say no to a shuttle launch, but only one can say go.

We thought we were ready to go on Friday, April 10, but as we got down to the last few seconds some computer glitch came up and we had to recycle for forty-eight hours. Of course, that put us on the twentieth anniversary of Yuri Gagarin's flight.

Bright and early on the morning of the twelfth we got down to the last ten seconds. At T-minus six, the three main engines on the orbiter fired up, one after another. At zero those big solids lit, and off she went.

Everything was smooth on the ride up, separation of the solids and all that. The only real blip came once John and Crip were in orbit. They opened up *Columbia*'s payload bay doors and saw that a couple of the goddamn thermal tiles were missing from the engine pods on the back of the orbiter.

Those particular tiles weren't mission-critical. You could lose a few of them and still have protection during reentry. But if you lost them on the bottom of the orbiter, you were in for a bad day.

We had confidence in the work, but the shuttle was such a new bird that we also knew it was going to surprise us.

I was out at Edwards for the landing on the fifteenth, and I was really sweating it. The last we heard from John was the words "I mark," in answer to capcom Joe Allen's "We're all riding with you." At that time *Columbia* was over the Pacific somewhere near Wake Island, pitching up to forty degrees to let those black belly tiles take the brunt of reentry.

A long sixteen minutes. Then we heard John: "Hello, Houston, *Columbia* here. We're doing Mach ten-point-three at one eighty-eight. Our L over D is nominal." Mach 10.3 at 188,000 feet! I wished I could have been the one to say that.

Columbia passed over Big Sur on its way across California. John took manual control and did some S-turns. He found that *Columbia* handled pretty much the way it was supposed to.

At Edwards our first warning of *Columbia*'s arrival was a *boom-boom*—twin sonic booms. Then this little speck appeared in the sky . . . got bigger as it went around. John and Crip set up for landing and came straight in . . . gear down . . . dust on the runway. Roll to a stop.

Half an hour later John was outside doing a walk-around. He was about as happy as I've ever seen him.

One of my jobs was looking over the vehicle, which came through in good shape. A little damage to one of the control flaps and around the doors for the nose gear. Tile damage on the underbelly was mostly from pebbles kicked up after landing.

I was confident we could actually get a hundred missions out of one of these orbiters.

The second OFT mission, STS-2, had a crew of Joe Engle and Dick Truly. Good old Joe, who had lost the lunar module pilot seat on Apollo 17. We hoped for a five-month turnaround—still quite a ways from the two-week turnaround NASA management was promising.

One thing we had gotten concerned about was backwash right at liftoff. The exhaust from the solid rocket motors was higher than anyone had expected, and as it reflected off the ground and the launch tower it actually pushed some of the control surfaces on *Columbia* out of alignment. This was bad enough for the orbiter, and it was also likely to cause problems for scientific instruments in the payload bay.

The solution to this was to flood the base of the launch complex with water just before ignition. Getting that installed added a couple of weeks to the turnaround time.

No sooner did we have *Columbia* back on the pad on September 22 than some nitrogen tetroxide spilled over the nose of the orbiter. The stuff dissolved the material that bonded tiles to the skin of the orbiter, so we had to fix them right on the pad.

We were also running up against the fact that the shuttle was only going to be able to carry about thirty-eight thousand pounds into orbit, not the sixty-five thousand as planned.

STS-2 finally got launched, after another scrub, on November 12. This mission was supposed to last five days, but we developed problems with the auxiliary power units in the orbiter, and then some fuel cell problems. No one in Houston was interested in taking chances with a "used spacecraft," so we went to a minimum mission—fifty-four hours.

The rest of STS-2 went fine. During the landing at Edwards, I was flying a T-38 chase plane and called off the final feet to wheels-down for Joe and Dick.

STS-2 was my last mission as a NASA employee. I had actually retired from the agency back in 1980, prior to the first OFT flights. How it happened was kind of funny: I was having a conversation with Bob Thompson and asked him what his future plans were. He said he was retiring. "What the hell do you want to do that for?" I said.

He explained this new deal that was available to senior management with a certain number of years in the agency: you could retire now and start drawing retirement pay . . . then come back as a "retired annuitant" for a year or more at a lower level, serving "at the discretion of the NASA administrator." But your combined pay would be higher than your retirement pay, and you would also raise your future base for retirement pay. This sounded like too good a deal to pass up, and when I found out that Chris Kraft and Max Faget had taken advantage of it, I did, too.

The joke on poor Bob was that here he struggled for years on the shuttle program that was continually starved for funds and postponed, and then became manager of McDonnell Douglas Space Operations in Houston . . . going through the same damn thing with the Freedom Space Station.

(I had resigned from the Air Force with nineteen years, one short of retirement. Had I stayed on an extra year, all that would have counted toward "combined federal service" on a pension. Since I hadn't, it didn't . . . though I eventually wound up getting credit for some of that time.)

There were other factors besides the financial that got me out of NASA. For one thing, the agency had changed in the time I'd been there. When I started, it was small and more focused on a particular goal. The agency that I found in the late 1970s and early 1980s, which seemed to exist solely for the shuttle, wasn't much fun for me.

I had also held out the slim hope that I might get a shuttle flight of my own, but when I dropped a few hints about that around 1980, it was clear no one was interested.

I also think the leadership was better in the 1960s, with James Webb, who was still the best NASA administrator, followed closely by Thomas Paine. I never had any problems with James Fletcher, who was a real straight shooter. He'd come in in 1972. But the new administrator picked by the Reagan people in 1981 was James Beggs, and he was a real horse's ass.

He was a political appointee in the worst sense of the word—he had some experience with NASA in the past, but he was really there for ideological reasons, to both commercialize NASA and militarize it. To that end he brought in Hans Mark as head of manned spaceflight. Mark was a weird guy, also ex-NASA, who had been undersecretary of the Air Force in charge of spy satellite programs from 1977 to 1979. There we'd be in flight readiness reviews, and suddenly these two wienies would fly in from Houston to "take charge."

After I'd been a retired annuitant for a year, I asked whether Beggs wanted to continue me and was told no. So that was it.

28

RACING

My retirement from NASA was official on February 27, 1982. I had logged 7,164 hours of flying time including my last T-38 flight on February 23. For the first time in thirty years I no longer had planes to fly, and naturally I wanted to keep doing it. So I looked around for something I could afford that would also be fun and challenging.

Finding something I could afford became somewhat easier thanks to a good friend.

Henri Landwirth had survived a Nazi concentration camp and come to the U.S., where he ran the Holiday Inn at the Cape. I got to know him during the Mercury days, and we stayed in touch over the years. Henri came to my NASA retirement party and right in the middle of it, knowing how much I wanted to buy a toy plane of my own, handed me a check! I was so touched and surprised I damn near didn't know what to say.

I had another friend up in Oak City named George Budde who was into air racing. Even before I left NASA he took me up to the Corvallis, Oregon, Air Expo in August 1981 to be grand marshal of a parade. That's where he got me started looking around for a plane to buy. I got close to buying one, but then some old character heard about this and came up to me, saying, don't buy that plane. He knew of a better one just sitting in a trailer down in Van Nuys, California.

It was a Williams 17, an all-metal plane built by John Paul Jones, a sheet metal worker. In fact, there were only two ever made: the other one, a bit bigger (it was customized for a bigger pilot), was owned by Bob Downey, who later crashed it at Reno. This particular Williams had the tail number 21 and was nicknamed the *Stinger*.

The trick in buying Formula One planes is that you generally have to buy them before you fly them. I guess too many of them have gotten broken while being flown around the block.

But Bob Hoover warned me that one thing I had to be sure of was that I fit in the plane. It was a good point, because John Paul Jones was smaller than me. The Williams turned out to be a tight fit—I always wound up flying a bit hunched over, because there wasn't a lot of room for my shoulders—but the plane's advantages made up for the tight squeeze.

So I didn't get to fly it until I bought it. It was a squirrelly little S.O.B., but I got the hang of it. And had her towed up to Wenatchee, Washington. This was April 1982.

Formula One racing has a set of strict rules concerning, for example, engine size and wing area. There's a lot of room for improvement within those rules—they've stayed the same for forty years, but people are flying twenty percent faster than they did back then. The planes themselves are pretty damned small; they don't take up much more room than a car. What you wind up with is eight of these guys looping around a three-mile oval in speeds in excess of three hundred miles an hour.

In my first race I made a typical rookie mistake. I was sitting there trying to set my watch to get the green flag and was still sitting there when everybody else took off. But I wound up second in the silver championship with a time of 7:08.89 and an average speed of 201.45 miles per hour. The top three were separated by only two seconds. I was hooked.

Over the next eight years I raced forty-three times in eighteen different meets. Best showing was fifth in the gold in San Diego.

I thought I would be able to fly five races a year, but with all my other commitments, I never managed to get over two or three. It was fun, though. One time it would be Cleveland, the next Salinas. Reno was every year.

When I started, there were two different Formula One organizations. I was put in charge of one of the procedures committees, but there were constant fights with the other organization. Finally we merged.

In case I thought I was just going to have fun racing, I got elected president of the International Formula One in 1985 and damned if they didn't keep reelecting me until 1989.

Formula One never got too big. It needed strong sponsors, which it never had. To be fair, there wasn't much money to be made in it. As we said, it was a great way for a millionaire to go broke.

Fun as it was to fly a Formula One plane, you still had to know what you were doing. Midair collisions weren't too common. The thing most likely to kill you was mechanical problems, or good old pilot error.

There were a few guys flying Formula One who didn't have any business doing it. A guy named Errol Jonstad lost three planes. He was a Pan Am pilot who tried flying three different kinds of aircraft without being especially proficient in any of them. He came to Reno one year and was getting coached by radio, which is against the rules. During one of our heats he was going from the number five to number six pylon and apparently lost number six. Instead of turning in he turned out, right in front of a woman who had no place to go but up. I thought she was going to hit me, but she didn't.

Then, damned if he didn't do the same thing on the next lap.

He ended up dying in a plane crash, running out of gas while qualifying for a race. I hated to see someone get killed, but I was sure glad to have him out of racing.

One year we lost three people. One guy testing a new plane decided to pull a buzz job, a fast low-level pass, but his tail wouldn't take it and broke off.

In July 1990 I came closer to wiping out than I had in a long time. It was at New Braunfels, Texas, a Formula One heat. Two of the other pilots were a doctor from Virginia named Rocky Jones and an astronaut named Hoot Gibson. Hoot was a Navy pilot who'd come into the program in 1978 and had flown the shuttle three times by then and was training for a fourth, scheduled for launch in the summer of 1991. I knew him best from racing.

Anyway, we were all in the fourth lap, right over a cornfield, when suddenly it was like the air was full of birds. Then I realized it wasn't birds but pieces of airplane: Jones had collided with Hoot, clipping off a good chunk of the wing on Hoot's little Cassutt racer. It was a plain old midair collision.

Jones went right into the ground and was killed. Hoot managed to land safely, quite a trick and a testament to how good a pilot he was. The rest of us hadn't suffered any damage.

There was some fallout at NASA a couple of days later, though. It turned out that there was a rule forbidding astronauts assigned to crews to do any "recreational" flying. (NASA had instituted the rule in the summer of 1989, when Dave Griggs, another shuttle pilot, got killed doing off-duty aerobatics.) Hoot had broken this rule and it cost him: Don Puddy, the guy who succeeded George Abbey as director of flight crew operations, kicked Hoot off his shuttle flight and grounded him for a year.

He stuck it out, though, and it didn't seem to wreck his career. He commanded a fourth flight, STS-47, in the fall of 1992—and right after that got named chief of the Astronaut Office, my old job.

Over the years Reno turned out to be a great place for a family gathering.

My brother Dick would come up from Los Gatos, California where he was living. And my sisters, Bev and Marie, would come out from Wisconsin. Bobbie and I would drive up there with the plane in a trailer and bunk down in a Winnebago. It was a great way to see old friends and still get some flying.

Other Voices

JOHN "DUSTY" DOWD

I met Deke through air racing. It's not strange to think that he would get into that, given his adventurous nature.

He never acted like he was Mr. Famous Astronaut, though other people sometimes treated him that way. (Jim Miller and I had a photo finish in a race which Deke also flew in, and in the paper there was a picture that said, "Astronaut Flies in Race.")

He helped me push a racer out at San Marcos, Texas, early on. My wife, Sue, hardly said a word because she was so in awe of this astronaut. Hell, I told her, he drinks beer and was brought up on a farm. How bad can he be?

In 1985 at Reno there was a heat getting ready to go out. Now, this was one of Deke's first races, and I was a little leery of flying with an astronaut. I figured he would be a hot dog.

It was about forty-five degrees and it was raining. I wasn't going out, and when I heard the heat start, I couldn't believe anyone else was. Deke got down in the starting line. I just stayed in the hangar and listened to them take off.

Four or five came by. No Deke. A few moments later he came taxiing back to the hangar.

"What happened?" I asked.

"I couldn't see a goddamn thing, couldn't keep the canopy from fogging."

A Formula One racer cockpit is about the size of a Mercury capsule—it's very cramped and there's no heat or anything. At the start of a race, he shut it off. He wasn't going to be pressured by anything into flying something that wasn't safe.

Another time I gave him hell because he wouldn't put a checklist in the aircraft. So there was a time in Cleveland when he and Bobbie had come in late and someone else had checked out the airplane. He took off and got about halfway down the runway and realized he'd forgotten to turn the fuel on. With his reflexes, he was able to get it on and keep flying, and proceeded to qualify.

When he got back to the hangar, he came up to me and asked if I had any tape.

"What for?"

"Thought I'd make a goddamn checklist."

29

SPACE SERVICES

Even before I officially retired from NASA, I'd been thinking about what I would do after that. I had been approached by a couple of big contractors such as Lockheed and the Aerospace Corporation about some kind of relationship. Under NASA rules, you weren't allowed to talk to contractors about possible jobs without the written consent of the NASA administrator. I got the consent and had a couple of conversations.

But I had set myself two rules about a new job. First, I didn't want to work for any of the big contractors, because that would be just like working for the government. And I had had my fill of that.

Second, I didn't want to leave the Houston area. Most of the big NASA contractors were in southern California or on Long Island, so that pretty much left them out. (By this time I was a confirmed Texan. I was close to water and hunting, and though Houston weather had some drawbacks, it still beat winter in Wisconsin.)

It was possible that I would have signed on as a consultant with one of these firms, like Chris Kraft did, while remaining in Houston.

But prior to my retirement I was contacted by a Houston businessman named David Hannah, who had gotten my name from Kraft and Max Faget. Hannah had made some money in Texas real estate and was looking for new worlds to conquer. While he was on an airplane one day he ran across an article by Gerard O'Neill, a professor at Princeton University who popularized the idea of colonies in space. (O'Neill had been a candidate for the 1967 group of astronauts, but hadn't gotten selected.)

This got Hannah all fired up. He started trying to figure out how to make something like a space colony happen. The first step was obviously to improve access to space for the private citizen. Eventually he hit upon the idea of creating a company that would launch payloads for customers for a profit. He incorporated it as Space Services in 1981 and assembled a team to build a small rocket called the Percheron. That July they had tried to launch from a site on Matagorda Island, off the Texas coast. Unfortunately, Percheron blew up.

But Hannah was still optimistic. He got in touch with Hubert Davis of Eagle Engineering later that year. (Eagle was formed by a bunch of newly retired NASA guys, including Max Faget.) They advised Hannah to switch to a solid fuel motor, and damned if he didn't manage to locate a surplus Minuteman first stage.

Prior to my retirement I took a bit of leave to do some design reviews for him. And when I left NASA for good, I became a consultant to Space Services. The entreprenurial approach appealed to me.

During 1982 I helped Space Services design and build a rocket called the Conestoga, and essentially wound up serving as program manager and range safety officer for that launch, on September 9, 1982, from Matagorda. We had a couple of hundred reporters around, but it was still a big change from the Apollo or shuttle way of doing things. Sally Chafer, who was married to one of the engineers on the team, did the countdown.

Conestoga took off with fifty thousand pounds of thrust, not much different from Al Shepard's Redstone, and shot up to an altitude of 196 miles. The payload was nothing—forty gallons of water—and landed 321 miles downrange.

But it was a big success, especially since we had done it with a team of about twelve people. The NASA director of commercial affairs at the time, James Rose, told me later that we had done it with one third the people for one fifth the cost, which was about $2 million. I don't know that I buy that: I'd say we did it for about forty percent of the cost and personnel that a bigger organization would have used. The money for the launch and our early operations came from a group of investors David had put together—a limited partnership of about fifty people. The launch had had adquate funding and had taken place on schedule.

Since it had gone okay, I was elected president of the company.

Other Voices

DAVID HANNAH

I got put in touch with Deke through Max Faget, and he was the key player in our Conestoga launch. I don't think Deke realized he was putting himself in for four years of one crisis after another as we tried to launch a commercial space enterprise. He could have walked away at any point and I wouldn't have complained. There were other things he could have done that would have made him rich and would have been much less work.

What struck me about Deke was his honesty and integrity. He wasn't always right—*he and I agreed that we had missed our best opportunity to get things going right after that Conestoga launch in late 1982, to ride that first wave of publicity and euphoria. I had wanted to move right then into raising money and getting investors, but Deke thought we should be more deliberate, get our technical house in order.*

Right or wrong, however, he was always there. When we got into the suborbital launch program and had the second one blow up, Deke stepped right in to find out what the problem had been and to make sure we fixed it. And he went everywhere, giving speeches, making presentations, doing whatever he could to make our dream of private spaceflight into something real.

It cost me my cash; it could have cost Deke his reputation. But we both hung in there.

I'm more philosophical than Deke is. During the many hours we spent together I told him I believed this project was our calling. "You're getting a little too religious for me," he would say. But he did agree that no one ever has an experience he doesn't need, and that even if we weren't successful in what we were trying to do, somebody else would come along and build on our work.

That alone made it worth doing.

Right after the Conestoga launch we had pretty ambitious plans, at one time envisioning as many as twelve commercial launches a year for a ten-year period. The goal was to get paying customers.

The U.S. government created an office of commercial space inside the Department of Transportation right about that time. The head of the office was Janet Dorn, and she just did a terrific job. Following the launch, we got licenses from the department for three more launches, and within thirty days of application. (I doubt nowadays they could find a way to license a launch within eighteen *months* of application.) It was for $40 million worth of launches.

I went to Los Angeles to talk to the Air Force Space Division people. They were planning a series of small satellites; they listened politely, then told me some of the facts of life. They couldn't get money for a payload unless the payload had a launcher attached to it. And the Conestoga was not a proven launcher, not until it had an orbital launch. And given the reality of Air Force procurement—such as getting a proposal together and getting it approved and funded—even if they could give us a go-ahead that day, it would be four to five years before we were really in business. So the Air Force wasn't an immediate market for us.

Nevertheless, thanks to the Conestoga launch, Space Services was getting phone calls from all over. The oddest one was from a couple of businessmen in Florida who had the idea of sending cremated human remains—cremains—into permanent orbit.

One of the guys was in the mortuary business and had come up with a process that would turn a body into a cremain massing four ounces rather than the usual four pounds. They were hoping to charge from $2,000 to $4,000 each. We could have sent three hundred pounds into a two-thousand-mile mile orbit, where they would stay for about sixty-five million years—that's as permanent as it gets. It came out to ten thousand cremains per launch at up to $4,000 per cremain. Do the math and you come up with about $40 million to be made per launch. I even made a couple of speeches at mortuary conventions to get this thing rolling.

But the mortuary guys ran into a couple of problems in Florida. First there were some flaky state regulations about transporting human remains that essentially would have required us to build a highway into space. (Cremains needed to travel on paved roads to the resting place, some damn thing like that.) There was some pressure from the competition—or maybe it was from a mortuary outfit that employed one of the businessmen. One of the original partners died. Anyway, eventually it all faded out. Too bad, too, because to this day I still get letters from people asking about that service.

The publicity from the ashes-to-orbit stuff really hurt us with NASA, however. It didn't matter that an ex-astronaut and ex-shuttle manager was involved; Space Services struck them as a flaky outfit. Not the kind a dignified organization like NASA should be involved with.

Another guy came along with a terrific idea: he wanted to put a small, commercial positioning satellite in orbit. The Air Force has its Navstar Global Positioning System (GPS) in orbit—that's eventually going to be a network of twenty satellites in a variety of orbits. The original idea was to provide pinpoint guidance for cruise missiles—which was demonstrated in the Gulf War—but it also has a lot of commercial applications. Nevertheless, GPS birds cost about $300 million each . . . and they are military. Our guy's idea was simpler.

It came about because he had a car stolen and got pissed off about it. He had worked in the satellite business, so he figured why not put a small, simple send-and-receive bird in geosynchronous orbit, where it would stay over the same region of the earth. You put a transmitter inside your car: if it's stolen, you beam your social security number or whatever up to the satellite, it looks for that transmitter . . . and locates it to within about eighteen meters within thirty seconds. And you get your car back.

Of course, we all started thinking of other commercial applications for this baby right away: rental car companies trying to keep track of their vehicles, train companies tracking boxcars, oil companies tracking their rigs.

We signed a contract. It was our only legitimate contract, because the guy gave me a dollar to make it binding. He was going to build the satellite out in Newport Beach . . . but he never found any money, either. But the commercial GPS is still a good idea that somebody is going to try.

We eventually got over 250 letters of intent from people who wanted payloads flown. The problem was nobody had any money. A guy at McDonnell Douglas came up with the perfect name for these projects: "DreamSats" The world's full of them.

I went through some personal changes during those years with Space Services. Bobbie Osborn and I got married in October 1983—Tom McElmurry was best man—and bought ourselves a house in a new development in League City, just across the lake from JSC. The corner we lived on was where Masters Drive met Admiral Way—named for one of the original developers, a guy named Admiral Alan B. Shepard.

I finally got a boat of my own, a Boston Whaler, which I kept at a slip a few hundred yards from the house. It made fishing a whole lot more convenient.

And for the first time in my life I began to read for fun. When I was flying for NASA, I had the T-38—never had to wait in an airport like I did flying commercial for Space Services. I got tired of re-reading newspapers after a while and started picking up novels. Larry McMurtry became one of my favorite writers—especially *Lonesome Dove*, his big book about two retired Texas Rangers taking a last trail drive.

Maybe I identified with those guys. Here I was, at an age where I could have been retired, still trying to launch rockets. I guess it was all those formative years I spent on the farm, all those chores morning and night; you learn that people are depending on you and it kind of sticks with you. Even flying, which was all I wanted to do for fifty years, had its own requirements. If you got lazy there, though, you could get killed.

Those were the key lessons. That's how I operated. Get the job done. Then go have fun.

* * *

On Tuesday, January 28, 1986, I was in my office when Bobbie called and told me the *Challenger* had blown up. The commander of that flight was one of my better friends in the new generation of astronauts, Dick Scobee.

I wound up on television several times in the next few days. Once I heard the weather conditions down at the Cape for the launch, I figured out where the problem was. The solid rocket motors simply weren't designed to be launched with ice hanging off them. We had developed those operational rules over the years: all you had to do was follow them and you'd be all right.

Not that anybody was talking about the shuttle as a perfect vehicle, or one that was a hundred percent safe. I never thought it was. I sure as hell wouldn't have said it was safe enough to fly congressmen and schoolteachers. It would do what it was designed to do with an acceptable amount of risk. They've had almost fifty flights and lost one; I guarantee you if we'd flown fifty Apollo missions, we'd have killed somebody. (We killed one crew on the ground as it was, and came damn close to losing another.)

Going into space isn't as safe as getting on an airliner. And it's never safe if you're ignoring flight rules.

During the years from 1983 to 1988 I must have talked to a hundred venture capitalists, bankers, brokers, and investors all over the world. The original partners were still financing this marketing—we had no launches and no hardware—and it was very patient of them. I even spent a couple of weeks in China in 1985.

I was getting a little frustrated. Finally a venture capitalist in Denver sat David Hannah and me down for a debrief on what our problems were. It was pretty illuminating.

For one thing, your typical venture capitalist is willing to spend from half a million dollars up to maybe $6 million, with most of them clustered at the smaller figure. Naturally, what we were looking for was a figure greater than $6 million.

Even if you got one of them to agree to give you money, it would have to be for a proven technology. Which the Conestoga rocket was not. Which, in fact, no rocket is. (Ariane, Delta, and Centaur all have failure rates of around five percent, which are basically unacceptable to bankers.)

Assuming you could get the money for the technology, the payback criteria were pretty brutal: you would wind up giving the original investor something like a thousand percent payback—in effect, they would wind up owning the company. It was clear to us that we didn't meet any of these criteria.

We even took a pass at Ross Perot, but he didn't make all that money by being wild with it.

But we didn't give up. We managed to romance Houston Industries, the parent company of Houston Power and Light—a private utility—into funding our operations for four years. Houston Industries paid dividends of a quarter of a billion dollars a year, so you can see that little Space Services wasn't much of a chunk compared to that. Nevertheless, our monitor at HI struck me as unaggressive and unimaginative—he never really had a feel for what we were doing, and he kept spoon-feeding us money . . . just enough to keep us crawling, not enough to let us run.

We kept crawling. At one point we hooked up with two other companies up near Washington, one of them a small satellite builder, the other for ground stations, and formed a partnership called Space America. Our plan was to do low-cost remote earth sensing, which is another idea someone is going to do—and make money on.

Right about this time, in 1983–84, NASA tried to turn the Landsat weather satellites over to a commercial operator and run it through the Department of Commerce. Since this was close to what we had planned anyway, we put in a bid, one of seven companies. We made it to the final three, then lost.

Our bid was one third the cost of our nearest competitor. We had taken our original cheap remote sensing proposal—price tag $50 million—and loaded it up with all the bells and whistles NASA wanted, and still had a bird that would cost $145 million . . . less than one third the cost of the one eventually selected. (And never flown: the whole project collapsed in mutual acrimony, and the last I heard, control of Landsat was about to be shifted to a new Department of Defense command.)

What cost us the Landsat bid? We didn't have a "proven" launcher.

Then there was the DARPA situation—the Defense Advanced Research Projects Administration. In 1983 they started planning a small satellite launch vehicle (SSLV), a cheap launcher to orbit things called Lightsats. We actually helped them get it going, since we had a Washington office at the time, with a couple of people who worked the congressional committees to get funding. There were nine companies that bid on SSLV; we were one of four finalists, then there was a management change at DARPA. Suddenly the Lightsat project went from procurement to a *study*. The four finalists were all given about $300,000 for their trouble . . . which was about a third of what we'd spent on our bid, thank you very much.

Six months later there was another management change at DARPA, and SSLV came back to life. Now they were gonna buy the hardware. We four finalists were fighting like hell for the bidding to be limited to us, but

DARPA decided to open it up again. Worse yet, since our original bids had been paid for with public money, they were public domain.

Suddenly there was a company none of us had heard of called Orbital Sciences in the mix. (They had been doing some other kinds of work for NASA on a transfer orbit stage, but were not in the launcher business prior to this.)

Orbital Sciences cooked up something called the Taurus launcher, which proposed to use their Pegasus winged upper stage atop a goddamn MX missile first stage for the SSLV. Our bid was for $15 million a launch; to no one's surprise, theirs undercut ours by a substantial figure, at $10 million.

They got the deal. And that was basically what caused Houston Industries to pull the plug on us. They didn't like dealing with the government as a rule, and this just confirmed it.

It all came to a crunch in the summer of 1990.

NASA had finally come up with a proposed set of small commercial launches it would buy, two or three of them, and NASA was looking for bidders. We spent $1.5 million on our bid and submitted it on June 15, 1990.

Houston Industries pulled its financing on July 1. We had four dozen employees: I had to call them all in and tell them we were out of money. For the next year or so I was the only employee of Space Services, along with my SSI controller, Joani Loveless—who was actually being paid by David Hannah.

The loss of our funding screwed us with NASA, since it put us in default. We got an official notice that our bid was rejected. (I heard later from friends at NASA that, technically, ours had been the best!)

A month later the Air Force came out with a request for proposal for a similar small launch program. I was sitting there by myself, and there was no way I could put together a bid, so that opportunity passed, too.

It was like having someone stake you to a long poker game . . . and when you've spent the evening losing money, he pulls you out just when you've finally got a winning hand.

Three to four months later, in early 1991, NASA proposed an orbital commercial flight called COMET. (Oddly enough, I'd discouraged them from putting that out for bid, figuring that was no longer the way to go about it. But NASA wanted to broaden the commercial space base.)

I couldn't let this one go by. So David and Joani and I took our original June 1990 NASA proposal and redlined the hell out of it, submitted it to NASA, and sold it. Of course, then NASA asked, where's your financing?

By that time we had managed to book a series of suborbital Conestoga launches. These were sounding rockets, small vehicles that would fire up to altitudes of two hundred miles or so with a package of scientific or technical experiments, then parachute back down. The payloads came from the University of Alabama in Huntsville, which was the clearinghouse for a consortium of sixteen other universities—the program was called Consort. By June 1990 we had had three Consort launches, with varying degrees of success, and a fourth was still on the books. I had promised the customers to make that happen even if I had to launch the damn thing myself.

By this time David had started talking to an engineering company called EER about taking us over. EER was involved in avionics and tracking systems and worked with DOD and NASA. They were in our business, and when the COMET deal came up, EER decided to acquire the company.

I won't go into dollar figures, but they got more in physical assets than they paid in cash. Essentially they got our chunk of the COMET program for free.

As of 1993, Orbital Science's Taurus system has yet to fly, though it was promised for 1991. Its projected cost is now $22 million a flight.

Orbital's Pegasus, though it had a couple of flights, is still frogging around. They originally sold it to customers at $6 million a flight, no risk. I figured it was one hundred percent underbid from the beginning—now it's $14 million a flight, if you can get one.

Orbital's technical setup is horrible, yet they're there. It's been a tough business: for example, Lockheed made some noise about getting into the SSLV business with a launcher based on its Poseidon missile. TRW under Dan Goldin was interested in launchers. When he went to NASA as administrator in 1991, that sort of faded. LTV, which made the Scout launcher for NASA and the Air Force for thirty years, has sort of gone belly-up. Boeing could have gotten into the market with a version of the Minuteman, but didn't.

Then you have E-Prime, a Florida-based company that tried to copy Orbital Sciences: they wanted to use the MX missile as the basis for a larger launch vehicle, something on the order of the Titan III, but at half the price. Of course, they had no authority—under the terms of the START treaty, you can't launch these ICBMs, no matter what they're carrying. They've since gone belly-up, too.

Thiokol, the solid rocket builders (remember the Shuttle SRB), built the MX and developed a satellite launcher version called the EG120. I don't know what's come of that.

Out in Camarillo, California, there's a company called AMROC—American Rocket—which was founded by a guy from the record business named George Koopman. AMROC planned to build a hybrid rocket motor and thus was always a bit confused: were they a motor company or a launcher company? They'd come around every now and then and try to sell me motors . . . then I would run into them at a conference, and find them trying to sell launchers in competition. I pointed out to them that I wasn't likely to buy motors from a competitor.

Poor George Koopman got killed in a car accident a couple of years ago, and AMROC one fizzled on the pad. I wouldn't count them out completely, though.

The company started by Max Faget, Space Industries, works with us on COMET, and it was just bought by a company in Indiana called ARVN. Max had already moved upstairs to chairman of the board, with Joe Allen as president. ARVN has come up with a four-man management team (something we tried at Space Services without notable success) that includes Joe and one of his people and two from the parent company. I suspect that the money guys are going to wind up running the thing. Max has already moved out of the company, to the board of directors.

We're building the COMET launcher out of existing components. Castor motors from Morton Thiokol in Huntsville, Alabama. (They fire nine of those every Delta launch. Those are the strap-ons to the Delta first stage.) Our upper stages are Star motors from Thiokol in Elkton, Maryland. These are as off-the-shelf as you can get in the aerospace business. You can't just walk in and buy them that way: you contract for them and then they start building. But the design is mature and that's exactly the path we were on from day one. We didn't want to invent anything.

Here's an example: In 1992 I was involved with General Dynamics in a proposal for a new engine concept that burns nitromethane. GD had the idea that maybe we could include their new engine in some of our proposals for SDI. I told them if I could figure out how to work it in, I would. One of the guys brought in a fuel injector he had bought at a hardware store for seven and a half bucks, just to show how cheaply you could put this new engine together.

Well, we tried the whole proposal on DARPA, and DARPA didn't have any money. So we went over to NASA, and Charlie Gunn at NASA jumped on exactly the same question I would have. "Goddamn it, you're telling me all these things are so simple and cheap, but you want me to give you thirty million dollars?" He turned to me and asked how much EER was spending to develop the whole vehicle, excluding the new engine. "About fifteen million," I said.

Here's General Dynamics, the second biggest aerospace company in the United States—whose chairman, by the way, is former Apollo 8 astronaut Bill Anders—whose stock has doubled in the past year. They're rolling in money, and they don't want to spend a nickel on a new concept like this. They're trying to get somebody else to cover it—little wienies like us, who haven't got any money to begin with! They should spent the money and develop the thing, then take it to the government and say, hey, we've got a whole new product.

Other Voices

KENT SLAYTON

After my mom died in February 1989, my dad and I grew closer than we had ever been in our lives. He had changed, and so had I. We were both able to talk about things, to be open with each other, in a way that was completely new. We always made sure we spent the holidays with each other. I was working in San Antonio, but we managed to stay very close.

The biggest surprise came when I found out he'd given away all his hunting trophies. He and Bobbie had gotten a dog named Bucky, and Dad really started to like the little guy. Remember, he had grown up on a farm; animals were for food or labor, they were never pets. Now that he had a pet, however, Dad couldn't bring himself to shoot anything.

Between 1989 and 1993 we launched six Consort missions out of White Sands. The last few used a launcher we called the Starbird.

Starbird is a two-stage vehicle using one motor from Thiokol and one from Bristol in Canada. We have those delivered from the factory to White Sands, where we integrate them right into the stack.

The payload comes from the Consort program. The guidance system we buy from Saab. Telemetry and recovery systems EER handles, since that's one of our areas of expertise. We do some things and subcontract out the rest. It's no different from the way Rockwell built the B-1 bomber.

We contract with the Army for range safety at White Sands—just write them a check in advance. Actually, our point of contact at White Sands is the Navy, because they have control over the sounding rocket pads going back forty years. The assembly building belongs to NASA, so that's another deal.

Dealing with NASA isn't too difficult, because they've done this kind of thing for years. But the Navy isn't really set up for commercial space. When we first approached them for authority to launch at White Sands, they wanted $500 million worth of liability insurance! For this little goddamn $1 million rocket! The Department of Transportation bailed us out on that and set the rate at $20 million, which was more like it. Because of that I ended up over at the Pentagon hassling with a bunch of Navy and civilian legal guys. I kept telling them, "Your argument's not with me. It's with the Department of Transportation. Why don't you guys get this sorted out?"

The first time we had to nail down a launch they insisted on having our money up front, so I went over to the Pentagon with a check for $300,000—and nobody could figure out who the hell had the authority to accept it!

It took two weeks to get that money transferred. Ultimately we ended up going to the Secretary of the Navy.

The first four launches at White Sands were done as Space Services. Beginning with the fifth it was EER, so we drew up a new memorandum of understanding, a contract between us and the Navy, just like the old one, with "Space Services" crossed out and replaced by "EER."

We handed it to the Navy office at White Sands, and they said it looked good to them. But it turned out that Navy authority for White Sands had been transferred to the facility up at China Lake in California.

Well, up at China Lake they had no idea what this was all about. That wound up right back at the Pentagon, too.

When we were acquired by EER, I went from being a company president to being director of a division about five levels down from the top.

Nevertheless, we got into serious business, with forty-five to fifty employees and two orbital COMET missions planned, with an option for a third. (We got some additional money for COMET from the movie industry. Columbia Pictures bought space on the side of the launcher to advertise an upcoming movie called *The Last Action Hero*, starring Arnold Schwarzenegger.)

And we also got a contract from the Strategic Defense Initiative Organization for the orbital launch of a twenty-five-hundred-pound satellite called MISTI. (SDI put two launches out for bid, a small one of five hundred pounds, and the big one, but announced up front it wouldn't put both with the same contractor. The small one went to a company called International Microspace, which has some severe technical and financial problems, and I wouldn't be surprised if EER eventually gets that one, too.)

There were seven suborbital Starbird launches on order, giving us a total of ten to twelve potential flights. It took me ten years getting to where I wanted to be with commercial space. It seems like a long time, but then, I spent ten years getting requalified to fly in space.

* * *

In the spring of 1991 I was a reasonably healthy international Formula One air race pilot with about fifty years' experience flying combat, test, operations, and air racing. I was smart enough to know I wasn't immortal, but fifty years of successfully dodging bullets left me rather casual about my near-term risk. I had enough racing friends getting dinged regularly to convince me they were more than holding up the casualty odds dictated by fate and I was relatively invulnerable. Then things changed radically.

After entering the 1991 Reno Air Races, I encountered a strange physical malady. Intermittently I would, for no obvious reason, lose my ability to tell up from down. (We call this "tumbling your gyros" in the flying business.) The immediate solution was to get on my hands and knees before I fell uncontrolled into some less benign attitude, like flat on my back or face.

These episodes occurred rarely and randomly, but it was clear only an idiot would put himself in control of a racing plane flying about twenty-five feet off the ground at max speed with a gaggle of competitors fighting for the pylons. I grounded myself and persuaded a good friend, Dusty Dowd, to race my airplane.

Dusty is an outstanding pilot, aircraft designer, builder, and friend. He worked on my airplane prior to Reno with unimaginable effort and hours, and succeeded in flying it to the number one position in the silver championship race and the winner's trophy.

From then on the Stinger sat in the trailer at his home in Kansas because I was unable to fly it, and Dusty had too many other jobs demanding his time. About that time I experienced blind spots in my vision, decrease in short-term memory, and increased stability and directional problems. A rerun by the medics uncovered that I had a malignant brain tumor entwined in the base of my brain and around the optical nerve.

The cause of my problems was now very obvious. The size and location of the tumor was such that surgery was a high-risk, low-probability-of-success option. A biopsy showed it was either anegleoma, which is difficult to treat, or lymphoma, which is generally controllable. The first input I had was that it was the bad stuff and I could expect six months longevity—maximum.

During the biopsy, which involved drilling a hole in my head while I was conscious, I experienced a weird perception that a black cloud was enveloping my brain. I was sure I was either dead or dying and it scared the hell out of me. I had been dodging lethal bullets, lead and steel, or accidents, for fifty years and had become very fatalistic—when your number is up, it's up. Suddenly, facing a finite end point put my thinking into a totally new perspective. I had too many things left to do and too many people dependent on me. I'm sure this is not a unique experience.

I made a resolution right there to get my act together, get my personal affairs in order, and create an ongoing plan for what's *really* important in the future. On the business side we were in the process of preparing two commercial rocket launches, and obviously the success of these was critical to our company's survival.

Mike Cassutt and I had started work on this book and I was deathly afraid if the brain tumor began to affect my memory, I would be useless to him. Fortunately, the Lord granted a stay of execution, which has held up so far. A review of the biopsy results showed a high probability of the lymphoma strain as opposed to the other and they had already started radiation treatments under that assumption. Upon completion of the first series, they concluded the tumor was really shrinking. I switched to a not yet fully defined chemotherapy protocol and am almost back to normal. I am confident the book will be finished with or without me from this point. Mike has my complete confidence in producing a quality product of interest to the reader.

On the personal side, Bobbie hates my travel schedule and I intend to minimize that radically while also maintaining better contact with my friends and relatives, both local and remote. We will continue to be kind to kids, animals, and people, work hard when we're working, and plan a maximum of play time for the family when we're not. Nothing new here for normal retired people, but I don't want to be retired—just downgrade my workaholic tendencies. I need to upgrade my spiritual life and increase my participation in charitable institutions, scholarship funds.

If there's a message in all this, it's that we all look up the barrel of a loaded gun regularly without adequately assessing the real potential for disaster. I don't recommend that you stop taking risks—even the turtle cannot proceed without sticking out his neck. But recognize that the good Lord has your number, and when it comes up, be prepared. You may not get as many chances as I've had.

In February 1993 we had our sixth Starbird launch. Like the Reno Air Races, the Starbirds turned out to be a nice way to see some old friends, such as Bill and Mariwade Douglas, who would drive down from Albuquerque. Bobbie came along, too, of course. There was a week of final checks, then the rocket would be moved over to the pad on one of the world's shortest railway systems.

The Starbirds took off fast—a few seconds and number six was out of sight going straight up. The recovery chopper headed out to pick up the payload, which would be coming down fifty miles farther into the desert.

It was a small operation—no more than a dozen people in an old blockhouse, one that had been in use for forty years. It was like Edwards . . . like NASA in the early days. Just a small group of people dedicated to getting the job done, and moving on to the next one. It was always fun to answer the phone while the smoke was still clearing, and hear Mike Bryant, our EER spokesman, asking how it had gone. "It went fine," I told him. "Nominal as hell."

ACKNOWLEDGMENTS

Deke Slayton died of cancer at his home in League City, Texas, on the morning of June 13, 1993. The following Saturday there was a public memorial service at the NASA Johnson Space Center, attended by four hundred friends, family members, and co-workers. Speakers at the memorial, organized by Alan Shepard, included John Glenn, Wally Schirra, Bill Dana, Alexei Leonov, Bob Thompson, Dusty Dowd, David Hannah, and NASA Administrator Daniel Goldin.

Deke and I began to work on his autobiography in the summer of 1991, conducting several hours of interviews over the next year. When medical reasons forced him to curtail his travel, we continued our conversations by telephone. This book is based on those conversations, supplemented by Deke's personal papers and previously published accounts. (Two books which were especially helpful must be noted here: *Manned Spaceflight* by David Baker and *Apollo: The Race to the Moon* by Charles Murray and Catherine Bly Cox.

However, it would not have been possible to finish this book without the reminiscences of Deke's friends, so my thanks to Joseph Allen, Dr. Charles Berry, Eugene Cernan, Dr. William Douglas, Bob Drew, David Hannah, James McDivitt, Dee O'Hara, Tom McElmurry, Harold "Bud" Ream, Thomas Stafford, Walt Williams, and Robert and Dot Thompson. (Some of the "other voices" were adapted from presentations at Deke's memorial.)

Bobbie Slayton, Kent Slayton, and Marie Madsen were also generous with their time and memories in what turned out to be a difficult year. Joani Loveless was also a lifesaver.

At NASA I was helped by Joey Kuhlman of the History Office and Becky Fryday of Media Services. Lois Morris of the Woodson Research Center at Rice University, and David Shayler of Astro Info Service, also provided valuable materials.

I must also thank our agent, Richard Curtis, and *Deke!*'s editor, Beth Meacham, who believed in this project from the beginning and made it happen.

—Michael Cassutt

INDEX